A Postcranial Guide to Domestic Neo-Natal and Juvenile Mammals

The Identification and Aging of Old World Species

Thomas Amorosi

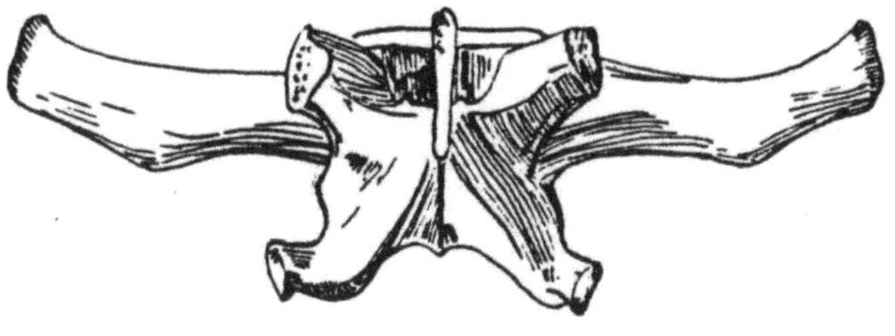

BAR International Series 533
1989

B.A.R.

5, Centremead, Osney Mead, Oxford OX2 0DQ, England.

BAR -S533, 1989: 'A Postcranial Guide to Domestic Neo-Natal and Juvenile Mammals'

ISBN 9780860546788 paperback
ISBN 9781407348278 e-book
DOI https://doi.org/10.30861/9780860546788
A catalogue record for this book is available from the British Library
This book is available at www.barpublishing.com

TO THE MEMORY OF MY DAD

NICK AMOROSI

WHO CONTRIBUTED SO MUCH
TO THE GRAPHIC QUALITY OF
THIS BOOK

ACKNOWLEDGEMENTS

Although a work of this nature is often conceived and written alone, the time between conception and the final manuscript reflects the interests of many people. A very special measure of gratitude is reserved for my Ph.D. advisor, Dr. Thomas H. McGovern (Department of Anthropology, Hunter College, C.U.N.Y.) as well as my M.A. advisors at the Department of Anthropology at New York University -- Dr. Randall K. White and Dr. C. Jean DeRouseau. I would also like to extend this thanks to Dr. Arthur E. Spiess (Maine Historic Preservation Commission), Eugene Boesch (Dept. of Anthropology, N.Y.U.) and Bobbi Brickman (Department of Science, Hunter College High School, C.U.N.Y.). If it was not for their many helpful disscussions and advise this work would not have been possible. Also a hearty thanks to Mr. Nicholas Amorosi (Dept. of Anthropology, American Museum of Natural History) who over the entire span of this project, provided many hepful suggestions concerning the illustrations in this manual, as well as the patience needed to look at yet another "bone drawing".

I would also like to acknowledge Dr. Daniel Bates, Dr. Francis P. Conant, and Dr. Gregory A. Johnson (Dept. of Anthropology, Hunter College) who provided "classes to teach" and research facilities which allowed me the needed time and place to bring this manuscript to a published form. Thanks are also due to Dr. David H. Thomas who found a place for me to work in the Department of Anthropology at the American Museum. Dr. Ian Tattersall (Dept. of Anthropology, A.M.N.H.) helped with interdepartment loans at the American Museum. Dr. Guy Musser (Dept. of Mammalogy, A.M.N.H.) obtained the needed permission to use the American Museum's Mammalogy collection as well as provided many helpful suggestions. Mr. Daniel Russell and Mr. Jean Augustian (Dept. of Mammalogy, AMNH) gave their time time and effort to help with the mammalogy collections at the American Museum.

Finally, I would also like thank Dr. Kathy Biddick, Dr. Gerald F. Bigelow, Dr. Juliet Clutton-Brock, Dr. Pam Jean Crabtree, Dr. Donald K. Grayson, Dr. Richard Meadow, Dr. Stanley J. Olsen, and Dr. Paul W. Parmalee.

May 1988 T.A.

CONTENTS

Chapter 1 THE PROBLEM 1
Methodological Concerns of Illustrating 3

Chapter 2 AGE DETERMINATIONS FROM BONE: A SURVEY OF METHODS 5
RELATIVE METHODS 6
Fusion of Epiphyses 6
Suture Closure 6
Problems and Limitations 7
Tooth Eruption, Wear and Crown Height Methods 11
Drawbacks to Dental Estimates of Age 12
Tables to Follow 14
INCREMENTAL METHODS 15
Historical Background 16
Tooth Cement and Dentine 17
Climatic Effects 19
Nutritional Effects 20
Reproductive Cycles 21
Bone 22
Horns 23
Selection of Specimens, Teeth 25
Selection of Specimens, Bone 25
Drawbacks to Incremental Methods 25
Comparison of Relative and Incremental Methods 28

Chapter 3 AGE DETERMINATIONS BASED ON BONE, TEETH AND HORN: TABLES OF AGING CRITERIA 30
Equus Fusion and Dental Data 33
Bos Fusion Dental Data 51
Capra Fusion and Dental Data 71
Ovis Fusion and Dental Data 81
Sus Fusion and Dental Data 91
Canis Fusion and Dental Data 103
Felis Fusion and Dental Data 113

Rattus Fusion and Dental Data 119
Grant's Method of Recording Mandibular Dental Wear for Bos, Ovis/Capra, and Sus 125
Sus Dental Wear Series 133
A Method To Assess Dental Wear for Rattus 141

Chapter 4 JUVENILE MORPHOLOGY 145
Plate I 149
Plate II 150
Axial Skeleton Plates III, Atlas 152
Plates IV, Epistropheus (Axis) 160
Plates V, Cervical Vertebrae 169
Plates VI, Thoracic Vertebrae 175
Plates VII, Lumbar Vertebrae 181
Plates VIII, Sacrum 187
Appendicular Skeleton: Thoracic Limb Plates IX, Scapula 192
Plates X, Humerus 205
Plates XI, Radius 217
Plates XII, Ulna 226
Plates XIII, Metacarpus. 235
Appendicular Skeleton: Pelvic Limb Plates XIV, Innominate (Pelvis) 245
Plates XV, Femur 256
Plates XVI, Tibia 268
Plates XVII, Fibula 278
Plates XVIII, Astragalus 283
Plates XIX, Calcaneus 289
Plates XX, Metatarsus 299
Plates XXI, Phalanges (Digits) 309
Contour Drawings of Bone Elements to Scale 315

Appendix A: MATERIAL CHOSEN FOR ILLUSTRATION 337

Appendix B: A LISTING OF OSTEOLOGICAL ATLASES, MANUALS AND KEYS 343

REFERENCES CITED 349

THE PROBLEM

Archaeologists today think of sites in terms of their human occupants, their subsistence behavior, and the animals that are exploited by that behavior. Increasingly archaeologists have begun to consider larger faunal samples essential in order to answer detailed paleoeconomic questions (Butzer 1975; Kenward and Engleman 1985 ; Higgs 1972, 1975; Levitan 1983; O'Connor 1985; Turner 1984) concerning hunting practices, animal husbandry, and marketing practices. As a result, there has a dramatic growth in research and literature using vertebrate archaeofaunas to derive information about past cultural activities (Binford 1981; Chaplin 1971 ; Frison and Stanford 1982; Higham 1967; Klein Cruz-Uribe 1984; Parmalee et al. 1972; Perkins 1973; Spiess 1979; Styles 1981; Wheat 1972).

There are numerous publications available to aid [1] the specialist with the identification of faunal remains, although the references cited here are no more than a window through which the reader can gain access to the vast literature concerning the identification of faunal remains. Each of these have their strong and weak points. None should be used without consulting a good comparative collection. There are manuals that deal exclusively with mammals, birds, reptiles, amphibians and fish (see appendix B for references). The manuals that pertain to mammals can be further subdivided into those with general mammalian osteology, combinations of cranial and post cranial keys, cranial remains, and those that pertain to postcranial material.

All of these deal exclusively with adult animals, with few useable illustrations of the bones of younger individuals. However, many archaeofauna contain immature bone fragments. The skeletal development in mammals entails morphological, as well as scalar change, and the

---

1. Bone manuals can only serve as an auxiliary guide for the identification of animal bone. Readers are stongly recommended to consult reference collections when ever possible.

relationship of adult to neo-natal and juvenile forms is not always straightforward.

During the subadult growth phase, neo-natal to juvenile mammalian bone experiences a dramatically visible change. Many of the skeletal elements have ossified from their cartilaginous pre-forms (Sinclair 1978), but the osteological landmarks found on the bone's surfaces are just beginning to take on their adult character. Without the use of osteological landmarks the differentiation of species can become difficult in some cases.

Growth phases can be defined by the following developmental characteristics:

> Neo-natal phase: The bone surface has a porous appearance consistent with primary bone deposition. Deciduous dentition has started to erupt or, in some cases, may already be present. Epiphyses have begun to form (Bass 1971 ; Schmid 1972: Stewart 1979).
>
> Juvenile phase: Deciduous teeth are mostly present. Replacement of deciduous teeth with the permanent set can occur. Epiphyses are starting to take on their adult character (Bass 1971; Schmid 1972; Stewart 1979).

The goal of this work is to compile information that will aid in the identification and analysis of domestic mammals within a neo-natal to juvenile age range.

Therefore the approach taken in this manual is similar to the practical "working group" approach outlined by Brown and Gustafson (1979) . Animals are selected for study based on their probability of being recovered from an archaeological context rather than by taxonomic classification. These animals can be grouped irrespective of taxonomic affinities (Brown and Gustafson 1979:IV). The definition of "working group" undoubtedly depends on the context of a particular project. The animals chosen for this study represent a type of "working group" useful for analysts working with domestic mammalian fauna.

The illustrations in this manual are designed to complement the work of Elisabeth F. Schmid (1972); portraying the adult forms of the domestic mammals -- horse, cow, sheep/goat, pig and dog. The domestic mammals selected for this particular manual represent European domesticates that have traveled with european colonists, primarily during the 9th to 19th centuries A.D. (Clutton-Brock 1981; Zeuner .1962). These species represent the overwhelming majority of specimens in Euro-american archaeofaunas, and normally cover the full range of domesticates whose culling pattern is of interest to paleoeconomists (Higgs 1972, 1975).

Illustrated in this work are the postcranial elements of the domestic horse (Equus caballus), domestic cow (Bos sp.), domestic sheep (Ovis aries), domestic pig (Sus scrofa), dog (Canis familiaris), cat (Felis catus L.), and the Norway rat (Rattus norvegicus). Although the taxonomic affinity of Ovis aries has been assigned to the illustrations of the sheep, it is recommended that the designation of Ovis/Capra be used. This is because of the close skeletal similarity between the domestic sheep, Ovis aries, and the domestic goat, Capra hirca (Boessneck 1970). It is also recommended that the taxonomic designations of Rattus sp. be used. While it is possible to differentiate the species of Rattus by their crania, it is impossible to do so postcranially (Brown and Twigg 1969:91).

The material illustrated here has incorporated as many neo-natal/juvenile skeletons and skeletal elements, either whole of fragmentary, as could be obtained (see appendix A). Numerous incomplete skeletons were utilized from the comparative collections at the Department of Anthropology, Bioarchaeology Laboratory at Hunter College at the City University of New York. Other whole skeletons were obtained from the collections in the Department of Mammalogy at the American Museum of Natural History or the author's collection.

Certain postcranial elements have been excluded from this manual. Ribs, coccygeal vertebrae, sternebrae, sesamoid bones, carpals and tarsals, are developing during this neo-natal to juvenile time phase. Most of these bone elements have ossified from their cartilaginous models and the osteological landmarks have just begun to take on their adult characters. Since the osteological landmarks needed to differentiate species have not as yet developed, the differentiation of species is difficult and highly inaccurate. An extensive comparative collection is mandatory for the identification of the adult forms of these bone elements. For this growth phase, it is doubtful if even such a comparative collection could be used to differentiate species.

Methodological Concerns Of Illustrating

The illustrative technique, the concern for an accurate reproduction of the illustration, the scale used, and the number of views per bone have varied from one manual to the next. Although this work is meant to complement Schmid's (1972), its format does not since the nature of the material presented requires a different format in order to be more comprehensive.

There are a number of illustrative methods that have been used for a manual of this nature. Each technique has its advantages and its compromises. It was decided that the illustrative method of pen & ink rendering was the best possible compromise for this manual. This will guarantee a compatibility of this work to Schmid's work. This also creates a compromise in that pen & ink rendering will not allow for the finer gradations of lights and darks that pencil or photography allow. However, the cost of duplicating a pen drawing is more economical and a photostated print of an ink drawing has the advantages of accurately reproducing well in published form.

The depiction of the scale of the bone element has varied from one manual to the next. Bone elements from different sized species have often been shown at the same scale (Olsen 1980, 1982). Another approach is to include a separate scale for each different size bone element (Brown and Gustafson 1979; Cornwall 1956; Schmid 1972). The direction taken by this manual is different from the previous manuals.

Two scales are used in connection with each illustration. An illustrator's scale is provided below a one-centimeter scale. The metric scale portrays the relationship a bone element has to an actual measurement. The Illustrator's scale provides information as to how the illustration was reduced or magnified. The advantage to this approach is that the reader is able to know how large or small an object is in relation to a unit of measurement. Also, a contour drawing of the bone element's actual size is provided in the bottom right hand corner of the illustration. This will allow for the comparison of the bone fragment to the actual size of the bone element.

Since this manual is concerned with young mammals, one other set of illustrations has been provided. These are the contour drawings found in chapter 4. Each plate is a comparison of one bone element's size in relation to the same bone of other species represented in this volume (i.e. the femora of each species). The bone elements are portrayed by a contour drawing of each element's natural size at a neo-natal or juvenile stage of development. The use of natural size is necessary in order to demonstrate this size relationship.

AGE DETERMINATIONS FROM BONE: A SURVEY OF METHODS

Accurate age determinations of faunal material is essential to the understanding of many aspects of prehistoric economy. Without reliable aging data, any assessment of hunting and/or husbandry strategies and the animal's economic significance is incomplete. A paucity of reliable aging information can preclude an understanding of many aspects of animal biology as well. Especially when the analyst must rely on scarce information about aging criteria in the skeleton. Unlike the mammalogist and zoologist, the faunal analyst can only rely on those methods that pertain to either whole or fragmentary bone. This problem is aggravted further by a lack of interdisciplinary cooperation. Olsen (1967:231) has drawn attention to the lack of cooperation between osteologists sharing information concerning bone identification and aging criteria with zooarchaeologists. Few mammalogists, or for that matter zoologists, are particularly interested in the problems faced by zooarchaeologists. The all too familiar attitude of "it's the archaeologists' problem" has not changed since Olsen wrote his commentary in 1967.

This chapter seeks to rectify one aspect of this problem providing technical aging data for those working with domestic mammals. It is hoped that these studies will provide the analyst with detailed information to better define age classes of domestic mammals which in turn can be applied to more complex questions pertaining to past economy. [2]

2. For a further background in studies concerning growth and development of bone, refer to Arey 1974; Lodge and Lamming 1968; Needham 1964; Simkiss 1975 and Vaughan 1980.

RELATIVE METHODS

Fusion of Epiphyses

Physiological events in an animal's life are often evident in its skeleton. Epiphyseal fusion is one such event commonly used for age determination. Growth in young mammals takes place in a cartilaginous region near the ends of long bones, vertebrae, sternebrae, and ribs. A pad of cartilage separates the metaphysis/diaphysis section from the epiphysis section of bone. Not all bones develop at the same rate, as growth can cease in different bones at different ages. It is possible to have a situation where the lower epiphysis fuses before the upper epiphysis in the same bone (Chaplin 1971 ; Morris 1972).

As a general rule, epiphyses subject to compressional forces fuse before those under traction (Morris 1972:83-85). This can be illustrated by the situation of an Equine radius fusing before the vertebrae. The epiphyses of the radius are under a weight bearing load or a compressional force; however, the body epiphyses of a vertebra are being pulled or under traction load. In a wide variety of mammalian species, the relative sequence of closure in various bones has essentially the same sequence as in Homo sapiens (Morris 1972:84). The sequence begins with the fusion of the distal humerus, followed by the fusion of the acetabulum, calcaneus, lesser trochanter, proximal ulna, proximal radius, femur head, distal fibula, proximal fibula, distal radius and ulna, proximal tibia, distal femur, proximal humerus, iliac crest, and vertebrae (Morris 1972:84; Cheverud 1981).

Epiphyseal fusion is of value to zooarchaeological studies because fusion sequences are indicators of relative age. Various scoring systems have been devised to indicate not only the stage of fusion ("solidly fused", "nearly fused", and so on), but a chronological age (Silver 1969; Chaplin 1971; Morris 1972; Wilson et al. 1982). An age profile can be constructed by plotting the percentage of individuals showing specified stages of epiphyseal fusion. If the sample is adequate, mortality curves can indicate different types of economic activities (cf. Legge 1981).

Suture Closure

Analogous to the fusion of the epiphyseal regions is the fusion of the bones of the skull. In mammalian species, this is a valuable method of age approximation. Fusion of bones in the skull can occur long after the permanent teeth erupt and the epiphyses have closed. This is especially true of domesticated mammals with long life spans (Chaplin 1971).

Problems and Limitations

Fusion data have been used for some time to age archaeological remains, but it was gradually realized that these methods have some serious drawbacks. For instance:

1. The timing of apparent bone fusion differs between gross anatomical specimens and radiological methods. In general, radiological estimates for domestic mammals tend to be younger than estimates derived from anatomical specimens (Chaplin 1971).

2. Methods of calculating the proportion of fused bone elements vs. unfused elements vary from one analyst to the next.

3. The nutritional status of the individual animal or a population of animals can delay or accelerate growth and cause variation from the standard fusion charts (Saxon and Higham 1968).

4. The closure times of epiphyseal fusion can vary from breed to breed and population to population and may have not been the same from the past to the present (Flannery 1969; Chaplin 1971; Watson 1978; Bull and Payne 1982).

5. Epiphyseal fusion is greatly affected by castration, the time of fusion apparently being delayed (Chaplin 1971; Noddle 1974; Watson 1978).

6. It has also been noticed that the younger, unfused elements are more vulernable to attrition of all kinds skewing the faunal sample in favor of the adult materials.

Furthermore, epiphyseal and suture closure times are often interpreted as points rather than ranges of time. When treated as a point, a given element in an individual is considered to fuse at the same time as the same bone element in another individual. Since bone fusion does not occur exactly at the same age in every individual, the use of an age range over which an element will fuse should be considered (Brickman 1984).

For example, consider a sample of proximal <u>Equus</u> femora, 20% of which are fused. If the proximal epiphysis is assumed to be fused by 3-3.5 years (Lesbre 1897; Brunni and Zimmerl 1951 ; Rooney 1963; Silver 1969; Schmid 1972), one would deduce that 20% of the horse femora represent animals that died at ages 3-3.5 yrs. or perhaps at ages over 3.5 yrs. old (Chaplin 1971; Watson 1978). However, the fused proximal femora sample indicate that 20% of the animals have died when over 3 yrs. old, but not necessarily over 3.5 yrs. The unfused portion of our imaginary sample indicate that 80% of the animals have died at ages under 3.5 yrs., but not necessarily under 3 yrs. Therefore, one is hard pressed to say that 20% of the horses represented in the sample died when they were over 3 yrs. old. What can be said, is at least 20% of the horse sample died at ages over 3 yrs. and 80% of the sample died under 3.5 yrs.. Alternatively, the whole sample could have come from animals that died between 3-3.5 yrs. The division of the sample into two age-classes on the basis of this fusion data provides little clear information about the age structure of the sample.

Watson (1978) used a similar sample of sheep distal metacarpals to demonstrate this same point. In his hypothetical sample, 40% of the metacarpals were fused. The conventional deduction is that 40% of the sheep died over 18-24 months (Schmid 1972; Silver 1969; Smith 1966), but as Watson suggests, these data do not necessarily indicate this situation. Instead, forty percent of the sheep died at ages over 18 months and 60% at ages under 24 months. Thus at one extreme, the sample could have come from animals killed between 18 to 24 months, at the other extreme, 40% died over 18 months and 60% under 24 months. Again, this provides little information about the population's age structure.

One way to minimize this uncertainty is to increase the number of age-stages (Watson 1978). The more bone elements used with known fusion dates, the more precise is the estimate of age distribution. For example, using a sample of Equine remains, Table 3-1 lists five bone elements of known fusion dates. All ages are based on Lesbre (1897), Tohra (1950), Brunni and Zimmerl (1951), Rooney (1963), Silver (1969), and Schmid (1972).

Table 2-1

Age in Months	Bone	Percentage of elements Fused	Unfused
9-15	Proximal Phalanx I	100%	0%
15-18	Distal Humerus	75%	25%
20-24	Distal Tibia	50%	50%
36	Calcaneus	5%	95%
42	Distal Radius	5%	95%

Based on Table 3-1, the following limits of the population dying under a certain age can be ascertained: 95-100% of the population died under 42 months, 50-95% under 36 months, 25-50% under 24 months, 25-50% under 20 months, 0-25% under 18 months, 0-25% under 15 months, and 0% under 9 months.

Similarly, the following limits can be deduced for the proportions of the population dying within given ranges: 0-5% died above 42 months, 0-5% died between 42-36 months, 0-45% died between 36-34 months (0-22.5% between 36-30 months and 0-22.5% between 30-24 months), 0-25% between 24-18 months, and 0-25% between 18-9 months (0-12.5% between 18-12 months and 0-12.5% between 12-9 months). Admittedly, once the proportion of individuals dying within any one of the age ranges is fixed, the possibilities for the other age ranges become restricted; however, variations in the age distribution can occur. Figure 2-1 demonstrates two different age distributions of which each is consistent with Table 2-1.

Watson (1978) demonstrates a similar example using a hypothetical sample of cattle. The sample consists of 4 elements, in which three age distributions are possible. The age distribution in Watson's example almost duplicate the results given above. By increasing the number of bone elements used, a more precise estimate of an age distribution can be made. Table 2-1 only uses a sample of 5 bone elements. If a larger sample of elements of known fusion dates were used, a more precise age distribution can be obtained.

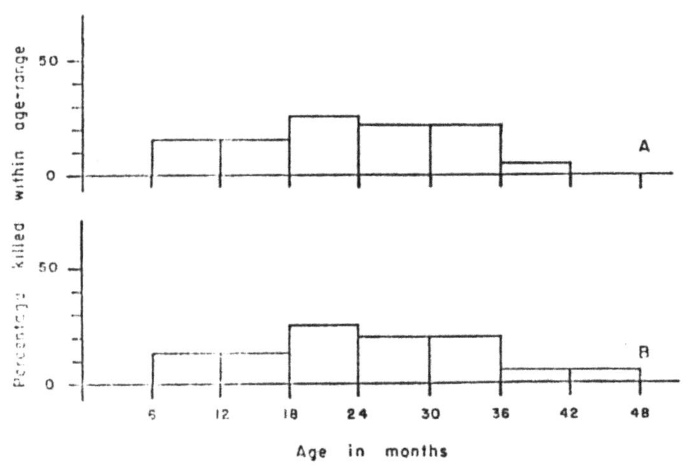

Figure 2-1, Age Distribution of Hypothetical Equus Sample.

Finally, the effects of sampling error on interpretation of fusion data must be considered. The serious problems of bone attrition, differential preservation, adequate sampling, and the interdependence of bone elements (Grayson 1977, 1978, 1984) must be resolved. The differential preservation and attrition of juvenile osteological material can vary. Unfused specimens are generally destroyed more rapidly than fused specimens (Noe-Nygaard 1977; Watson 1978; Gorden and Buikstra 1981). Sampling of faunal material are dependent on the recovery technique. For example, 1/2 inch wire mesh screen will not recover smaller animal species, such as Mus, nor the smaller bone elements of larger species (Jones 1982; Payne 1972, 1975; Struever 1968; Thomas 1969, 1972; Ziegler 1965). Fused specimens may be selected for the manufacture of bone tools and ornaments and thus may be missing from the faunal assemblage (Crabtree 1982a; Driver 1984; Payne 1972 a-c; Watson 1978).

Epiphyseal fusion as a method of age determination and demographic reconstruction is an important strategy in faunal studies, so long as potential skewing factors are considered. Epiphyses are often the most numerous age-related feature observable in an archaeofauna, far more common than complete tooth rows in most collections. The use of tooth eruption, wear, and crown height may be more reliable than epiphyseal fusion, but are not always possible.

Tooth Eruption, Wear and Crown Height Methods

The determination of age based on teeth is less problematic than the determination of age based on fusion data. Teeth, due to their durability, are usually abundant in faunal collections. With the exception of teeth from very young individuals, teeth from different age groups are equally durable (Klein et al. 1983). Destructive post-depositional processes are less likely to distort an age profile based on teeth, than one based on fusion data.

According to Klein (et al. 1983; also see Morris 1972; and Spinage 1973) there are three basic techniques to estimate age from teeth.

> "The first two methods are subjective comparisons of the states of eruption and wear in an archaeological specimen to eruption and wear rates in a series of specimens of known age. The third is the more objective measurement of a dimension that decreases with wear, the crown height of a tooth" (Klein et al. 1983:48).

Mammals generally possess a diphyodont dentition, where deciduous teeth are replaced by a permanent dentition. [3] The determination of the individual's age is based on a single visual inspection of the teeth to determine the developmental stage of the animal.

Dental eruption methods do not require special equipment or measurements. A specimen's dentition is examined and the state of eruption is scored for each individual tooth of both the deciduous or permanent sets. This data is compared to known dental eruption series and age estimates can then be made (Chaplin 1971:78-80; Morris 1972:78-82).

Tooth wear is one of the most widely utilized techniques used to determine age of mammalian species (Chaplin 1971; Wilson et al. 1982). Because of the constant attrition, after the permanent dentition is fully erupted due to use, tooth wear is usually assumed to be proportionate to an animal's age. The basis of these studies are animals of known age or wear sequence. These in turn are compared to archaeological specimens, and a wear or age sequence is constructed (Chaplin 1971:78-80; Morris 1972:78-82; Payne 1973).

The third method, age estimations based on crown heights of teeth, can approximate ages in high-crowned ungulate species. Crown height is, the distance between the occlusal surface and the line separating the enamel of the crown from the dentine of the root. Given that data points for an unworn crown and the age at which the crown is completely worn away are known, these data in turn can be used to calculate the age of the individual. Various adjustments for the sequence of deciduous to permanent teeth must of course be made (Levine 1982, 1983; Lewin 1983; Klein et al. 1983).

Drawbacks to Dental Estimates of Age

It has been gradually realized that dental methods also have drawbacks:

1. Poor nutritional health or disease can cause the eruption of teeth to be delayed (Grant 1975, 1978, 1982).

---

3. One such exception to this is the family of Trichechidae -- Manatees, whose teeth are constantly replaced as an adaptation to an abrasive diet (Lawlor 1979:262-263; Doming 1983).

2. Environmental stress, including life in marginal or unfavorable environments, can also delay the eruption of teeth (Grant 1978).

3. The actual hardness of the enamel and dentine can vary from individual to individual and species to species. Even the rate and nature of tooth attrition may vary from one individual to the next and can even vary regionally and between adjacent populations (Klevezal and Kleinberg 1969; Grant 1978; Stallibrass 1982).

4. The type of food and the degree of grit in the food will have an affect on dental attrition (Speth 1983:133-137).

5. Injury or disease to an area of the mouth will affect the other areas and cause differential eruption and wear.

6. Most of the dental eruption and wear data were developed for a particular population. Extrapolation of this data to other geographic populations or to different historical times can be a source of error.

7. The technique of comparing wear patterns to known series of patterns of wear involve subjective judgments on the part of the observer. Various researchers have found that, given the same set of data, different observers can produce different results (Keiss 1969; Low and Cowan 1963 ; Stallibrass 1982).

8. "The relationship between age and the height of the crown has not been established for many high-crown ungulate species. The basic problem is the scarcity of large samples of known-age dentitions (Klein et al. 1983:48)".

9. In the absence of a sound body of empirical information about crown heights, the determination of age requires assumptions on the nature and rate of crown attrition. Initially the assumption has been that the reduction in crown height is roughly constant through the life of a tooth. Therefore, the relationship between age and the decreasing crown height should be linear (Levine 1982; Klein et al. 1983). Also, the age at which a deciduous tooth is all but worn away is the same age at which its permanent counterpart will erupt. Although these assumptions do not account for difference due to the nutritional status of individuals or for differential attrition in individual teeth, they do

provide another measure of age by which to corroborate other estimates.

Tables to Follow

The tables in the following chapter are studies that are useful in defining dental wear for age parameters of Bos, Ovis/Capra, Sus, and Rattus (see tables 4-1 to 4-11). Included in these studies are Grant (1975, 1982). Bull and Payne (1982), and Amorosi (n.d.). Also included in the chapter are tables that pertain to epiphyseal and suture data, the development of horn cores, and tooth eruption charts (see tables 4-1 to 4-11).

The fusion and eruption data presented in chapter 4 are often inconsistent between authors. Epiphyseal fusion data estimates can vary as much as 2.5 years, for example, in the scapula of Equus. Similarly, as much as 19 months variation can be seen with Bos and Ovis. The overall variation apparent in these ages, however, is not very large. In some cases, different maturing breeds were used, as in Habermehl's (1975) comparison of early, middle, and late-maturing breeds of pigs. In others, there is a difference in the time periods in which species were studied. For example, the improved Landrace breed of pigs with which Reiland (1978) worked are from selected modern European commercial breeds. However, these data that Lesbre (1897) present on epiphyseal fusion was based on information available in the 19th century.

Brown (1902) seriously questioned whether improved husbandry practices did lead to significant differences in the rate of maturation. Brown's work was based on dental eruption for domestic animals in Britain during the late 19th and early 20th centuries. He focused his work on studies carried out between 1850 and 1902. Brown first accepted the possibility that the difference between early and late eruption dates were the consequence of improved animal rearing, but, by 1902, Brown concluded:

> "It did not first occur to the writer to question the truth of the assumption that the improvement of the various breeds of farm stock by selection and high feeding sufficiently accounted for the early appearance of the teeth; but now, after a further period of over half a century has elapsed, i.e. from 1850-1902, without any marked change having been noted in the development of the teeth - although breeders have been during the whole period steadily devoting their knowledge and energies to the cultivation of the various

> breeds of cattle, sheep and swine - it is impossible to avoid the conclusions that the original version of the development of the teeth was based on imperfect observation, or on the custom of one writer quoting in good faith from another" (Brown 1902:4).

Bull and Payne (1982) further feel that in the years between Brown's 1902 study and their (1982) no marked change has occurred. Both Brown, and Bull and Payne suggest that very early dates, mostly 19th century material, can be discarded as being unreliable.

Finally, some variation in tables 4-1 to 4-8 may be due to differences in the author's definition of the "moment" which an epiphysis or tooth will fuse or erupt (Bull and Payne 1982). This event is rarely defined by authors in their particular studies. Consider what is meant by the "moment of fusion" in an epiphysis. Such an event can be as early as when the epiphyses are no longer separate from the metaphysis/diaphysis region of the shaft or as late as almost complete fusion when the fusion line is still visible (Bull and Payne 1982). These two definitions can be as long as a year apart in some species. Lesbre (1897:8) provides a definition which refers to the final closure of the gap along the epiphyseal line:

> "On dit que noyaux d'ossification sont soudres lorsque tiut trace de cartilage de conjugaison a disparu, les vaisseaux de l'un s'etant reunis aux vaisseaux de l'autre. Avant de disparaitre, ledit cartilage, reduit a l'epaisseur de 1 a 2 millimeters, passe par une phase de calcification, et, a ce moment, les noyaux sout deja solidement soudes bien que non encore synostoses" (Lesbre 1897:8).

However, most faunal reports from archaeological sites have tended to consider fusion as the state in which epiphyses are no longer separate from the metaphysis/diaphysis. It is recommended that original studies be examined for the author's criteria if the variation in the timing of fusion or eruption dates is critical for the analyst's particular study.

INCREMENTAL METHODS

There is a growing literature concerning the use of counts of annual layers in tooth cement, dentine, and in the periostal region of bone for the assessment of age in animals (Andrews 1973; Bourque et al. 1978; Coy et al. 1982; Erickson and Seliger 1969; Evans n.d.; Gilbert 1966; Gorden 1982; Kay 1974; Klevezal and Kleinberg 1969; Langenau 1972; Low and Cowan

1963; Madsen 1967; Morris 1970; Ranson 1966; Saxon and Higham 1968; Scheffer and Peterson 1967; Sergeant and Pimlott 1959; Turner 1977). The literature, mostly American and Russian, is vast and tends to be redundant. This section discussess such methods, with special reference to problems pertaining to domestic species. I have attempted to incorporate some of the important Russian sources, but because of the language barrier, many Soviet studies are inaccessible. However, extensive bibliographies of Soviet studies can be found in Klevezal and Kleinberg (1969), Klevezal (1970), Morris (1972), Casteel (1976b), and Stallibrass (1982).

Historical Background

By the 1920s, it was generally known by biologists that there are more or less regular layers or lamina that relate to the growth of hard tissues in mammals. Klevezal (1970:2) describes one of the earliest incremental studies (Kler 1923), as a reliable document recording incremental growth in mammals. According to Klevezal, Kler considered it possible to analyze periodicity in the growth of mammals from periostal lamination of bone (Kler 1923, 1927). Kler thought these growth layers to be homologous to the appositional lamination of bones in fish. This Kler suggested could be used in age determination and growth analysis of mammals. During the 1930s and 40s, subsequent work established that growth layers were deposited in regular, seasonal cycles, and that lamina could be counted to give an absolute age of an individual (Laws 1952).

In the 1950s, mammalogists and wildlife management specialists shifted their emphasis towards studies of population age structure and aging individual animals. It was thought that monitoring the age structure of populations of various species will better manage natural resources. The most accessible source of such management data were sequences of tooth eruption and wear, established mostly for ungulate and other high quality fur bearing mammals. It was gradually realized that these methods have some serious drawbacks (see relative methods section of this same chapter). In particular, rates of tooth wear vary enough from individual to individual and population to population so that absolute ages cannot be assigned with confidence.

By the late 1960s and early 1970s there was a general dissatisfaction among biologists with tooth wear analyses. Most of the criticism was directed towards often questionable results of the comparison of field to known anatomical specimens. Age changes were not consistent and, even given the same set of data, different observers could produce different results.

Emphasis again shifted back toward incremental growth layers in teeth and bone that could provide absolute ages. Nearly every researcher who scrutinized incremental techniques found them to be highly accurate and relatively simple in the requirements of skill and equipment. These techniques further provided the researcher with the means to monitor age structures in game park animal populations through mortality profiles. Not only could the age of the animal be determined, but since layers are deposited seasonally, the outermost layers could indicate the season of death.

Annuli or annual incremental lines are annually deposited layers of secondary dentine or cementum in teeth. These growth rings are clearly visible under a microscope at low powers of magnification, usually between 10x-100x (Gorden 1982; Stallibrass 1982). The lamina are usually referred to as "annuli" in the literature, because the cross-section appears to form concentric rings. The underlying reason for the observed periodicity of the layering still remains unknown. However, the cyclic appearance of annuli might be associated with seasonality in general growth (Klevezal and Kleinberg 1969; Klevezal 1970; Stallibrass 1982).

Tooth Cement and Dentine

The same fundamental rules apply to tooth cementum and dentine, although they are distinct and dissimilar tissues (Sicker and Bhasker 1972). This discussion will normally consider both as a single phenomenon, but will separately consider them where appropriate.

Morris (1972:86-94) described both the dentine and cementum as being characterized by appositional growth, where newer material is formed and continually deposited around pre-existing material. Both the cementum and dentine are very stable structures, rarely subjected to normal remodeling and resorption, with exception of localized areas of damage (Stallibrass 1982). Cementum is deposited around the outside of the tooth root, providing a better anchorage for the tooth. While cementum is being deposited, dentine is deposited around the walls of the pulp cavity by odontoblasts. Environmental fluctuations can affect the rate of formation of the cementum and dentine (discussed later). Morris (1970:87) attributes environmental influences as a cause of the layering, rather than in the formation of a homogeneous structure.

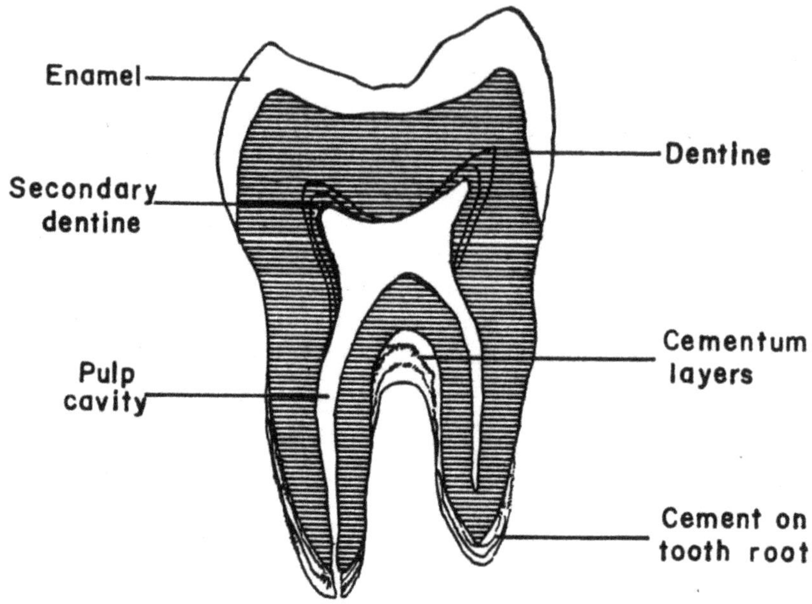

Figure 2-2, Cross section through a molar showing major structures and incremental lines in the cementum and secondary dentine (redrawn after Morris 1972:87).

The layering found in teeth has loosely been referred to as growth lines. However, these lines do not represent growth, but cessation of growth. Each annual layer consists of two zones, a broad and a narrow band, which will react differently to the intensity of staining reactions and different optical densities (Morris 1972:89; Stallibrass 1982:111-112). These differences are the result of periodic changes in the rate and nature of calcification of the growing dental tissues. These changes have been attributed to differences in seasonal feeding, hormonal activity, and vitamin D intake (Morris 1972).

There has been some confusion in the literature regarding which of the zones corresponds to which seasonal event, summer or winter (Klevezal and Kleinberg 1969; Morris 1972; Stallibrass 1982). The summer or winter bands can be described as translucent or opaque, or light or dark, depending on the methods of lighting and reaction to staining.

The summer band is characterized as being broad, due to the large number of active cementocyte cells. When this band is deposited, the presence of a large number of cells will reduce the content of calcium per unit volume (Sicker and Bhasher 1972; Stallibrass 1982:111-112). The winter band, in contrast, is narrow and primarily consists of calcium salts. It is unknown if deposition of minerals continues at a steady rate throughout the year or whether deposition continues at varying rates, sometimes stopping altogether. The resolution of this problem requires the examination of teeth from species taken throughout their yearly cycle and not killed during the legal hunting season. Without this precise interpretation, speculative assessments of age or season of death can only be general (Morris 1972; Stallibrass 1982:111-112).

Climatic Effects

It has been noticed by some researchers (Mayer and Bernick 1963; Morris 1972) that the contrasting clarity between bands will depend upon the degree to which the individual was affected by environmental factors. Seasonal growth differences are more likely to be pronounced in hibernating animals, where there can be an almost complete cessation of activity within the body during part of the year (Mayer and Bernick 1963). In non-hibernators, growth rates may be temporarily diminished by other seasonal factors, for example, a poor spring pasturage for grazing animals.

Reviewing the Russian literature, Klevezal and Kleinberg (1969) determined that the contrast between summer and winter bands are most distinctive in species that inhabit regions with marked climatic extremes. Species that live in temperate or tropical regions, where seasons are not strongly contrasted, have poorly defined annual bands. Klevezal and Kleinberg (1969) concluded that the degree of contrast between summer and winter bands is influenced by external factors such as seasonal variability.

The works of van Nostrand and Stephenson (1964) and Lockard (1972) on beavers and white-tailed deer, respectively, support this connection. These researchers found that the clarity of cementum layers can be positively correlated with the severity of seasonal changes. However, there is one problem as noticed by Lowe (1967) and Stallibrass (1982), it should be expected that the teeth from animals living in oceanic climates, such as coastal areas or islands, would have poorly defined summer and winter bands. This is due to the lack of severe seasonal contrasts in oceanic climates. But this does not seem to be the case. However, Lowe (1967), working with red deer from the Scottish island of Rhum, could not correlate the bands, summer and winter, with known ages of the red deer. Lowe suggests that the banding does not represent regular seasonal cycles. Rather, when seasonal variations are slight, the influence of other factors such as local variation in nutrition might be relatively increased and able to "overprint" any climatic-induced pattern.

Nutritional Effects

The causal relationship that diet may play in affecting the layering of cementum and dentine is difficult to separate from environmental factors. One is hard pressed to say if environmental factors (variation in temperature, precipitation, etc.) are the direct cause of fluctuations in deposition, or if deposition can be influenced by the availability and types of food. There have been very few studies in which a single causal agent has been examined.

Such studies have been carried out with the use of both domestic and non-domestic animals. Mitchell (1967), working with Scottish red deer, compared those had been fed extra fodder in winter to those that had been left to fend for themselves. Mitchell concluded that there is a higher proportion of the fed animals showing a poorly defined cementum layer than those that were not. Clearly, Mitchell's classification of a tooth section as "difficult" or "less clear" is subjective. However, these results do suggest that buffering by way of diet from periods of reduction in quantity and/or quality can obscure seasonal banding in cementum deposition.

Grue and Jensen (1976) and Grue (1976) produced a similar study on dogs and foxes that had been fed but not housed, fed and housed, or not fed and not housed. They concluded that the clarity of cement layers is influenced by both environmental and nutritional variations, but could not differentiate between the causal agents.

Additional studies involving comparisons of cementum and dentine layers from animals that are left out all year round, as opposed to those who are stalled, and from animals that are both fed and not fed supplemental foods are clearly desirable. Comparison of free-ranging versus stalled cattle, hill versus lowland sheep, free-ranging pigs versus penned pigs, could demonstrate variation in the clarity of layers according to different degrees of exposure to environment and according to different nutritional resources.

Another problem which must be further researched is the possible effect hormonal systems have upon the deposition of lamina in cementum and dentine. It is not well understood how an animals's metabolism is effected by changes in environmental factors such as temperature. Saxon and Higham (1968) mention that ewes from their New Zealand study show clear seasonal banding; the animals reduced their winter intake despite the provisioning of fodder at summer levels.

Another point of confusion in the literature is the direct cause of the winter band, especially as it relates to identifying the season of death of the individual. There is a general consensus (Klevezal and Kleinberg 1969; Klevezal 1970; Morris 1972; Stallibrass 1982) suggesting that the cause can be related to a reduction in the metabolism (see the comments of Low and Cowan 1963), a cessation of body growth, and the intake and/or the quality of food in winter (see Mitchell 1967; Grue and Jensen 1973; and Grue 1976, referred to above). Mitchell's (1967) work points to animals that died of starvation, having winter bands at the perimeter of the tooth sections. Though Mitchell does not mention the actual season of death of these animals, further investigations might be carried out to identify "starvation" or winter bands. Such results could have relevance for the zooarchaeologist trying to determine the cause of death of an individual animal.

Reproductive Cycles

Finally, another hormonal influence upon incremental growth with age may be the reproductive cycle of an animal species. Most notably in ungulate species, the male will go through periods of reduced feeding during the mating or "rut" season. During this phase, body growth will cease and the deposition of cementum and dentine will do likewise, forming narrow banded "rut" lines. The clarity between the signature of winter seasons and those of minor stress lines formed during a mating season can be a problem.

Mitchell's work (1967) with Scottish red deer showed that incremental lines associated with the rut in males occurred only in sexually mature individuals. However, only 25% of the males that Mitchell examined showed rut lines and their presence or absence could not be correlated to fertility, body size, or antler quality. Saxon and Higham (1968) demonstrated a similar lack of correlation in female sheep and known periods of pregnancy, lactation, and breeding. Yet, Mitchell's (1967) work did suggest that the clarity of cement layers tends to be less distinctive in females than in males. Castrates serve to make this matter more complex, since cement layers can be poorly defined and their rate of deposition can lag behind that of non-castrates (Aitken 1975). Further research on castrates and the sexual difference of males and females by the presence or absence of reproduction related lines needs to be further investigated.

Bone

It may not always be necessary to section teeth, since the destruction of the teeth for incremental methods can lessen the taxonomic value of the specimen. The orientation and thickness of periosteal bone layers can vary among species and also between different bones of the same species. Klevezal and Kleinberg (1969) summarize research on the peripheral region of long bones, which demonstrated that in some animals, the appositional bone forms a continuous sheath around the bone shaft, but in other bones or other species, layering occurs in discrete units. Morris (1972:92), in summarizing the work of Klebanova and Klevezal (1966), demonstrated the same: Klebanova and Klevezal found that the number of incremental layers in one bone was the same as in another belonging to the same animal. It was also noted that the number of incremental lines was in agreement with the known age of the animals investigated.

Theoretically, any bone structure in which periostal bone is overlaid without further resorptive remodeling, should show incremental lines. The function for the additional bone is to provide an extra means of support and strength for regions under stress (Morris 1972:91-94). For example, in mammals the mandible is subjected to considerable stress (torque), so that compensatory development of peripheral bone is sufficiently great as to be readily discernible. The mandible, therefore, provides an ideal structure in which to examine appositional growth layers.

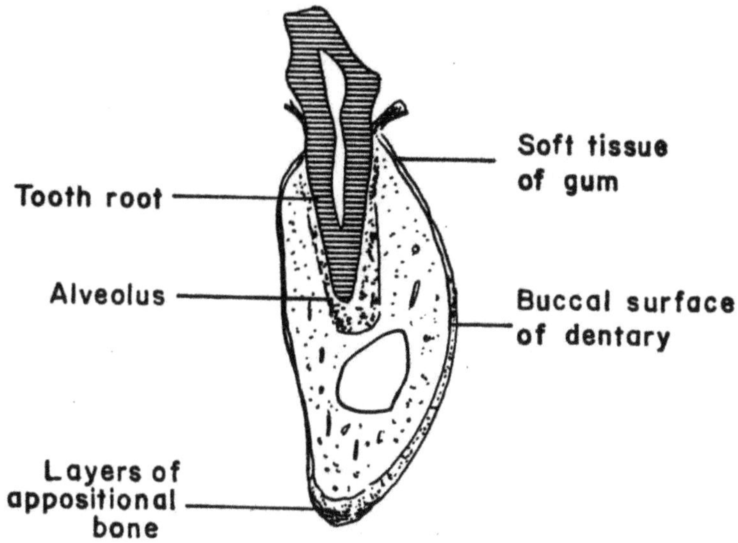

Figure 2-3, Cross section of a mandible showing tooth in situ and position of appositional bone layers useful for age determination (redrawn after Morris 1970).

Periosteal layering in long bones thus are of potential use to archaeological study. Incremental studies of bone have the advantage over the use of relative aging methods in that an absolute age can be given. There is continuous data provided unlike that of epiphyseal fusion which is a two state phenomena, either fused or unfused. Finally, bone fragments can be used to provide aging data, even in the absence of dental material and fragments that retain their epiphyses.

Horns

The horns of mammals are continuously growing structures (Caughley 1965; Morris 1972:94; Grigson 1982:25-35). As the horn grows, new material is added at the base of the horn. Seasonal fluctuations, diet, and reproductive cycles (discussed earlier) will influence the constancy of growth, the results of which are recorded as growth lines in the horn's keratin structure. Keratin is a hard, dense material that surrounds the core of the horn. Discontinous lines of growth are visible as thin regular lines interrupting a homogeneous area. Based on the recording of these lines, the age of the individual can be determined.

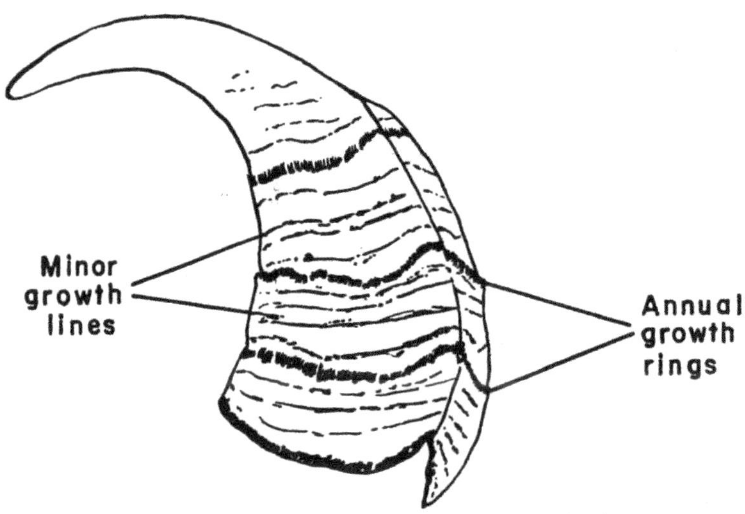

Figure 2-4, The keratin structure of a Thar, a wild asiatic ungulate, showing minor and major growth lines (redrawn after Caughley 1965; Morris 1972).

Grigson (1982:37-54), working with British post-medieval sites, devised a system for the aging of cattle by the surface texture and appearance of the horn core. Her system is based on the density of nutrient foramina found on the surface of the horn core. Horn texture is divided into six classes, the density of the foramina becoming less with age. Grigson (1982:39) also defines six classes of horn cores that demonstrate the size and state of development of the horn core. By scoring the horn core with the appropriate class, the horn core can be assigned to a relative age class (see tables 4-2F).

Grigson's study is included under incremental methods because this system is more of an attempt toward an incremental count than a relative aging technique. However, this system does fall short in that only relative ages can be assigned.

Selection of Specimens, Teeth

In many studies, the mandibular cheek teeth are the most preferred, because of their simple root structure. Tooth types that have a complete root structure, such as the upper molars, are avoided. This is because multiple root can cause an interlacing in the layering of cementum or dentine towards the pulp cavity (Gorden 1982). Incisors are often excluded, as they are rarely found in an archaeological context associated with the skull. Although canines, especially those of carnivores, can provide incremental data (Sauer 1966; Adams and Watkins 1967), they are rarely used for archaeological research. This is because the single root of the canine can easily come loose from the jaw after death and is often not recovered (Gorden 1982; Grant 1982). In addition, both the incisors and canines have frequently been used for ornamentation. Because of their decorative nature and value, these teeth are often transported from the source to different sites and held there for long periods of time (Ritchie 1980; Gorden 1982).

Selection of Specimens, Bone

The Utilization of periosteal bone for incremental work requires the careful selection of areas judged to contain complete sets of layers. Most of the research so far has concentrated on mandibles (Klevezal and Kleinberg 1969; Stallibrass 1982) because the mandible is subjected to considerable stress and the compensatory development of peripheral bone is sufficiently great to be readily discernible.

Drawbacks to Incremental Methods

In the previous sections, some of the drawbacks to incremental work were briefly mentioned. However, there are a number of other problems faced by researchers doing incremental studies. While some difficulties are not that severe and can easily be resolved, others require further investigation. Those difficulties that can easily be resolved are:

1. Problems are presented by differential preservation of bone and by the biases of sampling and recovery techniques and strategies. If these biases are not accounted for, age profiles provided by incremental methods will misrepresent the actual population present in the faunal sample.

2. Most incremental studies are potentially destructive to the specimen. The specimen is either sectioned or decalcifed, making it useless for gross morphological study (Lewin 1983:447).

3. The time and cost of doing an incremental study can become one of the more serious drawback. Few archaeological projects will allow an analyst the luxury of doing such a refined study. This is especially true when there are readily available relative aging techniques, that if properly used, can almost duplicate the information that incremental methods can without adding extra cost to an already limited budget.

The solutions to these problems are not difficult if resources and time are available. The first issue requires careful planning of research and excavation strategies, while the second can be resolved by making a reproduction (cast) of the taxonomic specimen. The last difficulty requires a firm commitment of time and funding.

While these difficulties can be resolved on the basis of current knowledge, others are more problematic and require further investigation:

1. Distinguishing the exact cause of growth layers in both teeth and bone is extremely difficult. The layering can be the result of a number of related factors, such as climate or even nutritional stress. Are the growth layers are representative of age or of the body's reaction to environmental and other stress factors? The interpretation of incremental structures needs to be based upon an understanding of the underlying causes of growth, preferably aided by known series of aged material (Klevezal and Kleinberg 1969; Morris 1972; Stallibrass 1982).

2. A source for frequent confusion is the occurrence of narrow, often multiple growth lines, usually in between the more distinct annual ones. Presumably, these growth lines are caused by minor short term disruptions in the animal's growth. Mitchell's (1967) work suggested that rutting behavior may be a possible cause. Scheffer and Peterson (1967) attribute multiple minor growth rings to the lactation patterns of Fur seal pups and concluded that these were correlated with regular periods of suckling from their mothers.

3. In certain carnivore species, the first growth line may not be formed until the second year of life. Linhart and Knowlton (1967), working with coyotes, noted that molars are especially prone to this, since molars do not erupt until a year or two after birth.

4. Dentine is deposited within the tooth, gradually filling in the pulp cavity. In many species, the pulp cavity can become occluded after a few years. Space inside the pulp cavity becomes an important factor when considering whether growth lines in dentine are more or less useful for age determination. In most mammalian species, dentine deposition is sharply weakened after attaining sexual maturity, although growth may continue in canine teeth for a long period of time (Morris 1972).

5. Periosteal bone is not as stable a structure as cementum and dentine. It is often subjected to resorption, especially in older individuals. The boundaries of layers can become indistinct, or even totally obliterated. There are other problems, the new material is not deposited evenly over the entire surface of any one bone. Patterns of deposition can vary between different bones of the same individual and between the same species (Morris 1972; Stallibrass 1982).

6. Banfield (1969) suggested that annual rings on antler pedicles of caribou might be suitable for age determination. However, as Morris (1972) pointed out, growth lines were indistinct because of the spongy texture of the bone in which they were formed.

7. What are the possible effects a non-annual breeding cycle may have on the formation of growth layers?

Most of these issues can be resolved through increased testing against known-age series and refinement of technique. Most authors who tested the accuracy of their own data by using material from individuals of known age, found a high degree of correlation between age in years and the number of growth layers. The unageable sections tended to be either badly prepared or made from disturbed layers. The resectioning of the same tooth or bone resulted in a more accurate age estimation date. Inaccurately aged sections were not very far from the expected mode, most cases showing an error of about a year (Stallibrass 1982). It seemed that the main cause of inaccuracy was due to the inexperience of the observer with incremental techniques.

While most of the published studies using incremental structures have been concerned with wild mammals, much of this discussion has been of studies that are relevant to that of domestic species. The focus has been that of looking at environmental, nutritional, and reproductive factors that can influence the formation of incremental structures.

At this time, the most useful function incremental studies can serve is to give more accurate estimates of season of death and age of the individual, especially if used in conjunction with the relative aging methods, such as bone fusion data or dental eruption or wear. Although there is still much confusion regarding the precise factors that influence the deposition of incremental structures, continued use of these methods will undoubtedly refine the technique and improve our knowledge of aging and the factors that influence it.

Comparison of Relative and Incremental Methods

There have been relatively few attempts to compare the results of relative aging methods to those of absolute techniques applied to the same archaeological material (Morris 1972). Even if a number of pilot studies were available, would such an exercise be fruitful? The biologist, who usually works with individuals of known ages, can use incremental methods to calibrate the relative aging techniques. But the zooarchaeologist has no such controlled sample. It is true that the use of incremental methods can be used to calibrate relative aging methods, but what is the rate of accuracy one could expect without a control population for contrast?

The question is not which techniques are better, but rather how these techniques can be used to complement one another. Especially, how can one use these techniques to gain the best possible aging criteria for a particular study of a particular archaeofauna. Ideally, the use of incremental work done on teeth can be used to calibrate such relative techniques as dental eruption and wear, the major requirement being that both the teeth and the epiphyseal regions come from the same animal. However, even relative methods can be used to calibrate one another, for example, the use of tooth eruption to correlate the rate of tooth wear. Only with such a calibration ooperation of aging techniques will age profiles become more meaningful.

British workers, in particular (Wilson et al. 1982), have been concentrating on aging and sexing studies of domestic animals. However, these works, for the most part, are only related to the horse, cow, sheep/goat and pig. A great deal of work is still needed, both for domestic and wild species.

AGE DETERMINATIONS BASED ON BONE, TEETH AND HORN: TABLES OF AGING CRITERIA

The following tables (3-1 to 3-11) pertain to age determination of osteological material. Complied from various sources, the best known and most widely used works on this subject are Lesbre (1897), Brunni and Zimmerl (1951), Habermehl (1961), Sisson and Grossman (1966), and Silver (1969). The zooarchaeological textbooks of Cornwall (1956), Ryder (1969 , and Chaplin (1971) contain some information on aging, but the data they is present are not useful for aging osteological material. Grigson (1974, 1975, 1976, 1978), Bullock and Rackham (1982), and Andrews (1982) are not listed here, since the data they contain cannot easily be incorporated without the loss of vital information. It is highly recommended that the reader review these sources for useful data on aging Bos and Capra. Work by Kroon (1929), Ellenberger and Baum (see Zietzschmann 1943), Zietschmann and Krolling (1955), Rudge (1972), and Benzie and Gill (1974) is not included here, since the author could not obtain copies of these studies. The data provided by Schmid (1972) are based on data complied after Cornwall (1956), Habermehl (1961), and Wolf-Heidegger (1961). The information presented in tables 3-9 to 3-11 pertain to methods for recording dental wear in Bos, Ovis/Capra, Sus and Rattus. These particular studies are efficiently simple, allowing for a quick method to record dental wear and some means to construct a relative age curve. Other works, Meadow (1978) and Deniz and Payne (1982), were found to be cumbersome to use, but nevertheless should be considered.

There are numerous inconsistencies in the data pertaining to fusion and eruption. This is due to the varying definitions of the "moment" of bone fusion or tooth eruption. Tooth eruption dates are given by Sisson and Grossman (1966) for Equus and Bos, but seem early compared with other data. Silver's (1969) bone fusion data for Bos and Ovis also seem early as compared to other sources. Silver's (1969) "18th century" dental eruption timing for Sus seems too late when compared with other sources. Bull and Payne (1982:55) suggest that Silver's data is improbable. They base their arguments on other archaeological specimens from the 18th century, which are consistent with modern sources. This would suggest that the 18th century data of Silver (1969) is unreliable. The reader is strongly urged to consult the original source for each author's criteria, if the variation in this data is crucial. For further discussion of inconsistencies found in these data, see discussion in chapter 2.

Equus Fusion and Dental Data

Table 3-1A, Cranial Fusion Ages in Equus. (Silver 1969)

Bone	Centre of Ossification	Fusion Age	Notes
Occipital	Supraoccipital 2 x exoccipital	Supra. with exocc. at 18 mos.	4 bones at birth, 1 in adult
	Basioccipital	Basi. with exocc. at 3-4 mos.	
Sphenoid	5 centres, one in each of 4 wings, 1 in body	5 yrs. (60 mos.) with occipital	Perpendicular & cribriform plates are cartilaginous at birth
Ethmoid	2 in each lateral mass, 1 in the perpendicular plate, a total of 5 centres		
Interparietal	2 main centres	W/ parietal at 3-4 yrs. (36-48 mos.)	
		W/ supraoccipital at approximately 5 yrs. (60 mos.)	
Parietal	Each has 1 centre	Parietals fuse across midline at about 4 yrs. (48 mos.)	
		Parieto-occiptal fusion at 5 yrs. (60 mos.)	
		Parieto-squamous fusion 12-15 yrs.	
Frontal	Each has a single centre	W/ each other or parietals only in old age age	

Table 3-1A, Cranial Fusion Ages in <u>Equus</u>.
(concluded)

Bone	Centre of Ossification	Fusion Age	Notes
Squamosal (squamous temporal)	Each has a single centre	W/ parietal at 12-15 yrs.	
Periotic (petrosal or petrous temporal)	At least 2 centres	At or soon after birth	Rarely fuses w/ adjoining bones
Pterygoid, Maxilla, Palatine, Zygomatic	One each (some may small secondary		May unite w/ adjoining bones in old age
Premaxilla	One centre	Fusion across the midline at about 4 yrs. (48 mos.)	
Nasal	One centre	Midline suture is unfused in old age	
Mandible	2 centres	The centres fuse before birth, the two halves at 2-4 mos.	

Table 3-1B, Deciduous Tooth Eruption in <u>Equus</u>.

Tooth	Schmid (1972)	Silver (1969)	Sisson & Grossman (1966)
Cental Incisor	Present at birth	Present at birth	Birth or first week
Lateral Incisor	Present at birth	3-4 weeks	4-6 weeks
Corner Incisor	Present at birtn	5-9 mos.	6-9 mos.
Canine	-	Rarely emerges from jaw	-
Premolar 1	-	Inconstant 6 mos.	Birth to first two wks.
Premolar 2	Present at birth	Present at birth	Birth to first two wks.
Premolar 3	Present at birth	Present at birth	Birth to first two wks.
Premolar 4	Present at birth	Present at birth	-

Table 3-1C, Permanent Tooth Eruption in *Equus*.

Tooth	Schmid (1972)	Silver (1969)	Sisson & Grossmam (1966)
Central Incisor	2.5 yrs. (30 mos.)	2.5- 3 yrs. (30-36 mos.)	2.5 yrs. (30 mos.)
Lateral Incisor	3.5 yrs. (42 mos.)	3.5-4 yrs. (42-48 mos.)	3.5 yrs. (42 mos.)
Corner Incisor	4.5 yrs. (54 mos.)	4.5-5 yrs. (54-60 mos.)	4-5 yrs. (54-60 mos.)
Canine	4.5 yrs. (54 mos.)	4-5 yrs. (54-60 mos.)	4-5 yrs. (54-60 mos.)
Premolar 1	vestigal	Inconstant 2.5 yrs. (30 mos.)	5-6 mos.
Premolar 2	2.5 yrs. (30 mos.)	2.5 yrs. (30 mos.)	2.5 yrs. (30 mos.)
Premolar 3	2.75 yrs. (33 mos.)	2.5 yrs. (30 mos.)	3 yrs. (36 mos.)
Premolar 4	3 yrs. (36 mos.)	3.5 yrs. (42 mos.)	4 yrs. (48 mos.)
Molar 1	1 yr. (12 mos.)	Present by 12 mos., but have a wide variation	9-12 mos.
Molar 2	2 yrs. (24 mos.)	2-2.5 yrs. (12-30 mos.)	2 yrs. (24 mos.)
Molar 3	3.5 yrs. (42 mos.)	3.5-4.5 yrs. (42-54 mos.)	3.5-4 yrs. (42-48 mos.)

Table 3-1D, Approximate Times of Appearance of Ossification Centres of the Thoracic Appendages of *Equus*.

Bone	Sisson & Grossman (1966) Time of Appearance (Day's Gestation)
Scapula	
proximal end -	9-12 mos. postpartum
body -	60-70 days
supraglenoid tubercle and coracoid process -	140-170 days
cranial part of glenoid cavity -	325 days to after birth
Humerus	
proximal end -	290-310 days
greater tubercle -	315-335 days
diaphysis -	60-70 days
distal epiphysis -	290-315 days
medial epicondyle -	315-335 days
Radius	
proximal end -	315-335 days
diaphysis -	60-70 days
distal epiphysis -	265-295 days
Ulna	
proximal epiphysis -	330 days - after birth
diaphysis -	65-75 days
styloid process (fuses to distal epiphysis of radius) -	330 days - after birth
accessory carpal bone -	250-290 days
Metacarpal III	
proximal epiphysis -	290-330 days
diaphysis -	70-80 days
distal epiphysis -	after birth
Metacarpal II & IV	
proximal epiphysis -	late in gestation
diaphysis -	90-110 days
distal epiphysis -	265-310 days

Table 3-1D, Approximate Times of Appearance of Ossification Centres of the Thoracic Appendages of *Equus*.
(concluded)

Bone	Sisson & Grossman (1966) Time of Appearance (Day's gestation)
Proximal Sesamoids	290-330 days
Proximal Phalanx (Phalanx I)	
proximal epiphysis -	290-320 days
diaphysis -	75-85 days
distal epiphysis -	265-310 days
Middle Phalanx (Phalanx II)	
proximal epiphysis -	290-335 days
diaphysis -	120-155 days
distal epiphysis -	310-330 days
Distal Seasmoid -	325 days - birth or soon after
Distal Phalanx (Phalanx III)	
proximal cap -	late in gestation
body -	85-135 days

Table 3-1E, Approximate Times of Appearance of Ossification Centres of the Pelvic Appendages of *Equus*.

Bone	Sisson & Grossman (1966) Time of Appearance (Day's Gestation)
Femur	
proximal epiphysis -	230-300 days
greater trochanter -	230-300 days
diaphysis -	60-70 days
distal epiphysis -	220-245 days
Patella -	325 days - birth
Tibia	
proximal epiphysis -	265-300 days
tibial tuberosity -	290-320 days
diaphysis -	60-70 days
distal epiphysis -	280-300 days
Fibula	
proximal epiphysis -	Soon after birth
diaphysis -	After birth
distal epiphysis (lateral malleolus of the Tibia) -	325 days - after birth
Calcaneus -	130-155 days
Talus (Astragalus) -	250-275 days
Other Tarsal Bones -	290-335 days

Table 3-1F, Epiphyseal Closure Times for the Thoracic Appendages of *Equus*.

Scapula

	Biciptal Tuberosity	Tuber Spine	Distal
Schmid (1972)	-	-	-
Silver (1969)	20 mos.	3 yrs. (36 mos.)	-
Rooney (1963)	-	-	10-12 mos.
Brunni & Zimmerl (1951)	-	-	9-12 mos.
Tohra (1950)	-	-	-
Lesbre (1897)	-	-	10-12 mos.

Humerus

	Proximal	Distal
Schmid (1972)	3.5 yrs. (42 mos.)	1.25-1.5 yrs. (15-18 mos.)
Silver (1969)	3-3.5 yrs. (36-42 mos.)	15-18 mos.
Rooney (1963)	3.5 yrs. (42 mos.)	15-18 mos.
Brunni & Zimmerl (1951)	3.5 yrs. (42 mos.)	15-18 mos.
Tohra (1950)	-	-
Lesbre (1897)	3.5 yrs. (42 mos.)	3.5 yrs. (42 mos.)

Table 3-1F, Epiphyseal Closure Times for the Thoracic Appendages of *Equus*.

(continued)

Radius

	Proximal	Distal
Schmid (1972)	1.5-1.75 yrs. (18-21 mos.)	3.5 yrs. (42 mos.)
Silver (1969)	15-18 mos.	3.5 yrs. (42 mos.)
Rooney (1963)	15-18 mos.	3.5 yrs. (42 mos.)
Brunni & Zimmerl (1951)	15-18 mos.	3.5 yrs. (42 mos.)
Tohra (1950)	-	-
Lesbre (1897)	15-18 mos.	3.5 yrs. (42 mos.)

Ulna

	Proximal	Distal
Schmid (1972)	3.5 yrs. (42 mos.)	-
Silver (1969)	3.5 yrs. (42 mos.)	Before birth to 2 mos.
Rooney (1963)	3.5 yrs. (42 mos.)	-
Brunni & Zimmerl (1951)	3.5 yrs. (42 mos.)	2-3 mos.
Tohra (1950)	-	-
Lesbre (1897)	3.5 yrs. (42 mos.)	2-3 mos.

Table 3-1F, Epiphyseal Closure Times for the Thoracic Appendages of *Equus*.
(continued)

Metacarpus III

	Proximal	Distal
Schmid (1972)	-	1-1.25 yrs. (12-15 mos.)
Silver (1969)	Before birth	12-15 mos.
Rooney (1963)	-	10-12 mos.
Brunni & Zimmerl (1951)	Before birth	15 mos.
Tohra (1950)	-	17-18 mos.
Lesbre (1897)	-	15 mos.

Phalanx I

	Proximal	Distal
Schmid (1972)	1-1.25 yrs. (12-15 mos.)	-
Silver (1969)	13-15 mos.	Before birth
Rooney (1963)	12-15 mos.	1 week postpartum
Brunni & Zimmerl (1951)	12-15 mos.	Before birth
Tohra (1950)	9-10 mos.	At birth
Lesbre (1897)	12-15 mos.	Some weeks after birth

Table 3-1F, Epiphyseal Closure Times for the Thoracic Appendages of *Equus*.

(concluded)

Phalanax II

	Proximal	Distal
Schimd (1972)	1-1.25 yrs. (12-15 mos.)	-
Silver (1969)	9-12 mos.	Before birth
Rooney (1963)	10-12 mos.	1 week postpartum
Brunni & Zimmerl (1951)	10-12 mos.	Before birth
Tohra (1950)	7-8 mos.	-
Lesbre (1897)	12-15 mos.	Some weeks after birth

Phalanx III

	Proximal
Schimd (1972)	1-1.25 yrs. (12-15 mos.)
Silver (1969)	No true epiphysis
Rooney (1963)	Last part of pregnancy

Notes based from Silver (1969):

Ulna - Interosseus radio-olecranon ligament ossifies at 3-4 yrs.(36-48 mos.).

Phalanx III - In old horses the lateral cartilagge may ossify to form "side bones".

Table 3-1G, Radiographic Epiphyseal Closure Times for the Thoracic Appendages of *Equus*.

Bone	Myers & Emmerson (1966) Colt	Myers & Emmerson (1966) Filly	Schmidt (1960)
Scapula			
distal	-	-	Before 18 mos.
Humerus			
proximal	27 mos.	26.5 mos.	-
distal	14 mos.	15 mos.	Before 12 mos.
Radius			
proximal	14 mos.	14 mos.	Before 12 mos.
distal	24 mos.	24 mos.	3yrs. (36 mos.)
Ulna			
proximal	27 mos.	30 mos.	-
distal	9 mos.	6.5 mos.	Before 4 mos.
Metacarpus III			
proximal	None seen after birth		-
distal	7 mos.	7.5 mos.	6 mos.
Phalanax I			
proximal	9 mos.	7.5 mos.	6 mos.
distal	1 mo.	Before birth	-
Phalanx II			
proximal	8 mos.	7.5 mos.	6 mos.
distal	None seen after birth		-
Phalanx III			
proximal	None seen after birth		-

Table 3-1H, Epiphyseal Closure Times of the Vertebrae and Sacrum of *Equus*.

Bone	Silver (1969)
Atlas	Wings not fused till 6 mos.
Axis	Epiphysis between body and odontoid open till 3-4 yrs. (36-48 mos.)
Vertebrae	1 - Body and arch fuse at or just after birth 2 - Bodies fuse with epiphyses at 5 yrs. (60 mos.)
Sacrum	1 - Body epiphyses may never fuse 2 - Unite with each other before uniting with body

Table 3-11, Epiphyseal Closure Times for the Pelvic Appendages of _Equus_.

Innominate

	Primary centers	Secondary Centers
Schmid (1972)	.75-1 yr. (9-12 mos.)	-
Silver (1969)	-	All fused at 4.5-5 yrs. (54-60 mos.)
Rooney (1963)	10-12 mos.	-
Brunni & Zimmerl (1951)	10-12 mos.	4.5-5 yrs. (54-60 mos.)
Lesbre (1897)	10-12 mos.	-

Femur

	Proximal	Distal
Schmid (1972)	3-3.5 yrs. (36-42 mos.)	3.5 yrs. (42 mos.)
Silver (1969)	3-3.5 yrs. (36-42 mos.)	3-3.5 yrs. (36-42 mos.)
Rooney (1963)	3-3.5 yrs. (36-42 mos.)	3.5 yrs. (42 mos.)
Brunni & Zimmerl (1951)	3 yrs. (36 mos.)	3.5 yrs. (42 mos.)
Lesbre (1897)	3-3.5 yrs. (36-42 mos.)	3.5 yrs. (42 mos.)

Table 3-11, Epiphyseal Closure Times for the Pelvic Appendages of *Equus*.
(continued)

Tibia

	Proximal	Distal
Schimd (1972)	3.5 yrs. (42 mos.)	2 yrs. (24 mos.)
Silver (1969)	3-3.5 yrs. (36-42 mos.)	20-24 mos.
Rooney (1963)	-	-
Brunni & Zimmerl (1951)	3.5 yrs. (42 mos.)	2 yrs. (24 mos.)
Lesbre (1897)	3.5 yrs. (42 mos.)	2 yrs. (24 mos.)

Fibula

	Proximal	Distal
Schimd (1969)	-	2 yrs. (24 mos.)
Silver (1969)	Doubtful 2-3 yrs. (24-36 mos.)	Fuses with tibia 1-3 mos.
Rooney (1963)	-	2 yrs. (24 mos.)
Brunni & Zimmerl (1951)	3.5 mos. (42 mos.)	3 mos.
Lesbre (1897)	-	3-5 mos.

Table 3-II, Epiphyseal Closure Times for the Pelvic Appendages of <u>Equus</u>.
(concluded)

Calcaneus

	Proximal
Schmid (1972)	3 yrs. (36 mos.)
Silver (1969)	3 yrs. (36 mos.)
Rooney (1963)	3 yrs. (36 mos.)
Brunni & Zimmerl (1951)	3 yrs. (36 mos.)
Lesbre (1897)	3 yrs. (36 mos.)

Metatarsus

	Proximal	Distal
Schmid (1972)	-	1-1.25 yrs. (12-15 mos.)
Silver (1969)	Before birth	16-20 mos.
Rooney (1963)	-	-
Brunni & Zimmerl (1951)	-	-
Lesbre (1897)	-	-

Notes:

Fibula - lateral malleolus is separate at birth to 3 mos. in the foal (Silver 1969).

Innominate - Acetabulum is fused with the pubis at .75-1 yr., 9-12 mos. (Schmid 1972).

Bos Fusion and Dental Data

Table 3-2A, Deciduous Tooth Eruption in *Bos*.

Tooth	Schmid (1972)	Silver (1969)
Central Incisor	Present at Birth	Present at Birth
Middle Incisor	Present at Birth	Present at Birth
Lateral Incisor	Present at Birth	Present at Birth
Canine	Present at Birth	Present at Birth or in first 2 wks.
Premolar 1	-	Occasional
Premolar 2	Birth	Birth to 3 wks.
Premolar 3	Birth	Birth to 3 wks.
Premolar 4	Birth	Birth to 3 wks.

Table 3-2B, Permanent Tooth Eruption in *Bos*.

Tooth	Schmid (1972)	Silver (1969)	Miller & Roberston (1947) Commercial Crossbred	Miller & Roberston (1947) Ranch Cattle
Central Incisor	1.75 yrs. (21 mos.)	14-25 mos.	20-24 mos.	22-24 mos.
Middle Incisor	2.25 yrs. (27 mos.)	17-36 mos.	30 mos.	30-36 mos.
Lateral Incisor	3 yrs. (36 mos.)	22-40 mos.	36 mos.	42 mos.
Canine	3.75 yrs. (45 mos.)	32-48 mos.	42-48 mos.	54-60 mos.
Premolar 1	Absent	Occasional but lost 3 yrs. (36 mos.)	-	-
Premolar 2	2.25 yrs. (27 mos.)	24-30 mos.	30 mos.	-
Premolar 3	2 yrs. (24 mos.)	18-30 mos.	30 mos.	-
Premolar 4	2.75 yrs. (33 mos.)	28-36 mos.	36 mos.	-
Molar 1	.5 yrs. (6 mos.)	5-6 mos.	6 mos.	-
Molar 2	1.25 yrs. (15 mos.)	15-18 mos.	15-18 mos.	-
Molar 3	2.25 yrs. (27 mos.)	24-30 mos.	24 mos.	-

Table 3-2C, Tooth Eruption and Wear Stages of *Bos* based on Schneider (1958), Degerbol (1970), Streitferdt (1972), Boessneck & von den Driesch (1975), and Grigson (1982).

Tooth	Age
M 1 not yet erupting	Less than 6 mos.
M 1 erupting	6-12 mos.
M 1 in wear, M 2 not yet erupting	6-18 mos.
M 2 erupting	18 mos.
M 2 erupted, M 3 not in wear	18-30 mos.
M 3 erupting	28-30 mos.
M 3 erupted, P 4 erupting	30-36 mos.
M 3 slightly worn	more than 36 mos.
M 3 medium worn	more than 36 mos.
M 3 strongly worn	more than 36 mos.

Table 3-2D, Stages of the Development of Bovine Cheekteeth Row after Higham (1967)

Stage	Dentition Development	Estimated Age
1	Deciduous undeveloped or erupting	Foetal
2	Deciduous in late eruption, unworn	Birth-3 wks.
3	Deciduous in early wear, M 1 unerupted	1-4 mos.
4	M 1 in primary eruption	5-6 mos.
5	M 1 in secondary eruption	6-7 mos.
6	M 1 in tertiary eruption	7-9 mos.
7	M 1 in primary wear, M 2 unerupted	8-12 mos.
8	M 2 in primary eruption	15-16 mos.
9	M 2 in secondary eruption	16-17 mos.
10	M 2 in tertiary eruption	17-18 mos.
11	M 2 in wear, M 3 unerupted	18-24 mos.
12	M 3 in primary eruption	24 mos.
13	M 3 in secondary eruption	24-30 mos.
14	M 3 in tertiary eruption	30 mos.
15	M 3 in primary wear, PM 2 & 3 in primary eruption	30-31 mos.
16	M 3 in primary wear with cusp 3 untouched by wear, PM 2 & 3 in secondary eruption	31-32 mos.
17	M 3 in late primary wear, PM 2 & 3 in secondary eruption	32-33 mos.
18	M 3 cusp 3 just in wear, PM 4 in primary eruption	36 mos.

Table 3-2D, Stages of the Development of Bovine Cheekteeth Row after Higham (1967).

(concluded)

Stage	Dentition Development	Estimated Age
19	M 3 cusp 3 in primary wear, PM 4 in tertiary eruption	38 mos.
20	M 3 cusp 3 in early secondary wear, PM 4 in tertiary eruption	40 mos.
21	PM 4 in primary wear	40-50 mos.
22	PM 4 in secondary wear	50 mos.
23	PM 4 in tertiary wear	over 50 mos.

Table 3-2E, Skull Suture Closure in Domestic Cattle, *Bos taurus* after Grigson (1982).

Suture Closure	Age
Interparietal/supraoccipital with parietal	6 mos.
Supraoccipital/parietal (on the corner of the temporal fossa) with squamosal/petrous and frontal/parietal (on the occipital plane)	12 mos.
Basi-occipital/exoccipital with excoccipital halves	12-24 mos.
Basisphenoid/basioccipital	24-36 mos.
Temporal/exoccipital	36-48 mos.
Temporal/supraoccipital	48-60 mos.
Temporal/parietal (in the temporal fossa) with parietal/frontal (in the temporal fossa)	60-84 mos.
Lacrimal/frontal (in the orbit) with parietal and frontal (edge of the temporal fossa below horncore base)	84-120 mos.
Temporal/alisphenoid (in the temporal fossa) with zygomatic processes of squamosal and jugal maxilla/jugal	120-180 mos.
Maxilla halves with the Mandibular Symphysis	180-480 mos.
Frontal/lacrimal (on the face) with Nasal halves	Extreme Age

Table 3-2F, Age Classes of <u>Bos</u> horn cores based on Armitage (1982).

Stage	Decription	Age Class
1	Soft "spongy" bone, light in weight. Many large channels formed by nutrient foramina. Surface is pitted by numerous minute pores. Stage of actively growing bone.	Juvenile
2	Soft porous bone. Channels leading into nutrient foramina are less conspicuous than in stage 1. Numerous minute pores present. Stage of actively growing bone.	Sub-adult
3	Porous bone. Nutrient foramina appear mostly as large spherical or elliptical pits which are arranged closer together than in stage 2. Numerous minute pores are present. Stage of growing bone.	Young Adult
4	Bone becoming more compact but surface still feels rough to the touch. Appearance of nutrient foramina as in stage 3, but more scatttered. Many minute pores present.	Adult
5a	Compact bone with scattered nutrient foramina. Only very few minute pores present. Growth of bone is complete at this stage.	Old Adult
5b	Smooth, compact bone. Very few nutrient foramina and minute pores are present. Stage of senescent bone.	

Table 3-2F, (continued).

Figure 3-1, Categories of surface texture and appearance of the horn core (after Armitage 1982).

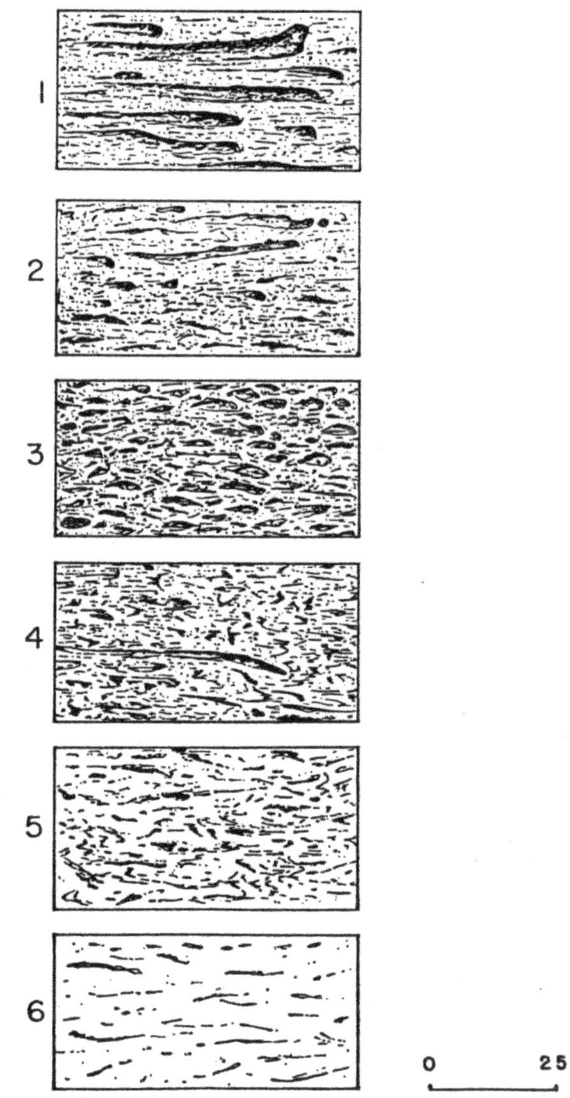

Table 3-2F, (concluded).
Figure 3-2, Schematic drawings of horn cores showing the five age classes defined by Armitage (1982).

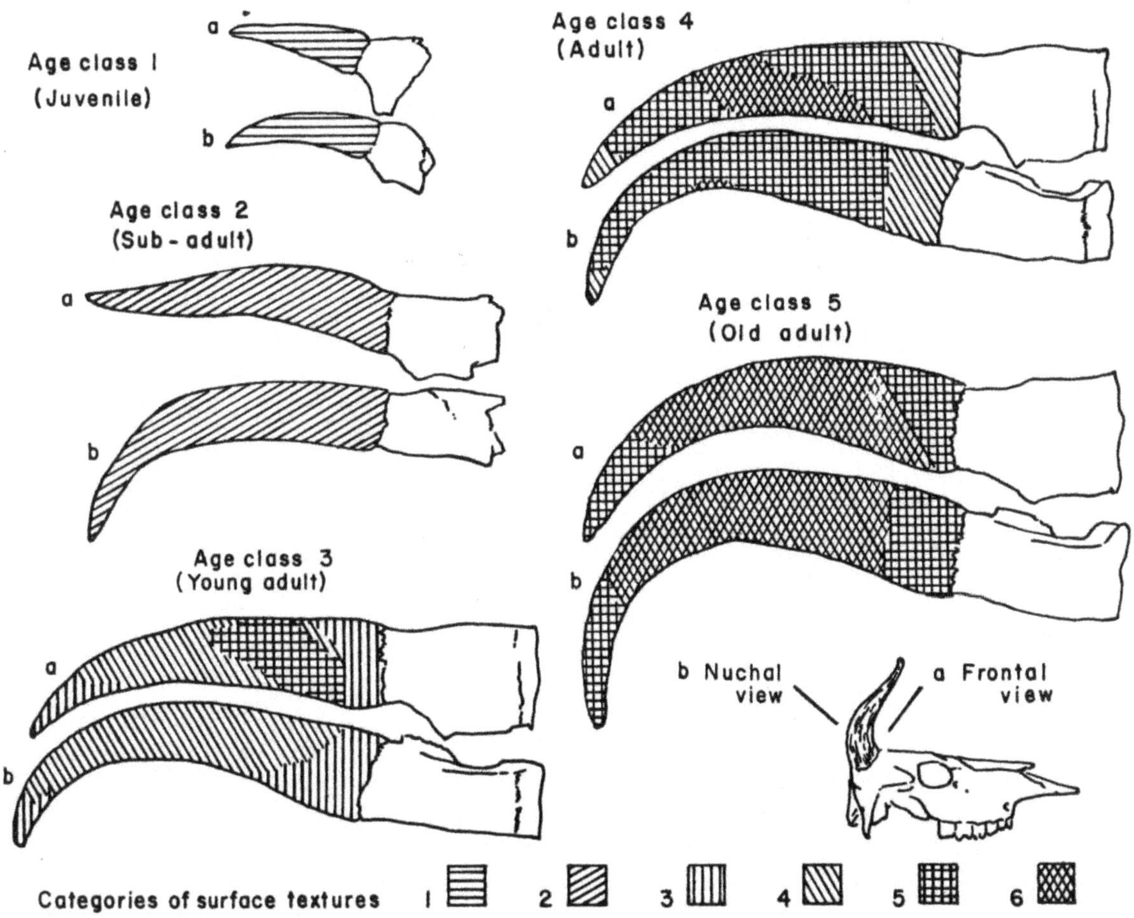

Table 3-2G, Epiphyseal Closure Times for the Thoracic Appendages of *Bos*.

Scapula

		Biccipital Tuberosity
Schmid (1972)	-	-
Silver (1969)	-	7-10 mos.
Habermehl (1961)	-	7-10 mos.
Brunni & Zimmerl (1951)	7-10 mos.	-
Lesbre (1897)	7-10 mos.	-

Humerus

	Proximal	Distal
Schmid (1972)	3.5-4 yrs. (42-48 mos.)	1.5 yrs. (18 mos.)
Silver (1969)	3.5-4 yrs. (42-48 mos.)	12-18 mos.
Habermehl (1961)	3.5-4 yrs. (42-48 mos.)	15-20 mos.
Brunni & Zimmerl (1951)	3.5-4 yrs. (42-48 mos.)	15-20 mos.
Lesbre (1897)	3.5-4 yrs. (42-48 mos.)	15-20 mos.

Table 3-2G, Epiphyseal Closure Times for the Thoracic Appendages of *Bos*.
(continued)

Radius

	Proximal	Distal
Schmid (1972)	1-1.5 yrs. (12-18 mos.)	3.5-4 yrs. (42-48 mos.)
Silver (1969)	12-18 mos.	3.5-4 yrs. (42-48 mos.)
Habermehl (1961)	12-15 mos.	3.5-4 yrs. (42-48 mos.)
Brunni & Zimmerl (1951)	1-1.25 yrs. (12-15 mos.)	3.5-4 yrs. (42-48 mos.)
Lesbre (1897)	1-1.25 yrs. (12-15 mos.)	3.5-4 yrs. (42-48 mos.)

Ulna

	Proximal	Distal
Schmid (1972)	3.5-4 yrs. (42-48 mos.)	3.5-4 yrs. (42-48 mos.)
Silver (1969)	3.5-4 yrs. (42-48 mos.)	3.5-4 yrs. (42-48 mos.)
Habermehl (1961)	3.5-4 yrs. (42-48 mos.)	3.5-4 yrs. (42-48 mos.)
Brunni & Zimmerl (1951)	3.5 yrs. (42 mos.)	3 yrs. (36 mos.)
Lesbre (1897)	3.5-4 yrs. (42-48 mos.)	3.5-4 yrs. (42-48 mos.)

Table 3-2G, Epiphyseal Closure Times for the Thoracic Appendages of *Bos*. (continued)

Metacarpus

	Proximal	Distal
Schmid (1972)	-	2-2.5 yrs. (24-30 mos.)
Silver (1969)	Before birth	2-2.5 yrs. (24-30 mos.)
Habermehl (1961)	Before birth	2-2.5 yrs. (24-30 mos.)
Brunni & Zimmerl (1951)	-	2-2.5 yrs. (24-30 mos.)
Lesbre (1897)	-	2-2.5 yrs. (24-30 mos.)

Phalanx I

	Proximal	Distal
Schmid (1972)	1.5-2 yrs. (18-24 mos.)	-
Silver (1969)	Before birth	1.5 yrs. (18 mos.)
Habermehl (1961)	20-24 mos.	-
Brunni & Zimmerl (1951)	20-24 mos.	-
Lesbre (1897)	20-24 yrs.	-

Table 3-2G, Epiphyseal Closure Times for the Thoracic Appendages of Bos.

(concluded)

Phalanx II

	Proximal	Distal
Schmid (1972)	1.5-2 yrs. (18-24 mos.)	-
Silver (1969)	Before birth	1.5 yrs. (18 mos.)
Habermehl (1961)	15-18 mos.	-
Brunni & Zimmerl (1951)	1.25-1.5 yrs. (15-18 mos.)	Before birth
Lesbre (1897)	1.25-1.5 yrs. (15-18 mos.)	-

Phalanx III

	Proximal	Distal
Schmid (1972)	1.5-2 yrs (18-24 mos.)	-
Silver (1969)	No true epiphysis, partly ossified at birth.	-
Habermehl (1961)	-	-
Brunni & Zimmerl (1951)	-	-
Lesbre (1897)	-	-

Note: The distal epiphysis of the Ulna first fuses with the Radius before fusing to the diaphysis of the Ulna.

Table 3-2H, Epiphyseal Closure Times of Vertebrae, Sacrum, Costal Cartilage and Sternum of *Bos*.

Bone	Silver (1969)	Habermehl (1961)
Atlas	Wings are not fused till 6 mos.	-
Cervical		
body epiphyses	-	8 yrs. or more (+96 mos.)
Thoracic		
Anterior:		
body epiphyses	-	8-9 yrs. (96-108 mos.)
bony caps ossified	-	3-6 yrs. (36-72 mos.)
bony caps fused	5 yrs. (60 mos.)	6-7 yrs. (72-84 mos.)
Posterior:		
body epiphyses	-	9 yrs. (108 mos.)
bony caps ossified	-	2.5-5 yrs. (30-60 mos.)
bony caps fused	5 yrs. (60 mos.)	5-6 yrs. (60-72 mos.)

Table 3-2H, Epiphyseal Closure Times of Vertebrae, Sacrum, Costal Cartilage and Sterum of *Bos*.

(concluded)

Bone	Silver (1969)	Habermehl (1961)
Lumbar		
body epiphyses	-	2.5-4 yrs. (30-48 mos.)
bony caps ossified	-	1.5 yrs. (18 mos.)
bony caps fused	5 yrs. (60 mos.)	2-3 yrs. (24-36 mos.)
Sacrum	Body epiphyses may never fuse.	Completely fused by 8 yrs. (96 mos.)
Costal Cartilage	Cartilage will ossify or calcify commonly in old age.	-
Sternum & Sternebrae	Centres fuse early in last sternebrae of most ruminants in 2 parts till old age.	-

Table 4-21, Epiphyseal Closure Times for the Pelvic Appendages of <u>Bos</u>.

Innominate

	Primary Centres	Secondary Centres	Pubic Symphysis
Schmid (1972)	.5-.75 yrs. (6-9 mos.)	-	-
Silver (1969)	-	4.5 yrs. (54 mos.)	-
Habermehl (1961)	7-10 mos.	5 yrs. (60 mos.)	3-8 yrs. (36-96 mos.)
Brunni & Zimmerl (1951)	7-10 mos.	-	-
Lesbre (1897)	7-10 mos.	-	-

Femur

	Proximal	Distal
Schmid (1972)	3.5 yrs. (42 mos.)	3.5-4 yrs. (42-48 mos.)
Silver (1969)	3.5 yrs. (42 mos.)	3.5-4 yrs. (42-48 mos.)
Habermehl (1961)	3.5-4 yrs. (42-48 mos.)	3.5-4 yrs. (42-48 mos.)
Brunni & Zimmerl (1951)	3 yrs. (36 mos.)	3.5 yrs. (42 mos.)
Lesbre (1897)	3.5 yrs. (42 mos.)	3.5-4 yrs. (42-48 mos.)

Table 3-21, Epiphyseal Closure Times for the Pelvic Appendages of Bos.
(continued)

Tibia

	Proximal	Distal
Schmid (1972)	3.5-4 yrs. (42-48 mos.)	2-2.5 yrs. (24-30 mos.)
Silver (1969)	3.5-4 yrs. (42-48 mos.)	2-2.5 yrs. (24-30 mos.)
Habermehl (1961)	3.5-4 yrs. (42-48 mos.)	2-2.5 yrs. (24-30 mos.)
Brunni & Zimmerl (1951)	4 yrs. (48 mos.)	2-2.5 yrs. (24-30 mos.)
Lesbre (1897)	3.5-4 yrs. (42-48 mos.)	2-2.5 yrs. (24-30 mos.)

Fibula

	Proximal	Distal
Schmid (1972)	-	-
Silver (1969)	Fused with Tibia 2-3 yrs. (24-36 mos.)	Fused with Tibia 2-3 yrs. (24-36 mos.)
Habermehl (1961)	-	-
Brunni & Zimmerl (1951)	3.5 yrs. (42 mos.)	3 yrs. (36 mos.)
Lesbre (1897)	-	-

Table 3-21, Epiphyseal Closure Times for the Pelvic Appendages of Bos.
(concluded)

Calcaneus

Schmid (1972)	3 yrs. (36 mos.)
Silver (1969)	3-3.5 yrs. (36-42 mos.)
Habermehl (1961)	3 yrs. (36 mos.)
Brunni & Zimmerl (1951)	3 yrs. (36 mos.)
Lesbre (1897)	3 yrs. (36 mos.)

Metatarsus

	Proximal	Distal
Schmid (1972)	-	2-2.5 yrs. (24-30 mos.)
Silver (1969)	Before birth	2.25-3 yrs. (28-36 mos.)
Habermehl (1961)	-	2-2.5 yrs. (24-30 mos.)
Brunni & Zimmerl (1951)	-	-
Lesbre (1897)	-	-

Capra Fusion and Dental Data

Table 3-3A, Deciduous Tooth Eruption in *Capra*.

Tooth	Silver (1969)
Central Incisor	Birth
Middle Incisor	Birth
Lateral Incisor	Birth
Canine	1-3 weeks
Premolar 1	Usually Absent
Premolar 2	3 months
Premolar 3	3 months
Premolar 4	3 months

Table 3-3B, Permanent Tooth Eruption in *Capra*.

Tooth	Silver (1969)	
Central Incisor	15 months	
Middle Incisor	21-27 mos.	
Lateral Incisor	27-36 mos.	
Canine	36-40 mos.	
Premolar 1	Usually absent	
Premolar 2	17-20 mos.	30 mos.
Premolar 3	17-20 mos.	30 mos.
Premolar 4	17-20 mos.	30 mos.
Molar 1	5-6 mos.	
Molar 2	8-10 mos.	12 mos.
Molar 3	18-24 mos.	30 mos.

Note: The early dates refer to improved breeds, while the later dates refer to rough goats (Silver 1969:298).

Table 3-3C, Epiphyseal Closure Times for the Thoracic Appendages of Capra based on Noddle (1974).

Scapula

Domestic Female	9-11 months
Domestic Male	-
Feral	12 mos.
Feral Female	-
Feral Male	-
Castrate	-
Castrate Male	4 mos.

Humerus

	Proximal	Distal
Domestic Female	23-24 mos.	11 mos.
Domestic Male	-	12 mos.
Feral	48 mos.	12-13 mos.
Feral Female	-	-
Feral Male	-	-
Castrate	7 yrs. (84 mos.)	4-6 mos.
Castrate Male	-	-

Table 3-3C, Epiphyseal Closure Times for the Thoracic Appendages of <u>Capra</u> based on Noddle (1974).
(continued)

Radius

	Proximal	Distal
Domestic Female	4-9 mos.	33-40 mos.
Domestic Male	4-9 mos.	-
Feral	4-9 mos.	48 mos. +
Feral Female	4-9 mos.	-
Feral Male	4-9 mos.	-
Castrate	4-9 mos.	7 yrs. (84 mos.)
Castrate Male	4-9 mos.	-

Ulna

	Proximal
Domestic Female	24-31 mos.
Domestic Male	-
Feral	48-60 mos.
Feral Female	-
Feral Male	-
Castrate	5-7 yrs. (60-84 mos.)
Castrate Male	-

Table 3-3C, Epiphyseal Closure Times for the Thoracic Appendages of <u>Capra</u> based on Noddle (1974).

(continued)

Metacarpus	Distal
Domestic Female	23-30 mos.
Domestic Male	-
Feral	-
Feral Female	36 mos.
Feral Male	24 mos.
Castrate	23 mos.
Castrate Male	-

Phalanx I	Distal
Domestic Female	11 mos.
Domestic Male	12 mos.
Feral	24 mos.
Feral Female	-
Feral Male	-
Castrate	-
Castrate Male	15 mos.

Table 3-3C, Epiphyseal Closure Times for the Thoracic Appendages of <u>Capra</u> based on Noddle (1974). (concluded)

Phalanax II

Domestic Female	9-11 mos.
Domestic Male	-
Feral	12 mos.
Feral Female	-
Feral Male	-
Castrate	-
Castrate Male	13 mos.

Table 3-3D, Epiphyseal Closure Times for the Pelvic Appendages of *Capra* based on Noddle (1974).

Femur

	Proximal	Distal	Third Trochanter
Domestic Female	23-24 mos.	23-24 mos.	15 mos.
Domestic Male	-	-	-
Feral	36 mos.	48 mos.	24 mos.
Feral Female	-	-	-
Feral Male	-	-	-
Castrate	5-7 yrs. (60-84 mos.)	60 mos.	23 mos.
Castrate Male	-	-	-

Tibia

	Proximal	Distal
Domestic Female	23-24 mos.	19 mos.
Domestic Male	-	-
Feral	36 mos.	24 mos.
Feral Female	-	-
Feral Male	-	-
Castrate	60 mos.	23 mos.
Castrate Male	-	-

Table 3-3D, Epiphyseal Closure Times of the Pelvic Appendages of *Capra* based on Noddle (1974). (concluded)

Calcaneus

Domestic Female	23 mos.
Domestic Male	-
Feral	-
Feral Female	48 mos.
Feral Male	36 mos.
Castrate	60 mos.
Castrate Male	-

Metatarsal

	Distal
Domestic Female	23-30 mos.
Domestic Male	-
Feral	-
Feral Female	36 mos.
Feral Male	24 mos.
Castrate	23 mos.
Castrate Male	-

Ovis Fusion and Dental Data

Table 1, Deciduous Tooth Eruption in <u>Ovis</u>.

Tooth	Schmid (1972)	Silver (1969)
Central Incisor	Present at birth	Birth to 1 week
Middle Incisor	Present at birth	Birth to 1 week
Lateral Incisor	Present at birth	Birth to 2 week
Canine	Present at birth	Birth to 3 weeks
Premolar 1	-	Birth to 6 weeks
Premolar 2	Present at birth	Birth to 6 weeks
Premolar 3	Present at birth	Birth to 6 weeks
Premolar 4	Present at birth	Birth to 6 weeks

Table 3-4B, Permanent Tooth Eruption in *Ovis*.

Tooth	Schmid (1972)	Silver (1969) Modern	Silver (1969) 1790
Central Incisor	1.25 yrs. (15 mos.)	12-18 mos.	18 mos.
Middle Incisor	1.75 yrs. (21 mos.)	18-24 mos.	30 mos.
Lateral Incisor	2.75 yrs. (33 mos.)	27-36 mos.	42 mos.
Canine	3.75 yrs. (45 mos.)	33-48 mos.	50 mos.
Premolar 1	Absent	Usually absent	-
Premolar 2	1.5 yrs. (18 mos.)	21-24 mos.	30 mos.
Premolar 3	1.75 yrs. (21 mos.)	21-24 mos.	30 mos.
Premolar 4	2 yrs. (24 mos.)	21-24 mos.	40 mos.
Molar 1	.5 yrs. (6 mos.)	5 mos.(upper) 3 mos.(lower)	6 mos.
Molar 2	2 yrs. (24 mos.)	9-12 mos.	18 mos.
Molar 3	1.75 yrs. (21 mos.)	18-24 mos.	3-4 yrs. (36-48 mos.)

Note: The modern figures are based on improved breeds, the 1790 figure is based on semi-wild, hill sheep (Silver 1969:297).

Table 3-4C, Epiphyseal Closure Times of the Thoracic Appendages in *Ovis*.

Scapula

	Biciptal Tuberosity	Tubercle
Schmid (1972)	-	-
Silver (1969)	6-8 mos.	-
Smith (1966)	-	5 mos.

Humerus

	Prox.	Distal	Tubero-osity	Medial Epipcondyles	Lateral Epipcondyles
Schmid (1972)	3.5 yrs. (42 mos.)	3 mos.	-	-	-
Silver (1969)	3-3.5 yrs. (36-42 mos.)	10 mos.	-	-	-
Smith (1966)	17 mos.	4 mos.	28 mos.	6mos.	4 mos.

Radius

	Proximal	Distal
Schmid (1972)	3 mos.	3.5 yrs. (42 mos.)
Silver (1969)	10 mos.	3 yrs. (36 mos.)
Smith (1966)	4 mos.	21 mos.

Table 3-4C, Epiphyseal Closure Times of the Thoracic Appendages in Ovis.

(continued)

Ulna

	Proximal	Distal
Schmid (1969)	3-3.5 yrs. (36-42 mos.)	3.5 yrs. (42 mos.)
Silver (1969)	2.5 yrs. (30 mos.)	2.5 yrs. (30 mos.)
Smith (1966)	21 mos.	26 mos.

Metacarpus

	Proximal	Distal
Schmid (1972)	-	1.66-2 yrs. (20-24 mos.)
Silver (1969)	Before birth	18-24 mos.
Smith (1966)	-	16 mos.

Phalanx I

	Proximal	Distal
Schmid (1972)	.5-.75 yrs. (6-9 mos.)	-
Silver (1969)	Before birth	13-16 mos.
Smith (1966)	10 mos.	-

Table 3-4C, Epiphyseal Closure Times of the Thoracic Appendage in *Ovis*.

(concluded)

Phalanx II

	Proximal	Distal
Schmid (1972)	.5-.75 yrs. (6-9 mos.)	-
Silver (1969)	Before birth	13-16 mos.
Smith (1966)	8 mos.	-

Phalanx III

	Proximal
Schmid (1972)	.5-.75 yrs. (6-9 mos.)
Silver (1969)	Partly ossified at birth, but not a true epiphysis.
Smith (1966)	-

Table 3-4D, Epiphyseal Closure Times of the Pelvic Appendages in *Ovis*.

Innominate

	Primary Centres	Secondary Centres
Schmid (1972)	.5 yrs. (6 mos.)	-
Silver (1969)	-	3.5 yrs. (42 mos.)
Smith (1966)	-	-

Femur

	Proximal	Distal	Trochanter Major	Trochanter Minor
Schmid (1972)	3-3.5 yrs. (36-42 mos.)	3.5 yrs. (42 mos.)	-	-
Silver (1969)	2.5-3 yrs. (30-36 mos.)	3-3.5 yrs. (36-42 mos.)	-	-
Smith (1966)	17-18 mos.	18-20 mos.	17-19 mos.	13 mos.

Tibia

	Proximal	Distal	Tubercle
Schmid (1972)	3.5 yrs. (42 mos.)	1.25-1.66 yrs. (15-20 mos.)	-
Silver (1969)	3-3.5 yrs. (36-42 mos.)	1.5-2 yrs. (18-24 mos.)	-
Smith (1966)	25 mos.	15 mos.	30 mos.

Table 3-4D, Epiphyseal Closure Times of the Pelvic Appendages in *Ovis*.
(continued)

Fibula

	Proximal	Distal
Schmid (1972)	-	-
Silver (1969)	-	Separate bone
Smith (1966)	-	-

Calcaneus

Schmid (1972)	3 yrs. (36 mos.)
Silver (1969)	2.5-3 yrs. (30-36 mos.)
Smith (1966)	15 mos.

Metatarsal

	Proximal	Distal
Schmid (1972)	-	1.66-2 yrs. (20-24 mos.)
Silver (1969)	Before birth	20-28 mos.
Smith (1966)	-	15 mos.

Table 3-4D, Epiphyseal Closure Times of the Pelvic Appendages in *Ovis*.
(concluded)

Phalanx I

	Proximal
Schmid (1972)	-
Silver (1969)	-
Smith (1966)	9 mos.

Phalanx II

	Proximal
Schmid (1972)	-
Silver (1969)	-
Smith (1966)	6 mos.

Sus Fusion and Dental Data

Table 3-5A, Deciduous Tooth Eruption in *Sus*.

Tooth	Schmid (1972)	Silver (1969)
Central Incisor	Birth	1-3 weeks (upper) 4-14 days (lower)
Lateral Incisor	3 mos. (12 weeks)	10-14 weeks (upper) 6-12 weeks (lower)
Corner Incisor	Birth	Birth
Canine	Birth	Birth
Premolar 1	-	-
Premolar 2	1 month	7-10 weeks (1.75-2.5 mos.)
Premolar 3	Birth	1-3 weeks
Premolar 4	Birth	1-4 weeks (upper) (.25-1 month) 2-7 weeks (lower) (.5-1.75 mos.)

Table 3-5B, Permanent Tooth Eruption in <u>Sus</u>.

	I1	I2	I3	C	P1	P2	P3	P4
Habermehl (1975)								
early breeds	11	12	6	6	4	12	12	12
middle breeds	12	16	9	9	6	14	14	14
late breeds	14	18	12	12	8	16	16	16
Reiland (1978) improved landrace, mandible	12	17	8	8	5	12	12	13
Schmid (1972)	12	15	9	9	6	12	12	15
Silver (1969)								
modern	12-17	17-20	8-12	8-12	3.5-6.5	12-16	12-16	12-16
18th century	30-36	30-36	6-12	12	-	24	24	24
Sisson & Grossman (1966)	12	16-20	8-10	9-10	5	12-15	12-15	12-15

Note: All ages are given in months.
Abbreviation: I1 = Incisor 1
I2 = Incisor 2
I3 = Incisor 3
C = Canine
P1 = Premolar 1
P2 = Premolar 2
P3 = Premolar 3
P4 = Premolar 4

Table 3-5B, Permanent Tooth Eruption in <u>Sus</u>.
(concluded)

	Molar 1	Molar 2	Molar 3
Habermehl (1975)			
early breeds	4 mos.	7 mos.	16 mos.
middle breeds	6 mos.	10 mos.	18 mos.
late breeds	8 mos.	13 mos.	20 mos.
Reiland (1978) improved landrace, mandible	4 mos.	9 mos.	17 mos.
Schmid (1972)	6 mos.	12 mos.	15 mos.
Silver (1969)			
modern	4-6 mos.	7-13 mos.	17-22 mos.
18th century	12 mos.	15-24 mos.	36 mos.
Sisson & Grossman (1966)	4-6 mos.	8-12 mos.	18-20 mos.

Table 3-5C, Epiphyseal Closure Times of the Thoracic Appendages of Sus.

Scapula

	Bicipital Tuberosity	Distal
Schmid (1972)	-	-
Silver (1969)	1 yr. (12 mos.)	-
Brunni & Zimmerl (1951)	-	1 yr. (12 mos.)
Lesbre (1897)	-	1 yr. (12 mos.)

Humerus

	Proximal	Distal
Schmid (1972)	3.5 yrs. (42 mos.)	1.5 yrs. (18 mos.)
Silver (1969)	3.5 yrs. (42 mos.)	1 yr. (12 mos.)
Brunni & Zimmerl (1951)	3.5 yrs. (42 mos.)	1 yr. (12 mos.)
Lesbre (1897)	3.5 yrs. (42 mos.)	1 yr. (12 mos.)

Table 3-5C, Epiphyseal Closure Times of the Thoracic Appendages of <u>Sus</u>.
(continued)

Radius

	Proximal	Distal
Schmid (1972)	1 yr. (12 mos.)	3.5 yrs. (42 mos.)
Silver (1969)	1 yr. (12 mos.)	3.5 yrs. (42 mos.)
Bruni & Zimmerl (1951)	1 yr. (12 mos.)	3.5 yrs. (42 mos.)
Lesbre (1897)	1 yr. (12 mos.)	3.5 yrs. (42 mos.)

Ulna

	Proximal	Distal
Schmid (1972)	3-3.5 yrs. (36-42 mos.)	3-3.5 yrs. (36-42 mos.)
Silver (1969)	3-3.5 yrs. (36-42 mos.)	3-3.5 yrs. (36-42 mos.)
Bruni & Zimmerl (1951)	3.5 yrs. (42 mos.)	3 yrs. (36 mos.)
Lesbre (1897)	3 yrs. (36 mos.)	3.5 yrs. (42 mos.)

Table 3-5C, Epiphyseal Closure Times for the Thoracic Appendages of *Sus*.
(continued)

Metacarpus III

	Proximal	Distal
Schmid (1972)	-	2 yrs. (24 mos.)
Silver (1969)	Before birth	2 yrs. (24 mos.)
Brunni & Zimmerl (1951)	Before birth	2 yrs. (24 mos.)
Lesbre (1897)	-	2 yrs. (24 mos.)

Phalanx I

	Proximal	Distal
Schmid (1972)	1-2 yrs. (12-24 mos.)	-
Silver (1969)	-	Before birth
Brunni & Zimmerl (1951)	2 yrs. (24 mos.)	Before birth
Lesbre (1897)	2 yrs. (24 mos.)	-

Table 3-5C, Epiphyseal Closure Times for the Thoracic Appendages of *Sus*.

(concluded)

Phalanx II

	Proximal	Distal
Schmid (1972)	1-2 yrs. (12-24 mos.)	-
Silver (1969)	1 yr. (12 mos.)	Before birth
Brunni & Zimmerl (1951)	1 yr. (12 mos.)	Before birth
Lesbre (1897)	1 yr. (12 mos.)	-

Phalanx III

	Proximal
Schmid (1972)	1-2 yrs. (12-24 mos.)
Silver (1969)	Partly ossified at birth, but not a true epiphysis
Brunni & Zimmerl (1951)	-
Lesbre (1897)	-

Table 3-5D, Epiphyseal Closure Times for the Pelvic Appendages in *Sus*.

Innominate

	Primary Centres	Secondary Centres
Schmid (1972)	1 yr. (12 mos.)	-
Silver (1969)	-	6-7 yrs. (72-84 mos.)
Brunni & Zimmerl (1951)	1 yr. (12 mos.)	-
Lesbre (1897)	1 yr. (12 mos.)	-

Femur

	Proximal	Distal
Schmid (1972)	3.5 yrs. (42 mos.)	3.5 yrs. (42 mos.)
Silver (1969)	3.5 yrs. (42 mos.)	3.5 yrs. (42 mos.)
Brunni & Zimmerl (1951)	3 yrs. (36 mos.)	3.5 yrs. (42 mos.)
Lesbre (1897)	3-3.5 yrs. (36-42 mos.)	3.5 yrs. (42 mos.)

Table 3-5D, Epiphyseal Closure Times for the Pelvic Appendages of *Sus*.
(continued)

Tibia

	Proximal	Distal
Schmid (1972)	3.5 yrs. (42 mos.)	2 yrs. (24 mos.)
Silver (1969)	3.5 yrs. (42 mos.)	2 yrs. (24 mos.)
Brunni & Zimmerl (1951)	3.5 yrs. (42 mos.)	2 yrs. (24 mos.)
Lesbre (1897)	3.5 yrs. (42 mos.)	2 yrs. (24 mos.)

Fibula

	Proximal	Distal
Schmid (1972)	3.5 yrs. (42 mos.)	2.5 yrs. (30 mos.)
Silver (1969)	3.5 yrs. (42 mos.)	2.5 yrs. (30 mos.)
Brunni & Zimmerl (1951)	3.5 yrs. (42 mos.)	-
Lesbre (1897)	3.5 yrs. (42 mos.)	2-2.5 yrs. (24-30 mos.)

Table 3-5D, Epiphyseal Closure Times for the Pelvic Appendages in *Sus*.
(concluded)

Calcaneus

Schmid (1972)	2-2.5 yrs. (24-30 mos.)
Silver (1969)	2-2.5 yrs. (24-30 mos.)
Brunni & Zimmerl (1951)	-
Lesbre (1897)	2-2.5 yrs. (24-30 mos.)

Metatarsus III

	Proximal	Distal
Schmid (1972)	-	2 yrs. (24 mos.)
Silver (1969)	Before birth	2.25 yrs. (27 mos.)
Brunni & Zimmerl (1951)	-	-
Lesbre (1897)	-	-

Canis Fusion and Dental Data

Table 3-6A, Deciduous Tooth Eruption in <u>Canis</u>

Tooth	Schmid (1972)	Silver (1969)
Central Incisor	1 month (4 weeks)	4-6 weeks
Lateral Incisor	1 month (4 weeks)	4-6 weeks
Corner Incisor	1 month (4 weeks)	4-6 weeks
Canine	.75 month (3 weeks)	3-5 weeks
Premolar 1	Absent	Absent
Premolar 2	1 month (4 weeks)	5-8 weeks
Premolar 3	.75 month (3 weeks)	5-8 weeks
Premolar 4	.75 month (3 weeks)	5-8 weeks

Table 3-6B, Permanent Tooth Eruption in <u>Canis</u>.

Tooth	Schmid (1972)	Silver (1969)
Central Incisor	4 mos.	3-5 mos.
Lateral Incisor	4.5 mos.	3-5 mos.
Corner Incisor	4.75 mos.	3-5 mos.
Canine	4.75 mos.	5-7 mos.
Premolar 1	5 mos.	4-5 mos.
Premolar 2	5.25 mos.	5-6 mos.
Premolar 3	5.5 mos.	5-6 mos.
Premolar 4	6 mos.	5-6 mos.
Molar 1	4 mos.	4-5 mos.
Molar 2	5 mos.	5-6 mos.
Molar 3	6 mos.	6-7 mos.

Table 3-6C, Epiphyseal Closure Times for the Thoracic Appendages of *Canis*.

Scapula

Schmid (1972)	-
Brunni & Zimmerl (1951)	6-8 mos.
Lesbre (1897)	6-8 mos.

Humerus

	Proximal	Distal
Schmid (1972)	1 yr. (12 mos.)	.5-.66 yr. (6-8 mos.)
Brunni & Zimmerl (1951)	1.5 yrs. (18 mos.)	6-8 mos.
Lesbre (1897)	13 mos.	6-8 mos.

Radius

	Proximal	Distal
Schmid (1972)	.5-.66 yr. (6-8 mos.)	1.5 yrs. (18 mos.)
Brunni & Zimmerl (1951)	6-8 mos.	16-18 mos.
Lesbre (1897)	6-8 mos.	16-18 mos.

Table 3-6C, Epiphyseal Closure Times for the Thoracic Appendages in *Canis*.
(continued)

Ulna

	Proximal	Distal
Schmid (1972)	1.25 yrs. (15 mos.)	1.25 yrs. (15 mos.)
Brunni & Zimmerl (1951)	15 mos.	15 mos.
Lesbre (1897)	15 mos.	15 mos.

Metacarpal III

	Proximal	Distal
Schmid (1972)	-	.5 yr. (6 mos.)
Brunni & Zimmerl (1951)	Before birth	5-6 mos.
Lesbre (1897)	-	5-6 mos.

Phalanx I

	Proximal	Distal
Schmid (1972)	.5 yr. (6 mos.)	-
Brunni & Zimmerl (1951)	5-6 mos.	Before birth
Lesbre (1897)	5-6 mos.	-

Table 3-6C, Epiphyseal Closure Times for the Thoracic Appendages of Canis.
(concluded)

Phalanx II

	Proximal	Distal
Schmid (1972)	.5 yr.	-
Brunni & Zimmerl (1951)	5-6 mos.	Before birth
Lesbre (1897)	5-6 mos.	-

Phalanx III

	Proximal
Schmid (1972)	.5 yr. (6 mos.)
Brunni & Zimmerl (1951)	-
Lesbre (1897)	-

Table 3-6D, Epiphyseal Closure Times for the Pelvic Appendages of *Canis*.

Innominate

	Ilium, Ischium, and Pubis	Acetabulum with Pubis
Schmid (1972)	-	.5 yr. (6 mos.)
Brunni & Zimmerl (1951)	6 mos.	-
Lesbre (1897)	6 mos.	-

Femur

	Proximal	Distal
Schmid (1972)	1.5 yrs. (18 mos.)	1.5 yrs. (18 mos.)
Brunni & Zimmerl (1951)	1.5 yrs. (18 mos.)	1.5 yrs. (18 mos.)
Lesbre (1897)	1.5 yrs. (18 mos.)	-

Tibia

	Proximal	Distal
Schmid (1972)	1.5 yrs. (18 mos.)	1.25 yrs. (15 mos.)
Brunni & Zimmerl (1951)	1.5 yrs. (18 mos.)	14-15 mos.
Lesbre (1897)	1.5 yrs. (18 mos.)	14-15 mos.

Table 3-6D, Epiphyseal Closure Times for the Pelvic Appendages of *Canis*.

(concluded)

Fibula

	Proximal	Distal
Schmid (1972)	-	-
Brunni & Zimmerl (1951)	15-16 mos.	15 mos.
Lesbre (1897)	-	-

Calcaneus

Schmid (1972)	1.25 yrs. (15 mos.)
Brunni & Zimmerl (1951)	14-15 mos.
Lesbre (1897)	14-15 mos.

Felis Fusion and Dental Data

Table 3-7A, Deciduous Tooth Eruption in *Felis*.

Tooth	Silver (1969)
Central Incisor	3-4 weeks
Lateral Incisor	3-4 weeks
Corner Incisor	3-4 weeks
Canine	3-4 weeks
Premolar 1	Absent
Premolar 2	5-6 weeks
Premolar 3	5-6 weeks
Premolar 4	5-6 weeks
Molar 1	Absent

Table 3-7B, Permanent Tooth Eruption in *Felis*.

Tooth	Silver (1969)	Walker (1982)
Central Incisor	3.5-5.5 mos.	3 mos.
Lateral Incisor	3.5-5.5 mos.	-
Corner Incisor	3.5-5.5 mos.	4 mos.
Canine	5.5-6.5 mos.	5 mos.
Premolar 1	Absent	-
Premolar 2	4-5 mos.	-
Premolar 3	4-5 mos.	-
Premolar 4	4-5 mos.	-
Molar 1	5-6 mos.	-

Table 3-7C, Epiphyseal Closure Times in the Thoracic Appendages in *Felis*.

Bone	Curgy (1965)	Walker (1982)
Scapula		
coracoid process	251 days	-
tuber scapulae	251 days	-
Humerus		
head, lesser and greater tubercle	251 days	-
trochlear	140 days	-
epicondyle	140 days	-
distal end	251 days	18 mos.
Radius		
proximal	251 days	-
distal	341 days	20 mos.
Ulna		
distal	-	15 mos.
Metacarpus		
proximal	200-251 days	-
distal	200-341 days	-
Phalanges	180-251 days	-

Table 3-7D, Epiphyseal Closure Times in the Pelvic Appendages in *Felis*.

Bone	Curgy (1965)	Walker (1982)
Innominate		
ilium, ischium and pubis	251 days	-
ischium and pubis	140 days	-
Femur		
head	251 days ?	-
lesser and greater trochanter	251 days ?	-
distal	251 days ?	15 mos.
Tibia		
proximal	341 days ?	18 mos.
distal	341 days	-
Fibula		
proximal	435 days	-
distal	435 days	-
Calcaneus	480 days ?	12 mos.
Metatarsus	435 days ?	-

Rattus Fusion and Dental Data

Table 3-8A, Tooth Eruption in <u>Rattus</u> (4).

Tooth	Navia & Narkates (1980)	
Incisor	Continuously erupting, enamel becomes increasingly yellow at about 21 days after birth.	
	Maxillary	Mandibular
Molar 1	16 days	16 days
Molar 2	17-18 days	17-18 days
Molar 3	32-34 days	32-34 days

Notes: The rat is a monophyodont rodent, having only one set of teeth.

Navia & Narkates suggest that a one day difference occurs between the mandibular molar and the maxillary molar.

----

4. For a discussion concerning the development of molar teeth in <u>Mus</u> see Cohn 1957.

Table 3-8B, Epiphyseal Closure Times of the Thoracic Appendages in *Rattus* (5).

Bone	Dawson (1925)	Dawson (1927)
Scapula	-	Coracoid Proc. appears by 120 days, fused by 151 days.
	-	Dorsal Angle appears by 5 mos. (181 days), fusion occurs approx. 1000 days (2 yrs. 270 days), but never fuses completely.
Humerus		
head	1135-1270 days (3 yrs. 40 days- 3 yrs. 175 days)	-
distal extremity	31-42 days	-
medial epicondyle	130-158 days	-
Radius		
head	83-92 days	-
distal extremity	1135 + days	-
Ulna		
olecranon	940 days (2 yrs. 210 days)	-
distal extremity	1135 + days	

----

5. For a disccusion concerning lasped fusion see Dawson 1934 qnd 1935. Also see Johnson 1933 and Wirtschafter 1960 for the Appendicular development in *Mus*.

Table 3-8C, Epiphyseal Closure Times in Vertebrae, Sternum and Ribs of *Rattus*.

Bone	Dawson (1927)
Vertebrae	Cartilage is intact by 1000 days The epiphyses may never fuse.
Sternum & Sternebrae	With the exception of the caudal end of the Manubrium and sometimes the 2nd Sternebrae, epiphyses rarely develop.
Ribs	Cartilage is intact by 900-1000 days (2 yrs. 200 days-2 yrs. 270 days). The cartilage thereafter may never disappear. The tubercle of the rib never develops.

Table 3-8D, Epiphyseal Closure Times of the Pelvic Appendages in Rattus.

Bone	Dawson (1925)	Dawson (1927)
Innominate	-	Primary centres are fused at 3 mos. (92 days. Secondary centres (Iliac crest, ischial tuberosity, margin of ischiopubic rami) appears by 5 mos. (181 days), fusion occurs 1000 days (2 yrs. 270 days), but never completely fuses.
Femur		
head	1006-1091 days (2 yrs. 276 days- 2 yrs. 361 days)	-
greater trochanter	1006-1091 days	-
distal	1135 days (3 yrs. 40 days)	-
Tibia		
proximal	1135 days (3 yrs. 40 days)	
distal	92-98 days	
Fibula		
proximal	1135 + days	
distal	98-130 days	

Grant's Method of Recording Mandibular Dental Wear for *Bos*, *Ovis/Capra*, and *Sus*

Numerical values assigned to a mandible. (6)

Eruption	Numerical Equivalent
c	1
v	2
e	3
1/2	4
u	5
a	6
b	7
c'	8
d	9
e	10
f	11
g	12
h	13
j	14
l	15
k	16
m	17
n	18
o	19
p	20

----

6. The technique for recording wear patterns of mandibular teeth of cattle, sheep/goat and pigs was published by Grant (1975, 1982). This method has been applied to a variety of British sites, dated from the Bronze age to the eighteenth century. This set of tables summarizes Grant's method as would apply to a more varied sample.

The upper surface of a tooth is completely covered in enamel as the tooth first erupts. As the tooth wears, the enamel of the occlusal surface gradually reveals the dentine surface below. The shape of the enamel and dentine occlusal surfaces form clear and distinct wear patterns. As the tooth begins to wear, "islands" of enamel are left within the dentine. Gradually these islands will disappear, leaving an occlusal surface of dentine surrounded by a border of enamel. Finally, the tooth crown will wear down to the roots in the mandible (Grant 1975, 1982, Deniz and Payne 1982).

Figures 3-3 to 3-5 show the stages of wear observed on the last four teeth of the mandibular row. These stages are termed as Tooth Wear Stages (T.W.S.). They are shown from the earliest stage of wear, when only the enamel surface is worn and no dentine is exposed, (i.e. T.W.S. a), to when, in some instances only the roots of the tooth remain. Molars 1 and 2 are structurally very similar in nature, the wear stages for these teeth are identical (Grant 1982:91).

The period of time over which each wear stage lasts is variable according to Grant. Grant's T.W.S. g stage in Ovis/Capra molars, is a stage that persist for some time. However, the early stages shown in these charts are relatively short lived (Grant 1982:91).

Almost every worn tooth observed in a bone assemblage can be assigned to a T.W.S.. But occasionally there are some that fall between stages. Grant recommends that these teeth be assigned to those stages which resemble most closely the observed wear pattern. The key criterion is the anterior pillar or cusp which is generally considered as having most diagnostic wear pattern than any other pillar(s). This is because the anterior pillar is usually the first to erupt and consequently, the first to wear (Grant 1982:91-92).

Occasionally anomalies do occur, such as severe wear on the posterior pillar. In such a case Grant recommends that it is best to assign such a tooth to the T.W.S. that approximates the same amount of wear as the anomalous tooth. The anomaly can also be separately recorded (Grant 1982:92).

The early stages of wear on P4 of Ovis/Capra are not illustrated. The eruption and wear of P4 is difficult to record, since the tooth frequently impacts against M1. Grant (1982:95) suggests that this affects not only the angle and pattern of wear on the tooth, but also its rate of eruption.

Lastly, in regard to cattle molars, there are some occasional problems encountered with a precise assignment of wear stages. The height and wear of the accessory pillar can vary. However, Grant suggests that the inaccuaccy is not likely to be more than one or two T.W.S. stages (Grant 1982:95).

For erupting teeth, but not yet in wear, Grant (1982) suggests the use of the notation proposed by Ewbank et al. (1964). This method requires the recording of the pillar of the erupting tooth. These symbols are:

- c - perforation in crypt visible
- v - tooth visible in crypt but below head of bone
- 1/2 - tooth erupting through bone
- u - tooth almost at full height but unworn

The next wear stage would be Grant's T.W.S. a, as shown in the relevant dental illustration.

The observer then records each mandible across a chart with vertical columns ruled for each tooth. Although there are no T.W.S. for P2, P3, M2, and M3, presence, and details of anomalies can be recorded. Deciduous teeth can be distinguished from permanent teeth by the use of brackets around T.W.S., or a different color pen (Grant 1982:95).

Once the tooth wear and eruption for each mandible has been recorded, numerical values are assigned to each mandible, representing the total amount of wear on each molar tooth in the row (see chart below). Grant refers to this as the M.W.S. or Mandible Wear Stage. The M.W.S. can represent the T.W.S. of a single permanent molar, or two to three T.W.S.'s. Table 1 gives the numerical values for eruption and wear stages of teeth.

The M.W.S. can be obtained by totalling the numerical equivalent values of all the permanent molar teeth in the mandible. Grant does not include P4 and M4 in the M.W.S.. Generally, the higher the M.W.S., the older the individual at the time of death. Once all M.W.S. are calculated, it is easy to order the mandibles for a probable relative age.

Figure 3-3
Tooth Wear Stages in *Bos*

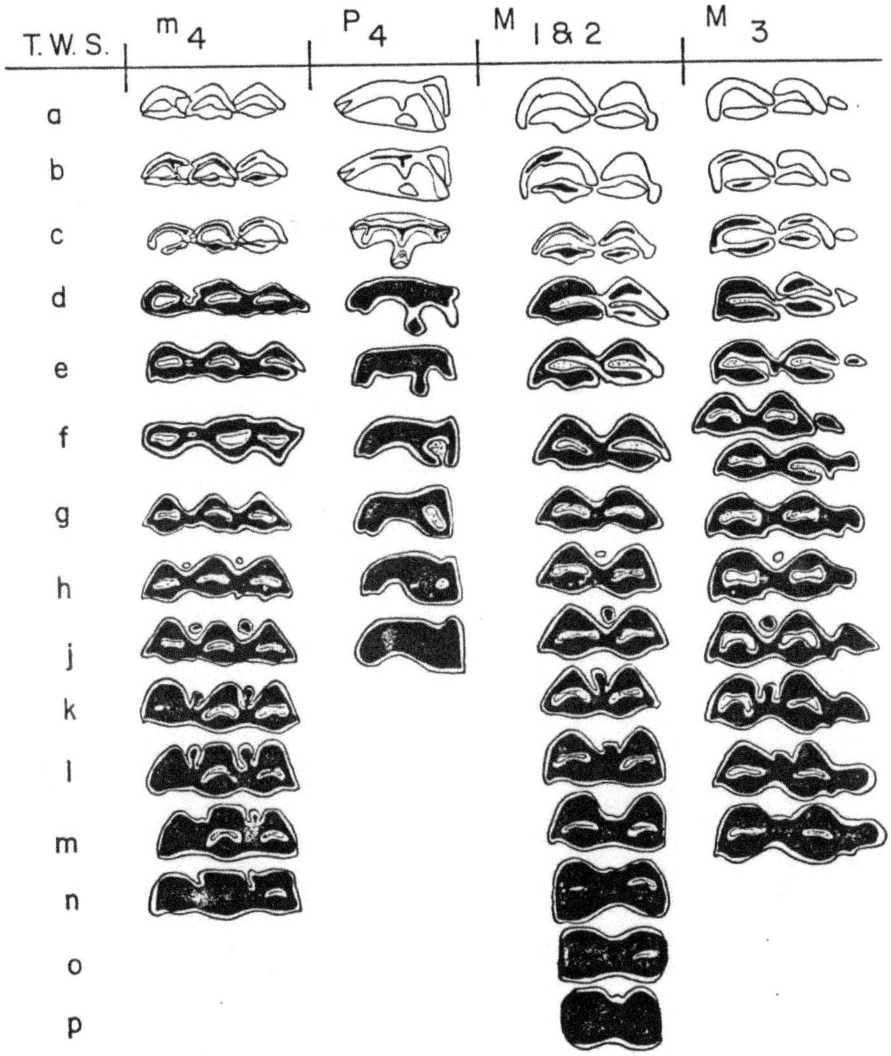

Figure 3-4
Tooth Wear Stages in *Ovis/Capra*

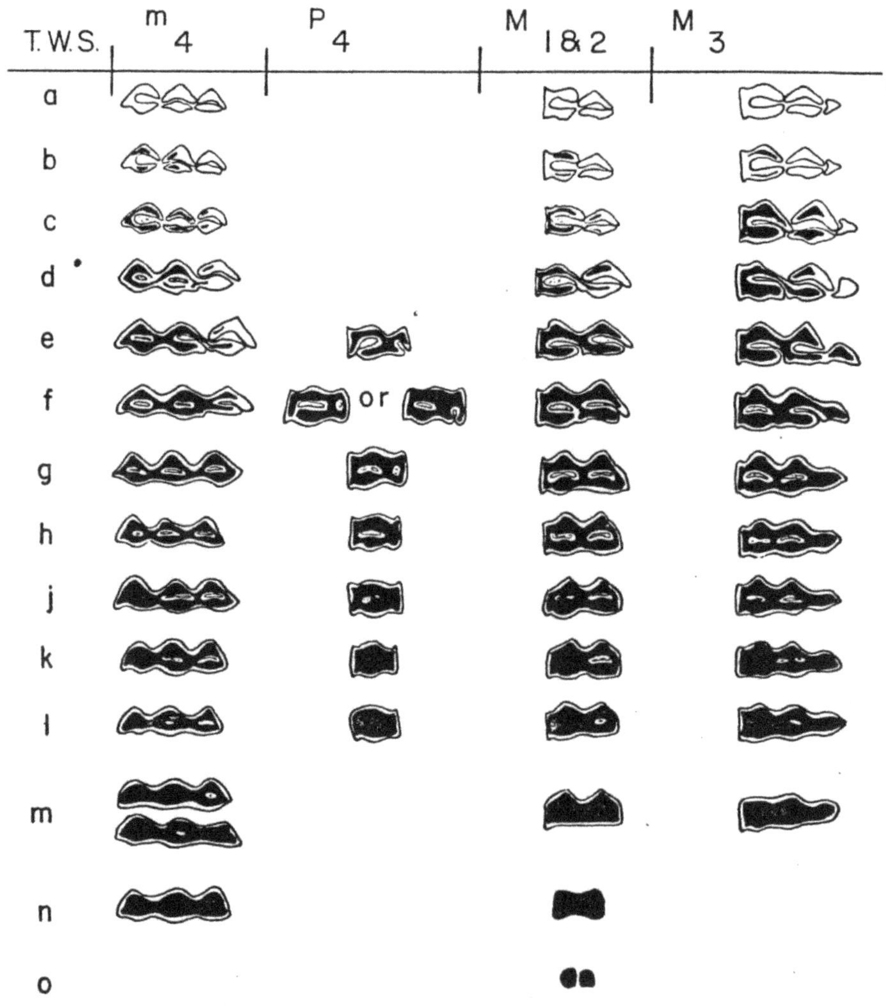

Figure 3-5
Tooth Wear Stages in <u>Sus</u>

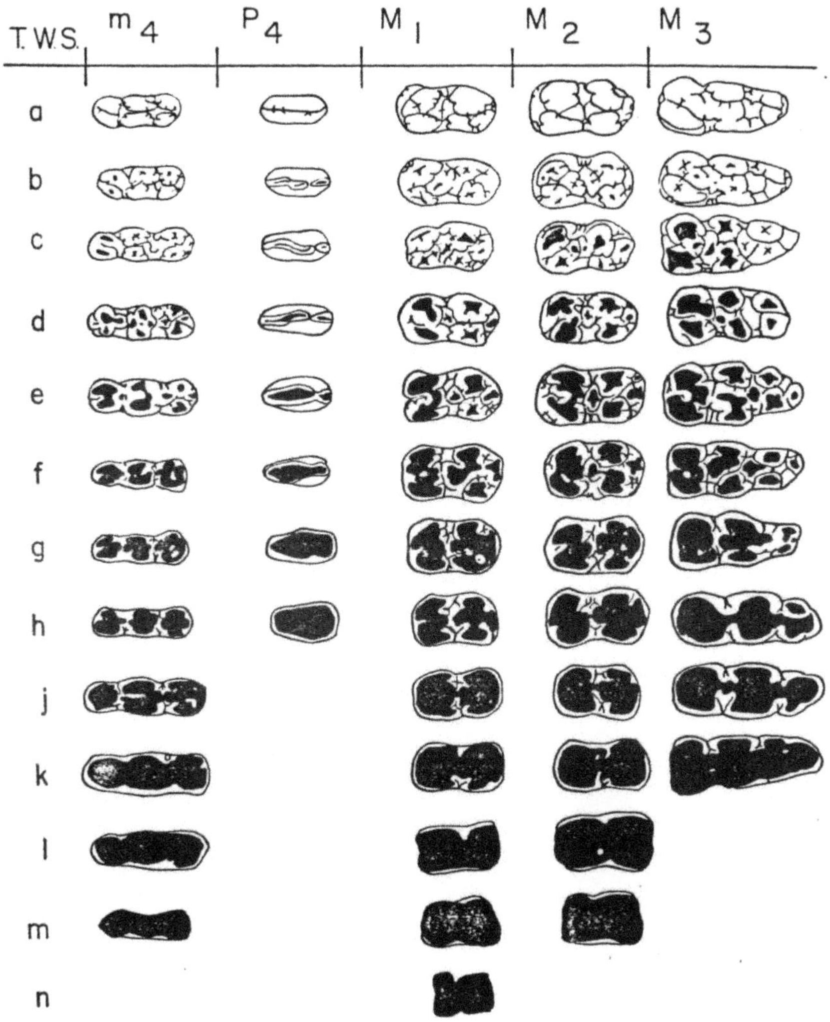

Sus Dental Wear Series

This dental wear series is based on a sample of 18 wild boar obtained in 1974-1975, in Kizilcahaman, north-central Turkey by Bull and Payne (1982). The sequence of tooth eruption agrees with known dental eruption data that can be found in tables 4-5. This sample is not based on animals of known age. Bull and Payne were able to establish wild boar births to a period of April to May. Bull and Payne feel that this will allow a developmental grouping into which the younger animals can be aged with confidence. The dental wear series is compiled into five groups:

Group 1 - Animals that have unreplaced milk molars and erupted second molars. These animals are presumably aged at 7 to 11 months.

Group 2 - This group has their P2-4 erupted, with the second molars in wear, and the third molars in the early stages of eruption. Bull and Payne age this group to be between 19 to 23 months.

Group 3 - These animals have their third molars, and these are starting to wear, aged between 31 to 35 months.

Group 4 & Group 5 - The remaining two groups are taken to be older than 35 months, with M3 in moderate to heavy wear.

Figure 3-6

Group 1 - Wear stages in lower and upper right dentition of Sus, 7-11 months (redrawn after Bull and Payne 1982).

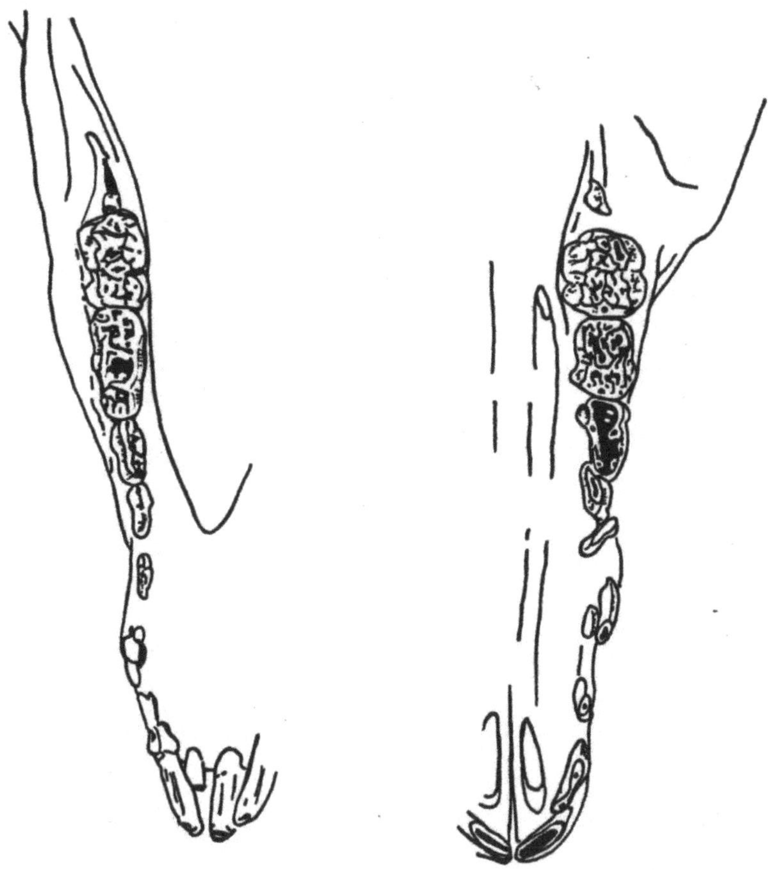

Figure 3-7

Group 2 - Wear stages in lower and upper right dentition of <u>Sus</u>, 19-23 months (redrawn after Bull and Payne 1982).

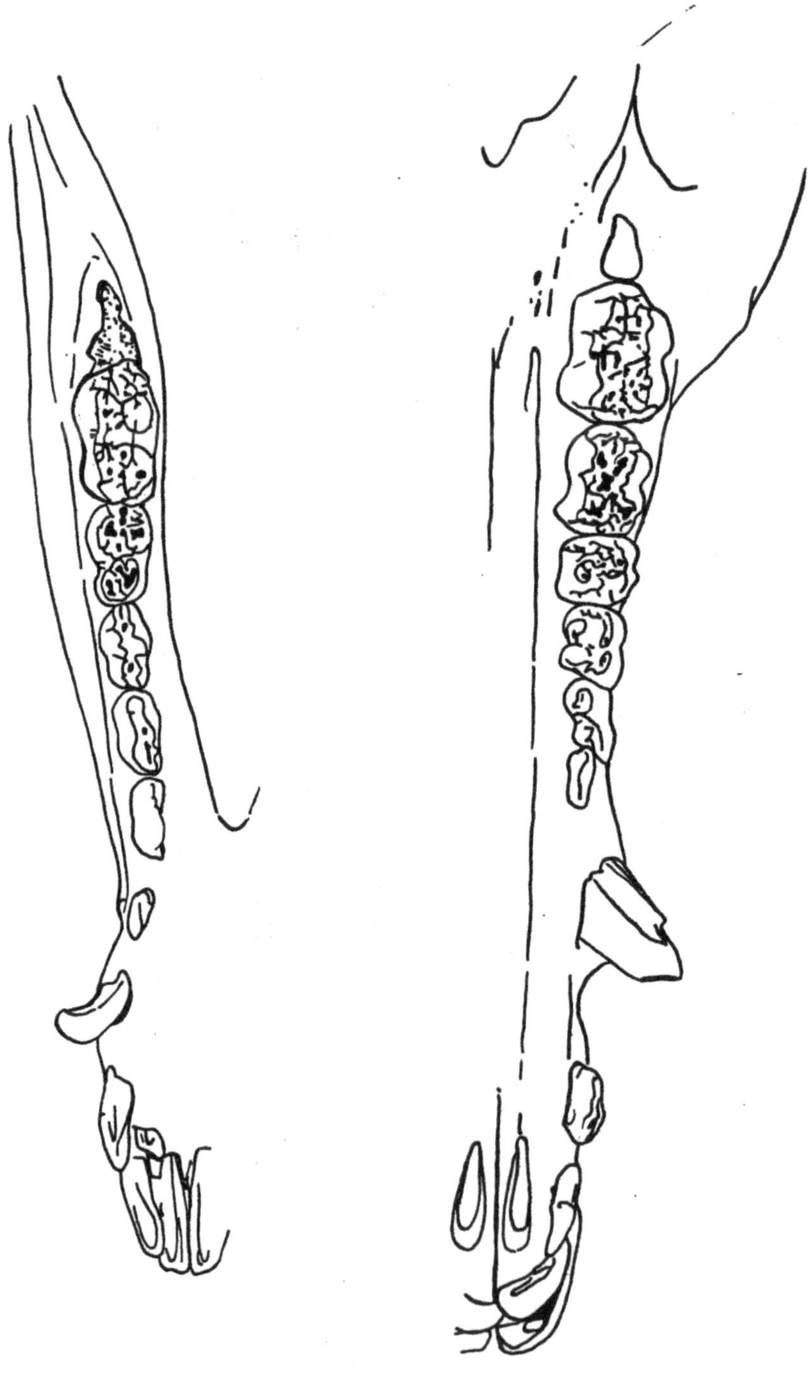

Figure 3-8

Group 3 - Wear stages in lower and upper right dentition of Sus, 31-35 months (redrawn after Bull and Payne 1982).

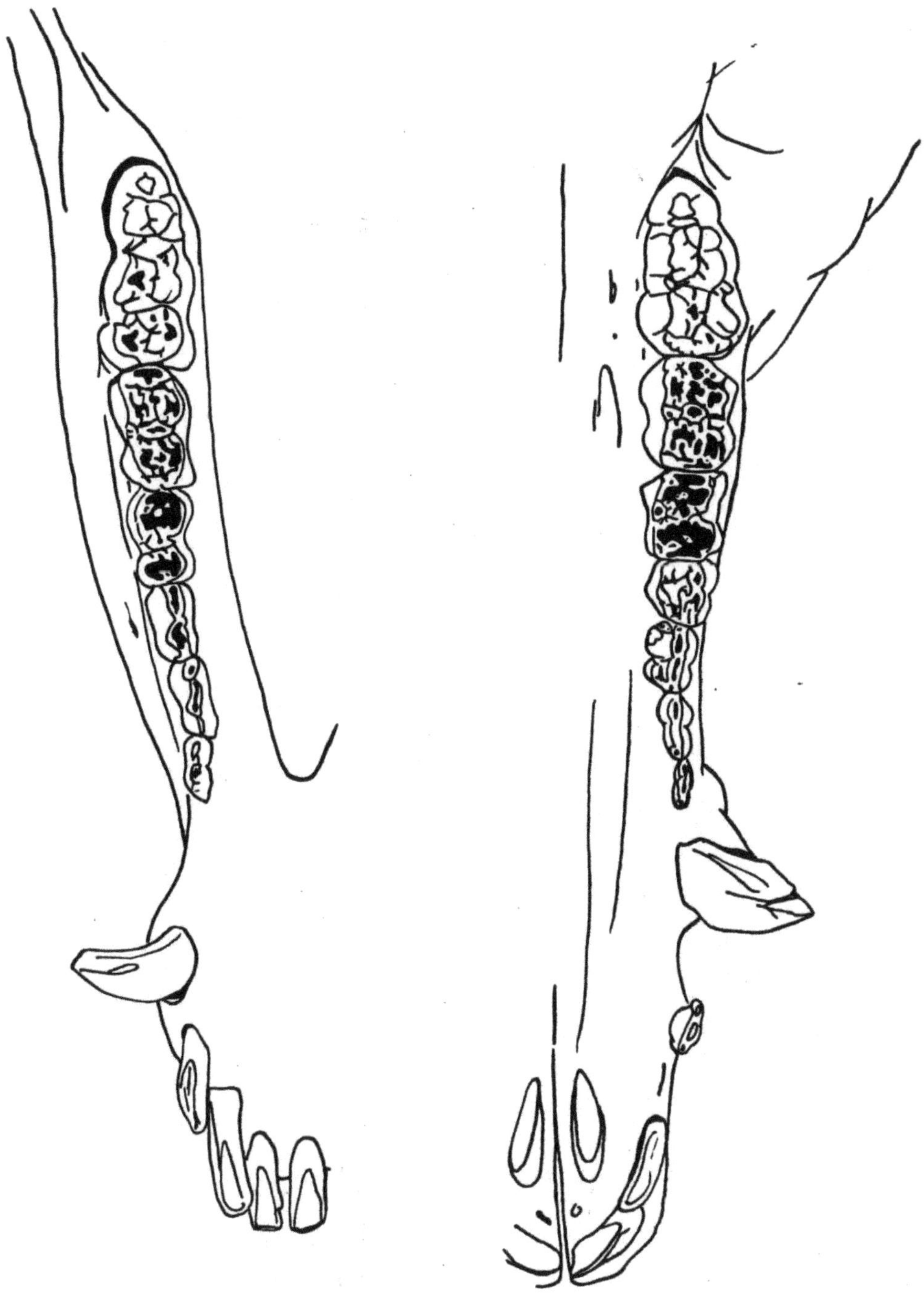

Figure 3-9

Group 4 - Wear stages in lower and upper right dentition of Sus, older than 35 months (redrawn after Bull and Payne 1982).

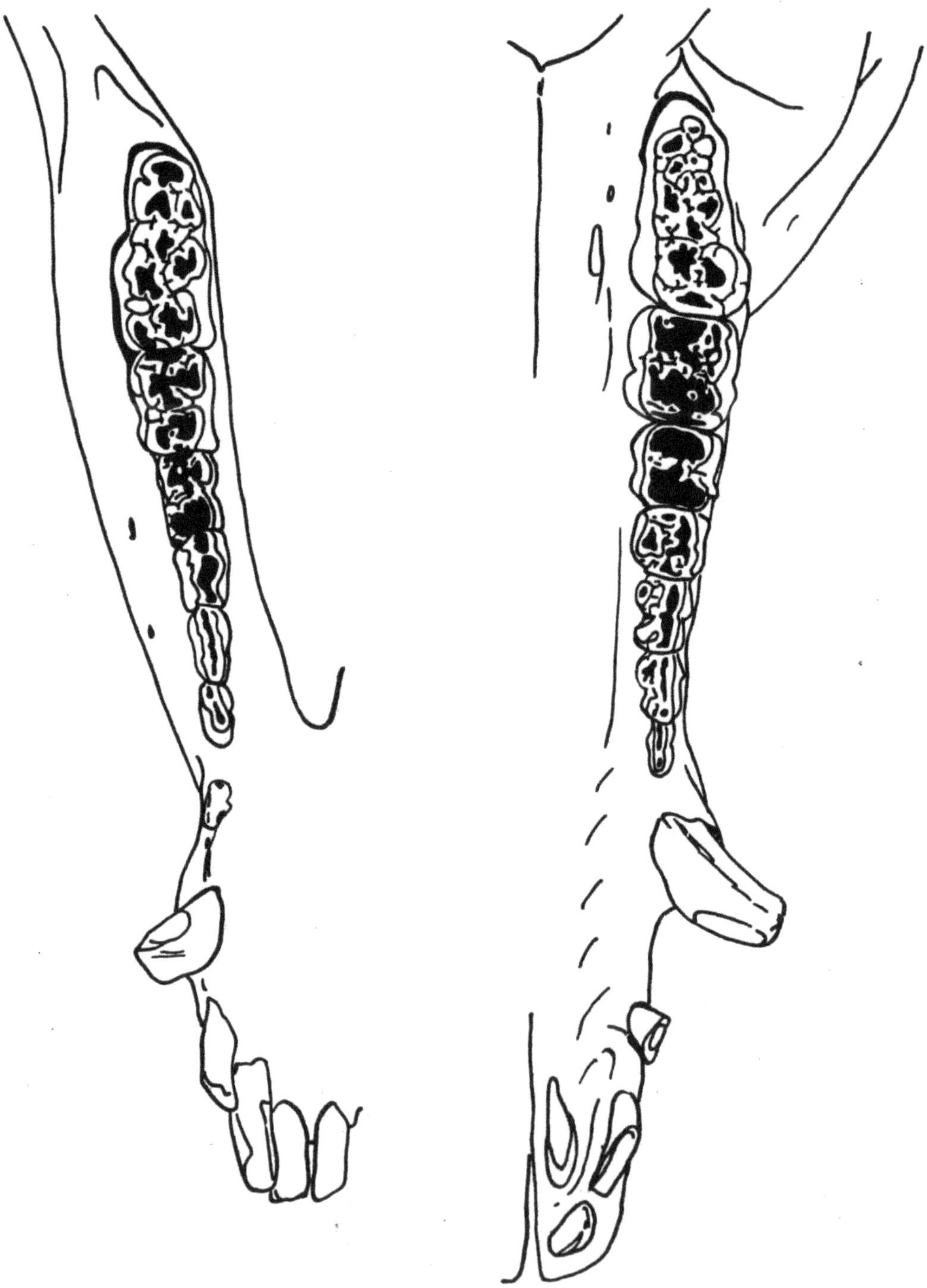

Figure 3-10

Group 5 - Wear stages in lower and upper right dentition of *Sus*, older than 35 months (redrawn after Bull and Payne 1982).

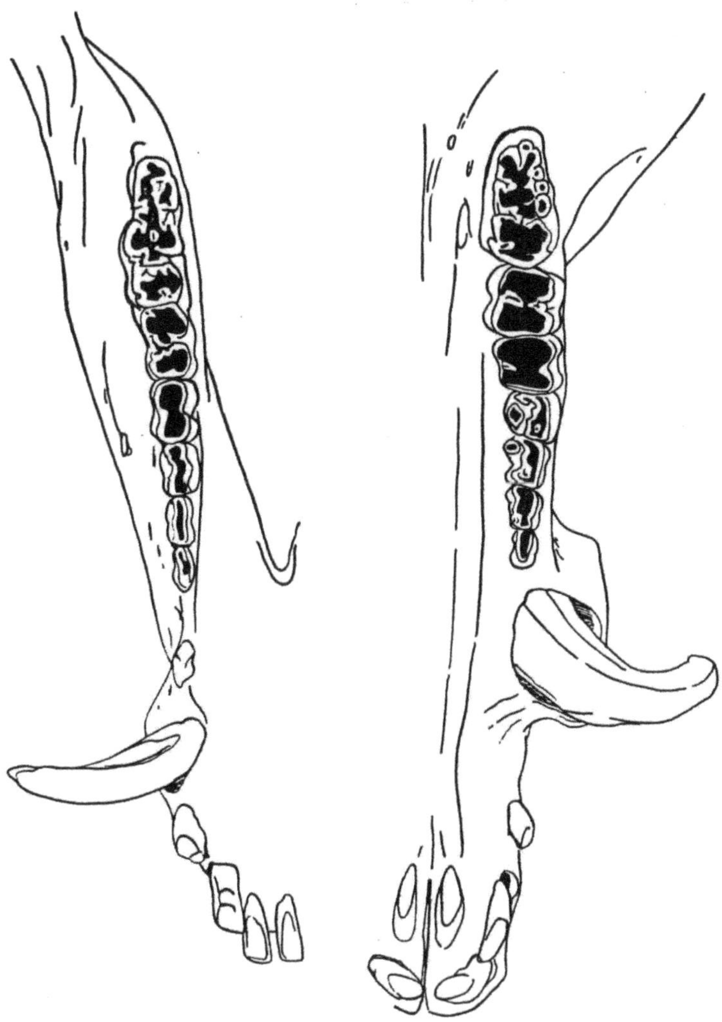

A Method To Assess Dental Wear for Rattus

The dental wear chart provided for *Rattus*, represents a generalized series of wear for molars 1, 2, and 3. This chart is based on a sample of 125 specimens of *Rattus norvegicus* from the 175 Water Street Site in New York City (Amorosi n.d. a). Although this chart does not provide a chronological means to age *Rattus*, it can provide the means by which a relative age curve can be constructed. Future research requiring a large collection of known aged specimens and observing their dental wear patterns can provide a means to chronologically age individuals.

The chart is used in a similar manner as the technique of Bull and Payne (1982). The specimen is compared to either the maxillary or mandibular row. Which ever grouping the specimen generally resembles is the relative age stage for the specimen. Once all specimens have been recorded, the ordering of mandibles for an age curve becomes easy.

Figure 3-11

Dental wear stages in _Rattus_.

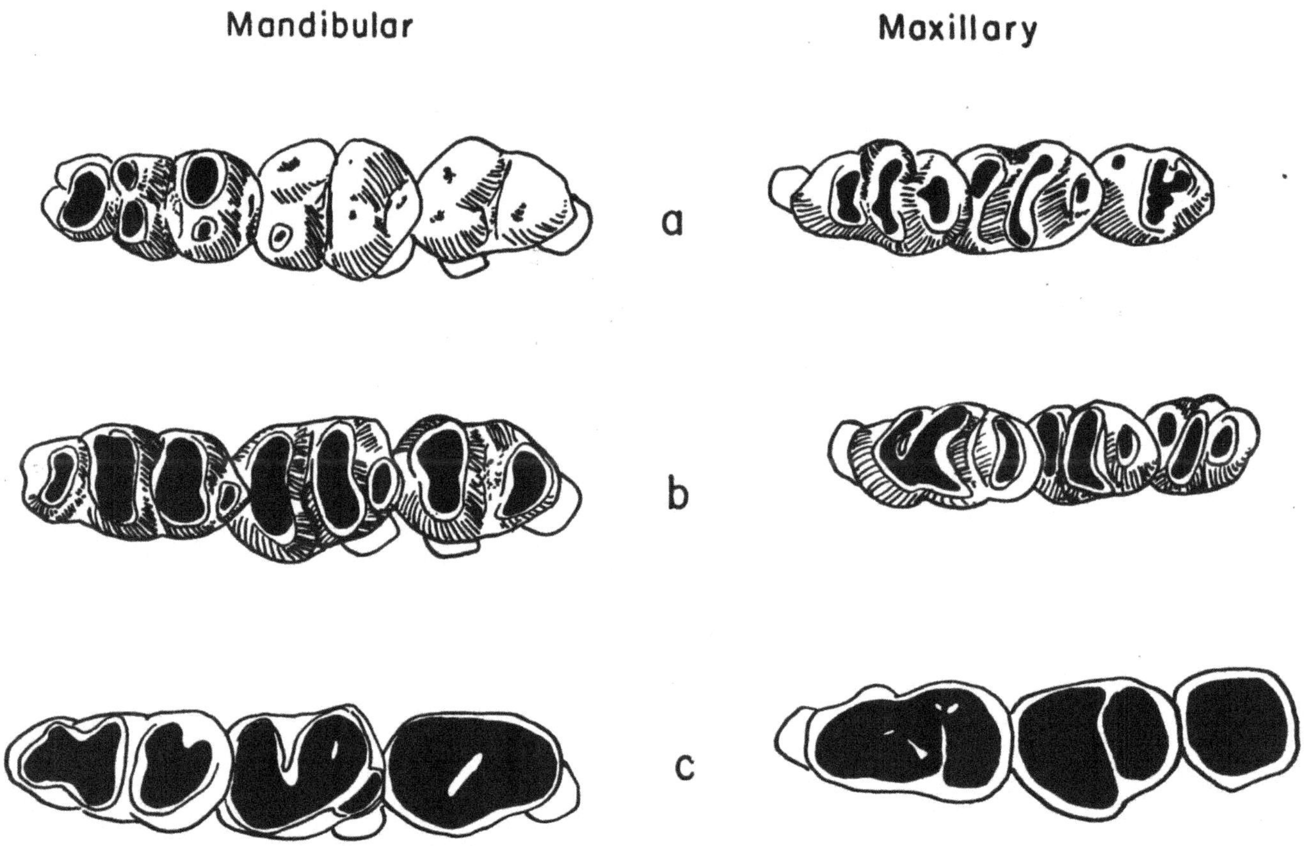

JUVENILE MORPHOLOGY

The illustrations in this chapter are organized into two sets of plates. The first set portrays bone in a detailed fashion, accompanied by some comparative anatomical descriptions. The second set is a comparison of a bone element's size across taxa. Both sets of illustrations were designed to aid in the identification of juvenile faunal remains, with the suggestion that more detailed anatomical texts and comparative collections also be consulted.

This manual is intended to be an introduction to the osteology of domestic mammals of neo-natal and juvenile ages. Yet, one of the most difficult problems of this work was to define neo-natal and juvenile in terms of skeletal anatomy. The stage of development which a bone has reached can be readily observed. As a result, a general ordering of osteological development has been noted by a number of authors (Bass 1971 Schmid 1972; Stewart 1979). These schemata can be complex (Stewart 1979) or simplistic (Schmid 1972). For example, such an arrangement of osteological development might be:

Stage	Development
Neo-Natal	The bone surface has a porous appearance consistent with primary bone deposition. Deciduous dentition has started to appear or can be present. Epiphyses have begun to form.
Juvenile	In this stage, deciduous teeth are mostly present. Replacement of deciduous teeth with the permanent set can occur. It is at this stage that epiphyses are starting to take on their adult character.
Preadult	The replacement of deciduous teeth

	with the permanent set may be nearly complete. The epiphyseal cap has formed completely, and started to fuse with the diaphysis of the long bone.
Young Adult	During this stage the permanent dentition is in occlusion. The epiphysis has fused with the metaphysis region of long bones. However, a visible epiphyseal line remains.
Mature Adult	At this stage teeth exhibit some degree of wear. Epiphyseal lines have disappeared forming one continuous surface. Areas of attachments for muscle and tendon on the bone are prominent.
Old Adult	Teeth are deeply worn or lost. If teeth are lost, resorption of alveoli have occurred. Degenerative joint disease is also present.

The most common mistake of those who use epiphyseal fusion and tooth methods, is to assume the terms neo-natal and juvenile imply the same chronological age across taxa. Yet, some species mature at much faster rates than others, for example a neo-natal rat matures at a faster rate as compared to a neo-natal cat. The crucial point is the definition of what is a neo-natal and juvenile has not attained a consensus over a wide range of species, and thus a comparison across taxa in these terms are severely hampered.

The illustrations in this manual were not designed to be a comparison across taxa. Instead, the illustrations demonstrate visually a stage of bone development. The developmental definitions that follow are thus broad and non-specific. Their utility lies in the fact that a stage of development can be discussed for all species, while recognizing the fact that chronological age can greatly vary between species.

Neo-natal bone can be characterized as having a porous appearance, consistent with primary bone deposition. Bone is not heavily mineralized, being light in weight, and somewhat chalky to the feel. The epiphyses of the long bones have started to form, and are characterized as round porous buttons of bone. The epiphyses are not proportional in size to the diaphysis, and generally have no noticeable osteological landmarks. The diaphysis is extremely porous towards the metaphysis region. The shaft starts to develop its adult characteristics, where some osteological landmarks can be determined. The flat bones, i.e. the scapula and innominate, are also starting to develop their adult characteristics. The illustrations of the horse, cow, sheep, and cat are representative of a neo-natal stage of development.

The bones of a juvenile are more mineralized than a neo-natal, in that the bone is dense, with a hard smooth surface. The epiphyses of long bones are proportional in size to the diaphysis, but have not begun to fuse to the diaphysis. Osteological landmarks can be determined on both the epiphysis and diaphysis surfaces. The adult characters become evident at this stage, not only on the long bones, but the flat bones and vertebrae (irregular bones) as well. The pig, dog, and rat are illustrative of a juvenile stage of development.

In each of the illustrations the epiphyses, shaft, etc., are portrayed as a unit. In reality, each anatomical specimen is composed of one to several units. For example, the innominate of the horse, while illustrated as one unit, is actually composed of three. In this way the relationship between the bone fragment and the whole bone should be made easier. For determination of which bone is composed of how many units, and at what age, see chapters 2 and 3.

The anatomical terminology used in this volume is based on the Greek and Latin names, invented by anatomists using classical words in their relative sense. These descriptions are of a general nature, and oriented for the identification of faunal remains, rather than as an exhaustive anatomical text. The reader is strongly advised to review Flower (1885), Romer (1949), Cornwall (1956), and Sisson and Grossman (1966, or Getty (1975 for the 5th edition) for comprehensive anatomical works on mammals. The anatomical terms used to describe osteological landmarks in this work are the same as those that appear in Schmid (1972) in order to ensure compatibility.

The smaller bone elements of the species represented in this volume that can pass through a 1/4 inch (3.18 mm) wire mesh were not included. The cat (Felis catus) and the rat (Rattus norvegicus) are the two species most affected by this bias. While most of the long bones are recovered, most of the vertebrae, metapodials, and phalanges do pass through the 1/4 inch wire screen. The atlas and axis have been included although both bone elements can pass through this size wire mesh. Both the atlas and axis represent the most diagnostic vertebrae of the cervical series, and are extremely useful in the differentiation of species. Certain other bone elements are not included in the illustrations because of their lack of availability for drawing. These include the sacrum, astragalus, and phalanx II of *Equus*, and phalanx III of *Sus*, metacarpal I and metatarsal I of *Canis*, and the astragalus of *Felis*.

Plate I

The purpose of this plate is to demonstrate the possible size ranges that young mammals might have in relation to a child of approximately 10 years. The animal types are generic in nature, with no intent of portraying any specific breed. The automobile (VolksWagen) in the background is a further illustration of scale in relation to the mammals.

Plate II

Directional nomenclature of the quadruped mammalian body.

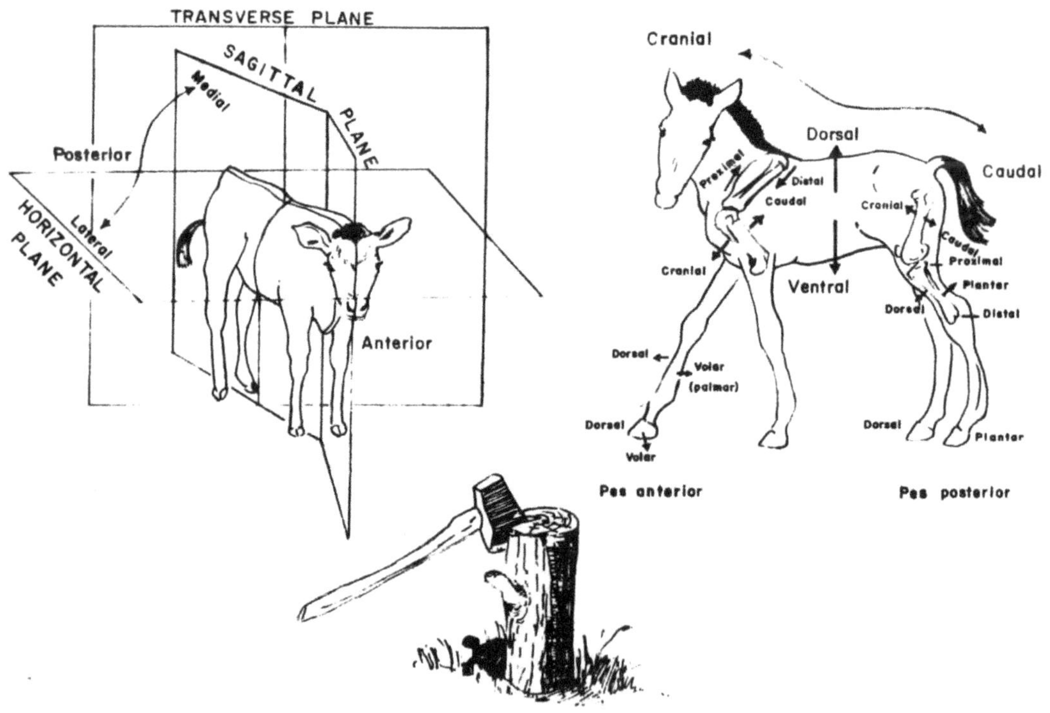

Plates III-XXII are organized into two sections: the Axial and Appendicular skeleton. The thoracic and pelvic limbs of the Appendicular skeleton are listed separately. The axial skeleton starts with the atlas and ends with the sacrum, as one would view an animal from the cranial position. The limbs of the Appendicular skeleton are organized from the most proximal bone to the most distal one, viewing an animal dorsally.

Axial Skeleton
Plates III
Atlas

There are three general functions that the atlas serves: 1) to protect the spinal cord, 2) support the head, and 3) to enable the head to turn. First, the atlas forms a ring of bone around the spinal cord protecting it. Secondly, the atlas has two deep articular fossae, facing cranially, that articulate with the occipital condyles of the skull. Last, the distal articular surface of the atlas will articulate with the cranial articular surface of the epistropheus (axis). Unlike the other vertebrae, the atlas has no centrum or body. The lateral processes serve as areas of muscle attachment. Schmid (1972:96) points out that the length of the atlas will correspond to the length of the animal's neck. Animals with short necks possess a low atlas, while those with a long neck possess a high atlas (as measured dorsal-ventrally on the atlas).

Views:
a - caudal
b - cranial
c - lateral
d - medial
e - dorsal
f - ventral

Landmarks:
1 - Foramen transverarium
2 - Tuberculum ventrale
3 - Ventral articular parts
4 - Process below the articulation
5 - Tuberculum dorsale
6 - Tuberculum ventrale
7 - Transverse processes
8 - Dorsal edge
9 - Indentation between the articulation and the Tuberculum ventrale
10 - Articulation ends above the ventral edge
11 - Rather low
12 - Transverse processes spreading more or less horizontally

Equus caballus

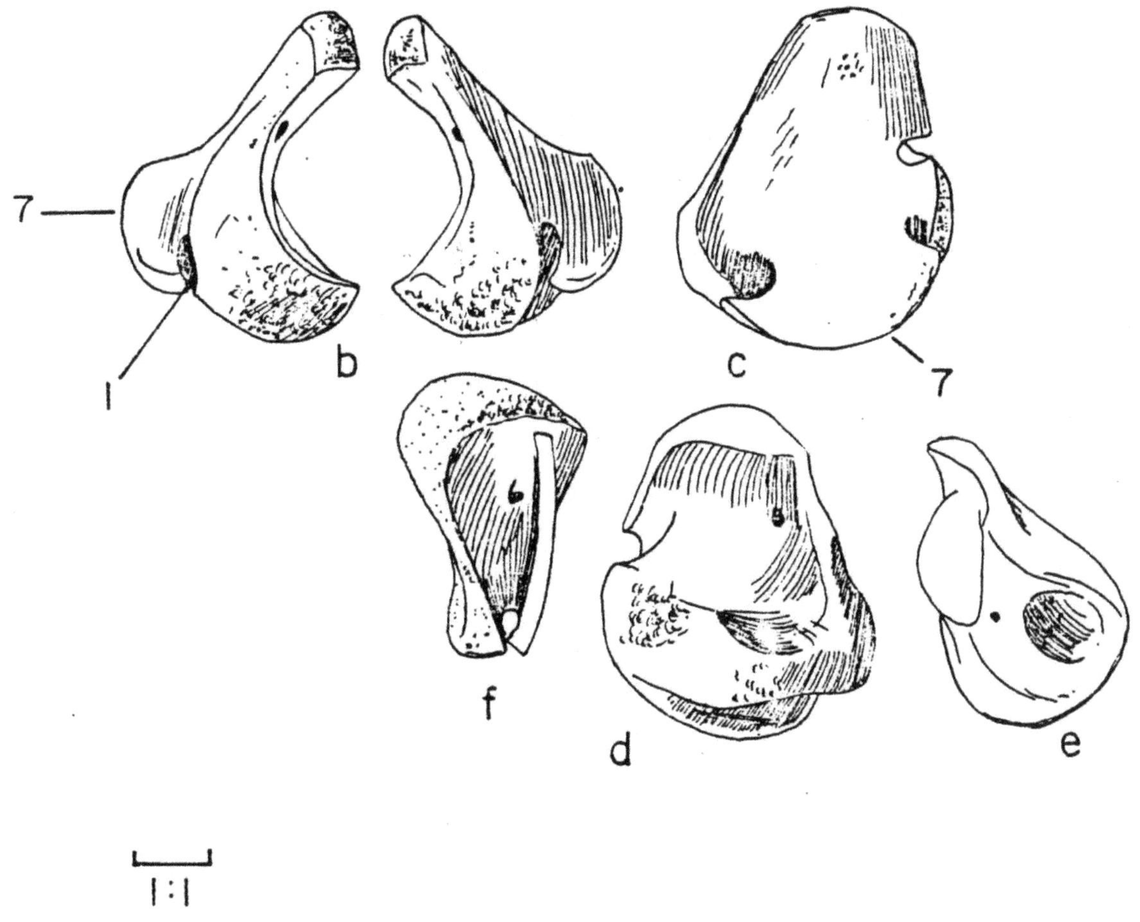

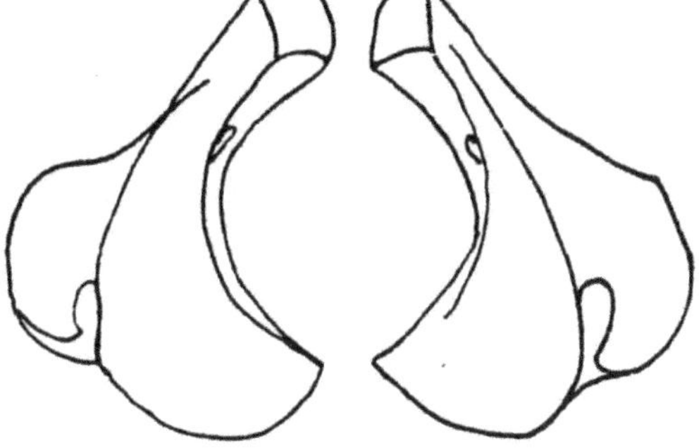

Bos sp.

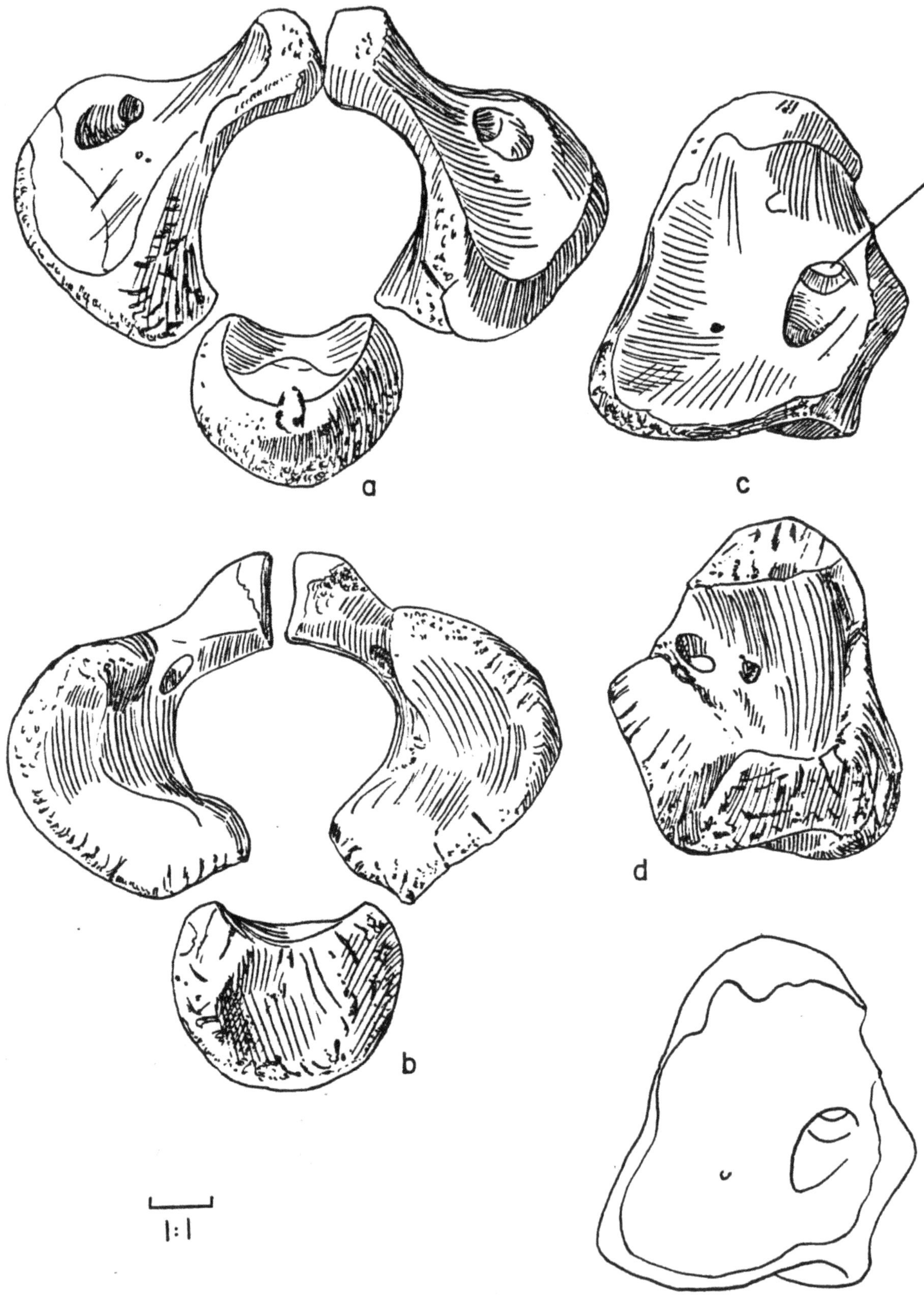

Ovis aries

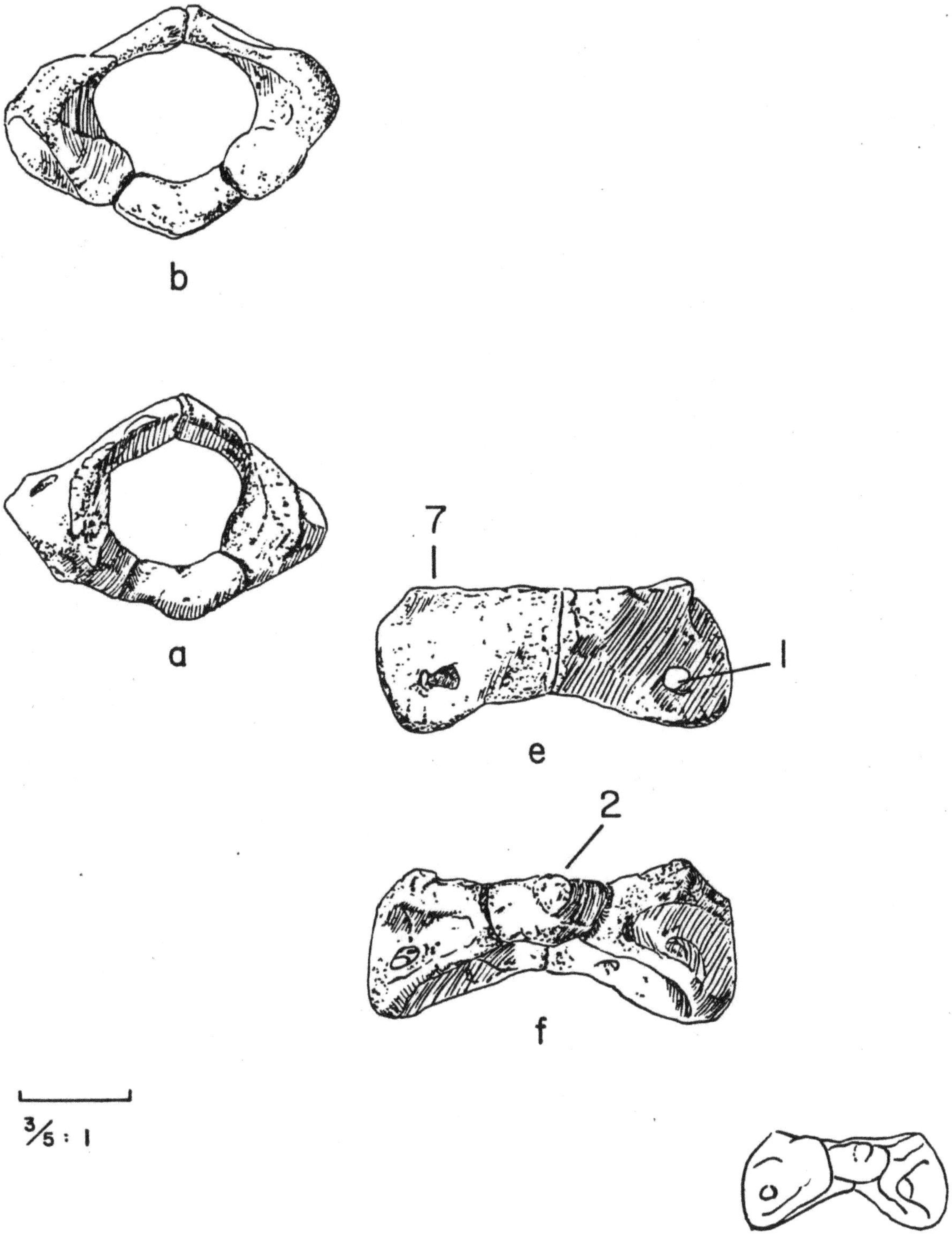

Sus scrofa

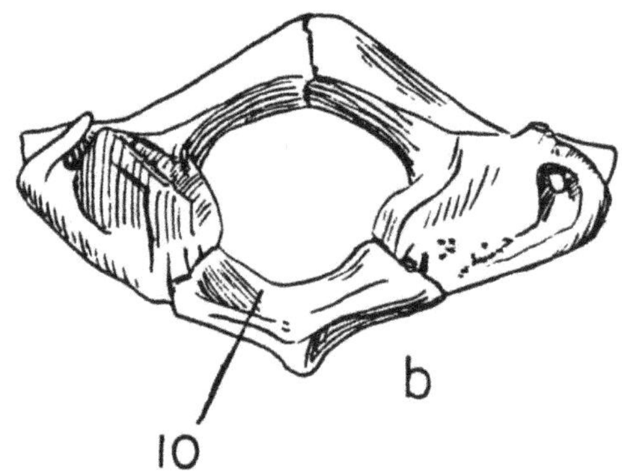

b

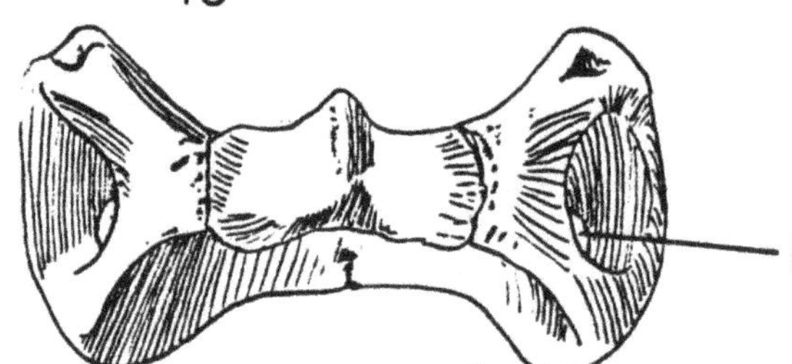

f

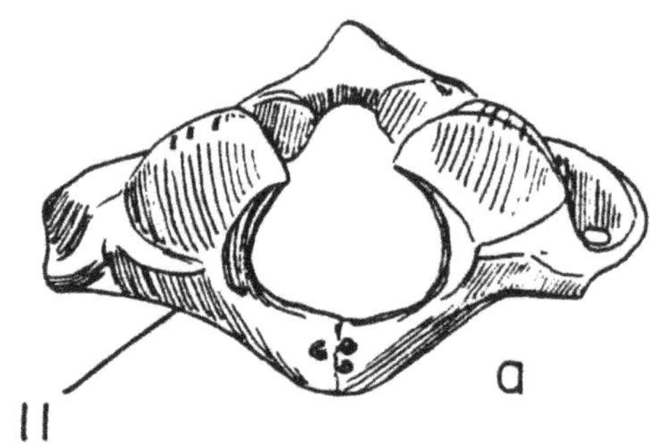

a

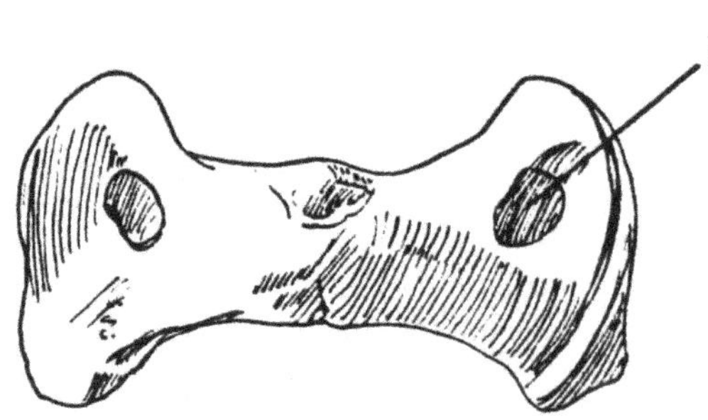

c

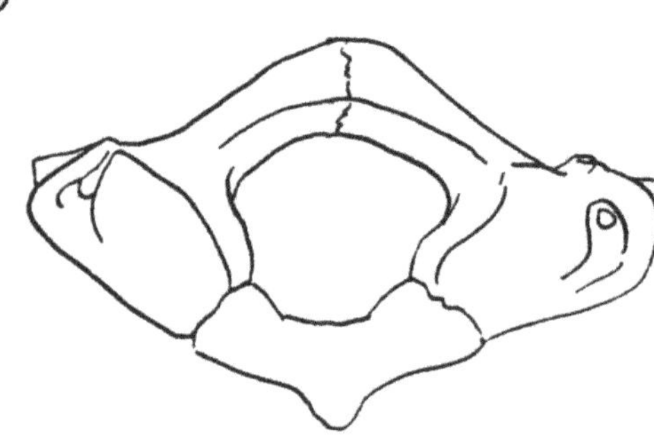

1:1

Canis familiaris

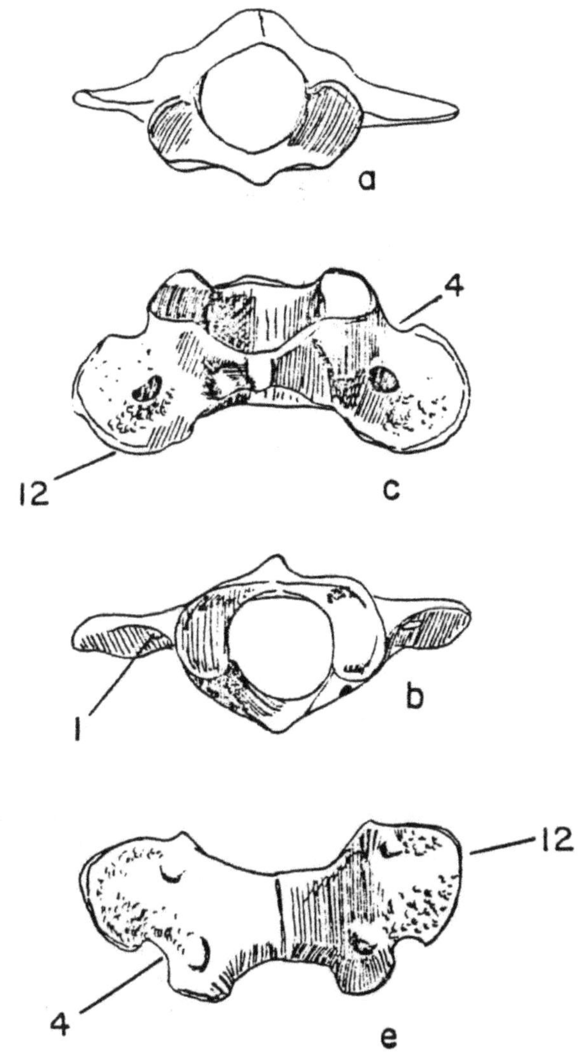

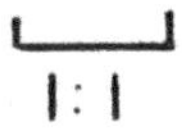

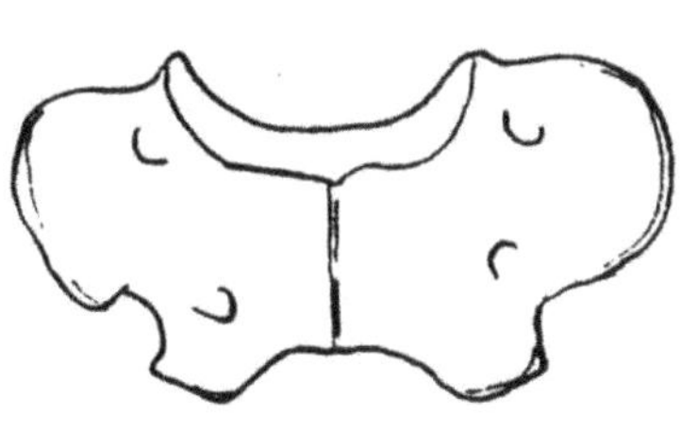

Felis catus L.

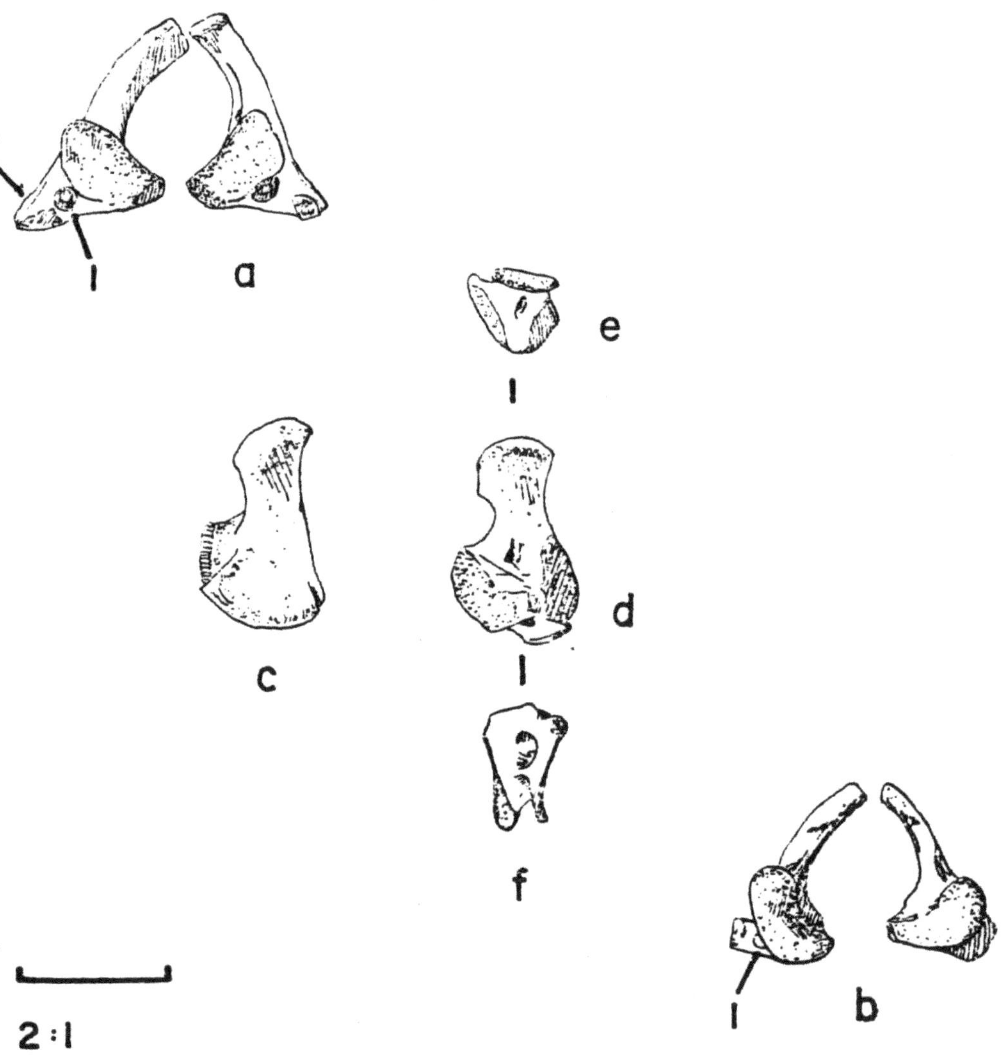

Rattus norvegicus

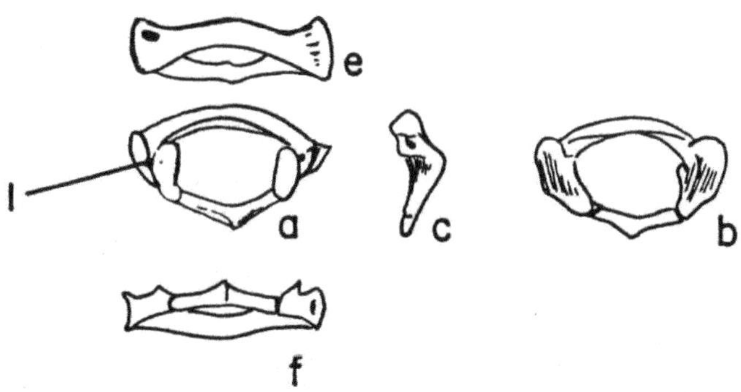

2:1

Plates IV
Epistropheus (Axis)

The axis is the vertebra responsible for the rotation of the head. The odontoid process (dens) will prevent any dislocation of the atlas to the spinal column. The shape of the odontoid process can be characteristic of the class of mammals. The odontoid process of carnivores, lagomorphs, certain rodents, and pigs tends to be cylindrical in shape. In ruminants, the process is spread out along the cranial articular surface of the epistropheus (Cornwall 1956; Sisson and Grossman 1966; Schmid 1972; Carter 1976).

Views:
a - caudal
b - cranial
c - lateral
d - medial
e - dorsal
f - ventral

Landmarks:
1 - Odontoid process
2 - Foramen travsversarium

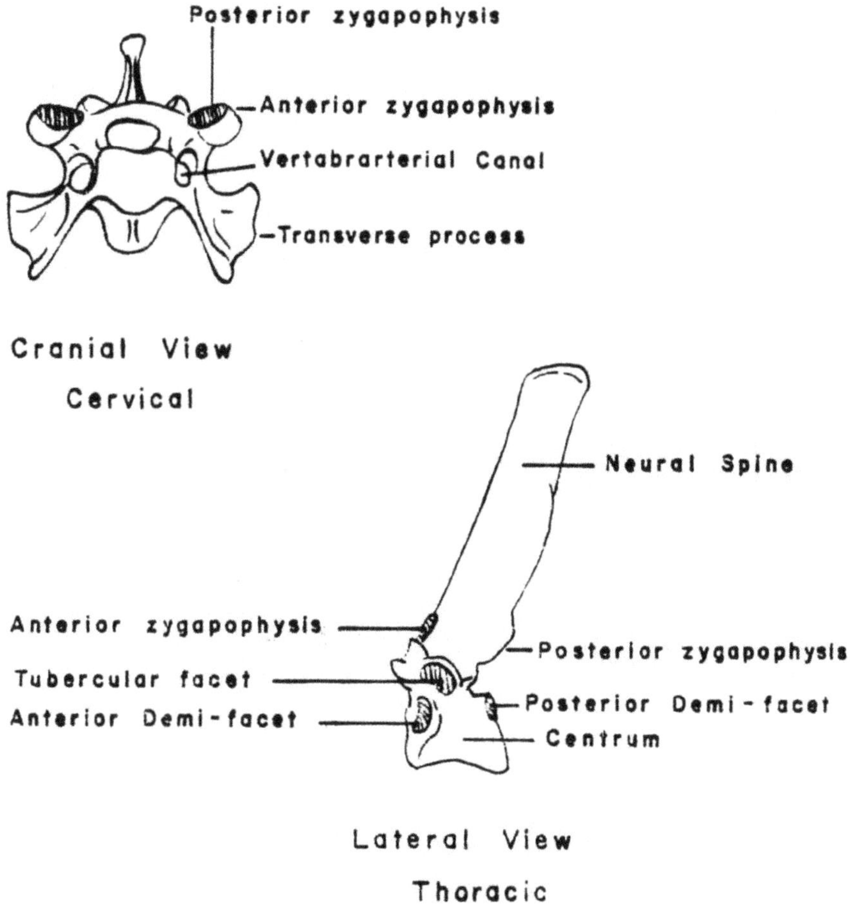

Figure 4-1, Landmarks as depicted on the cervical and thoracic vertebrae of Canis.

Equus caballus

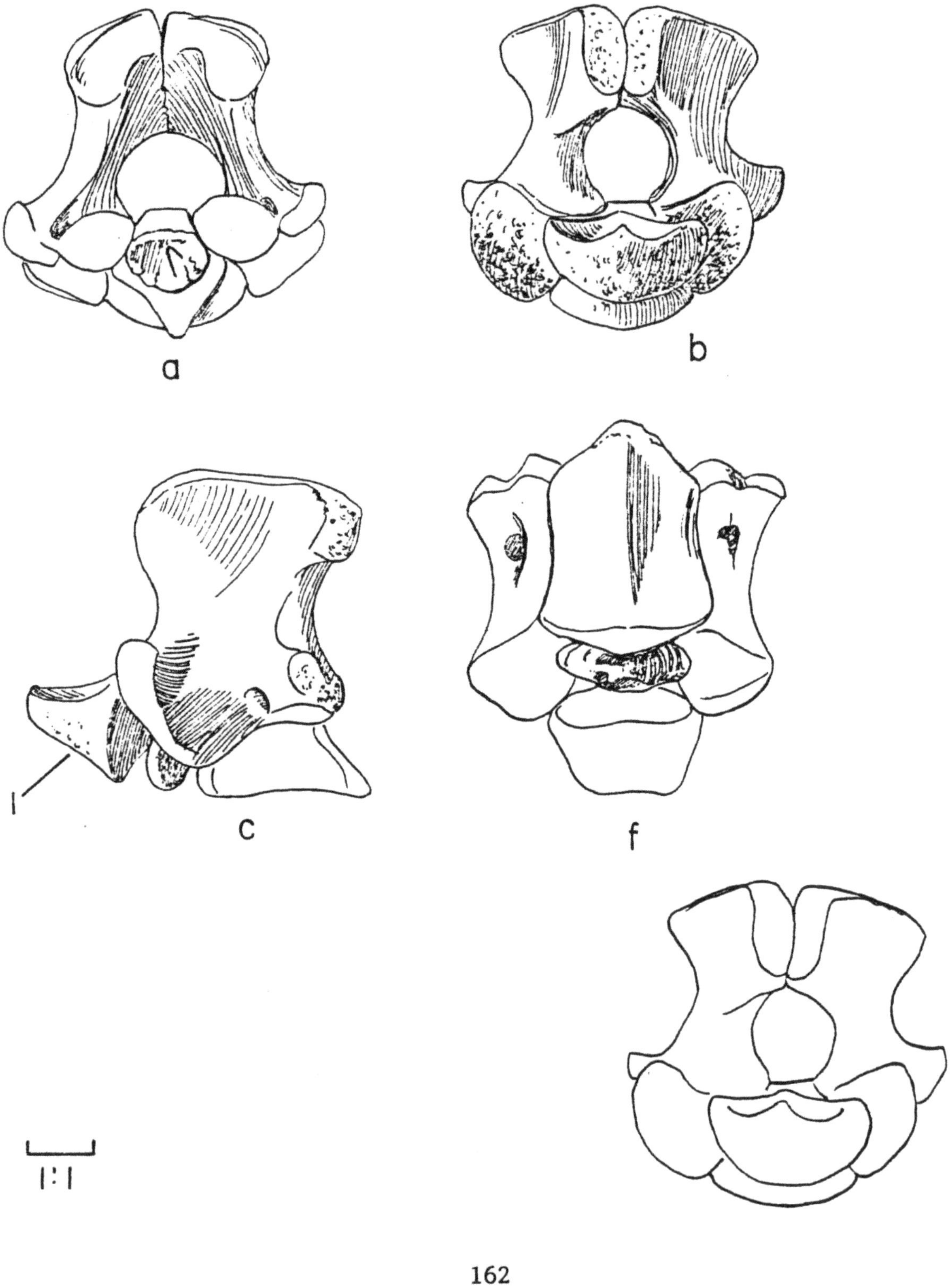

Bos sp.

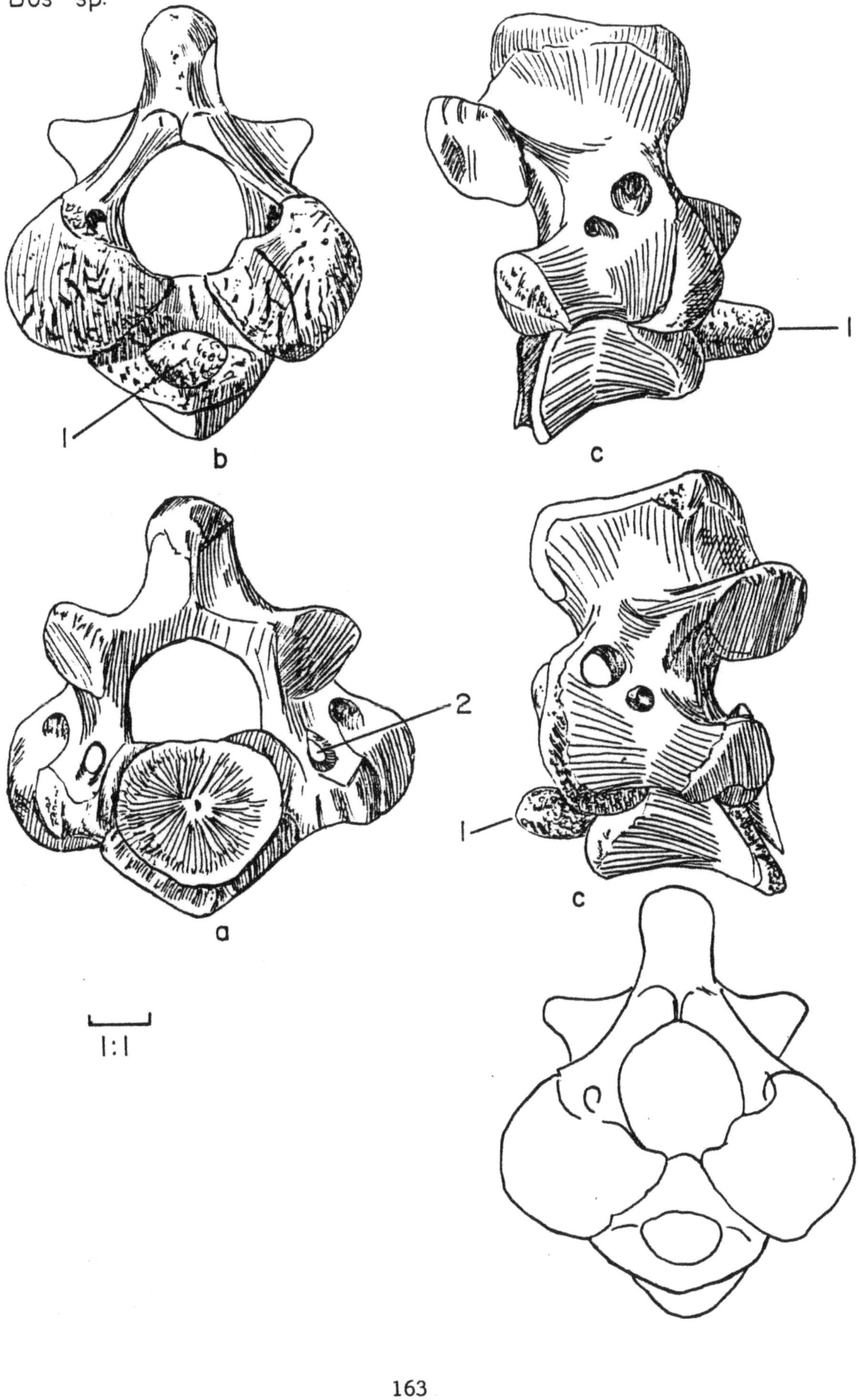

Ovis aries

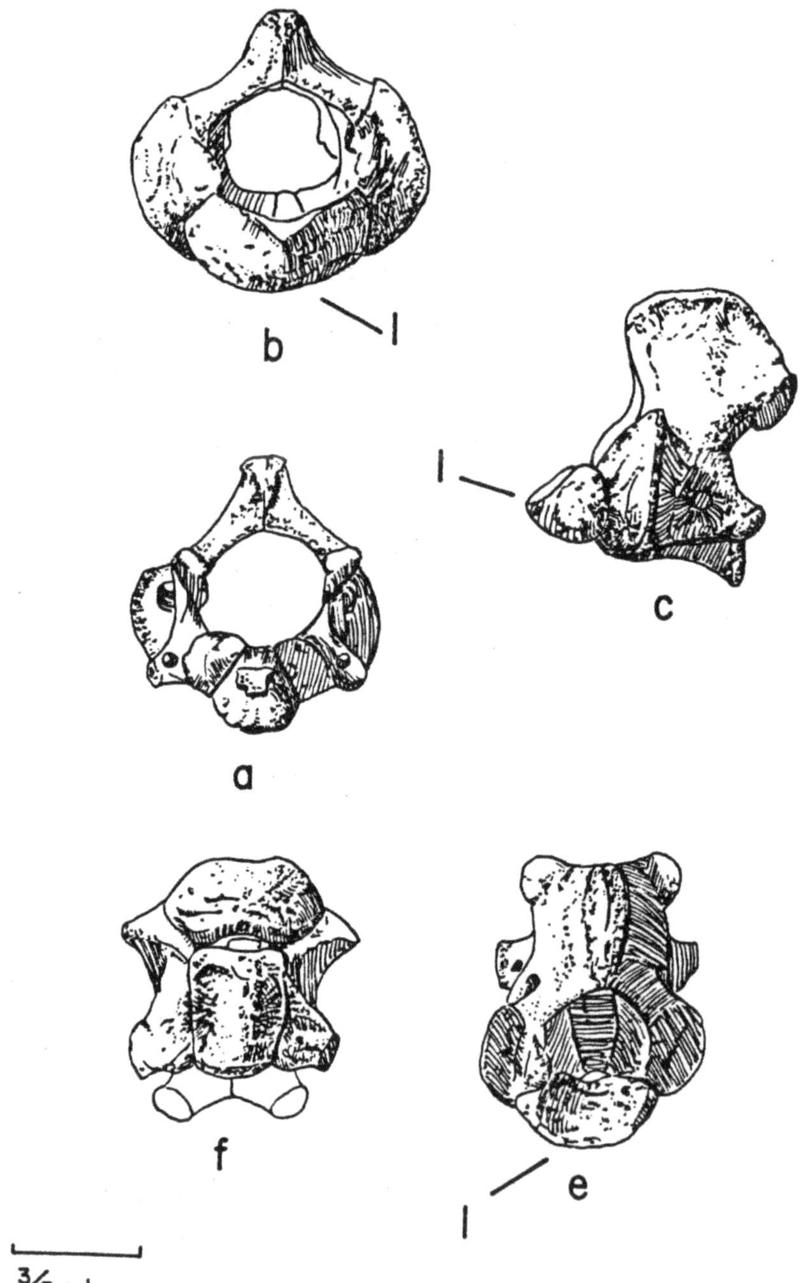

Sus scrofa

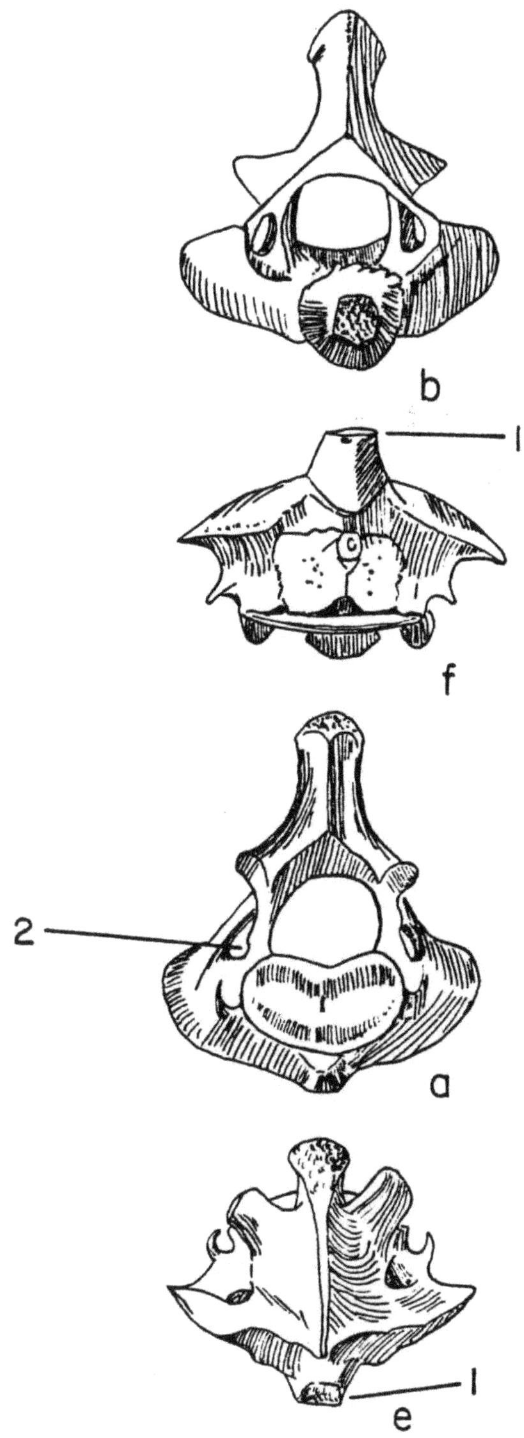

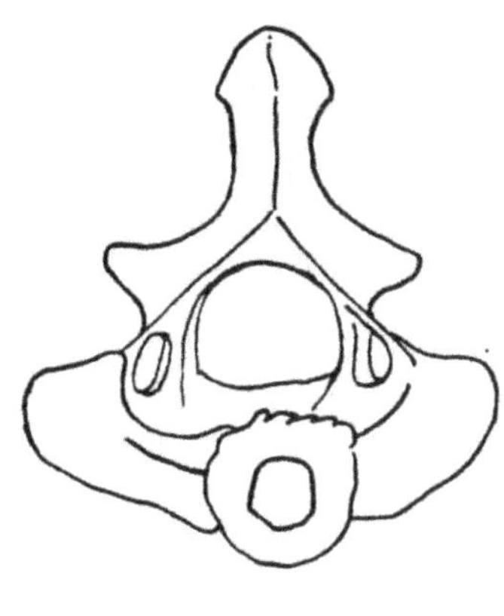

1:1

Canis familiaris

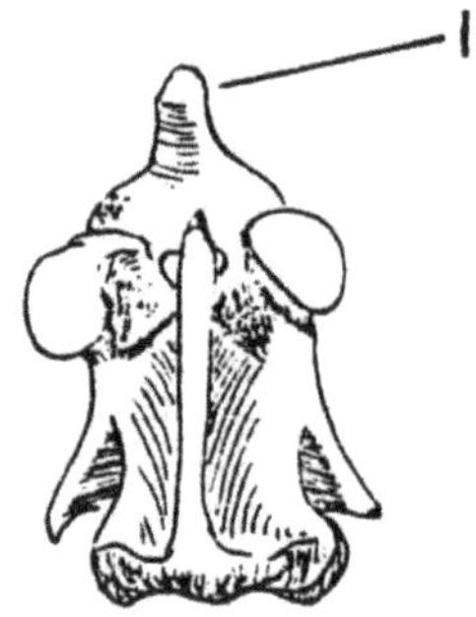

e

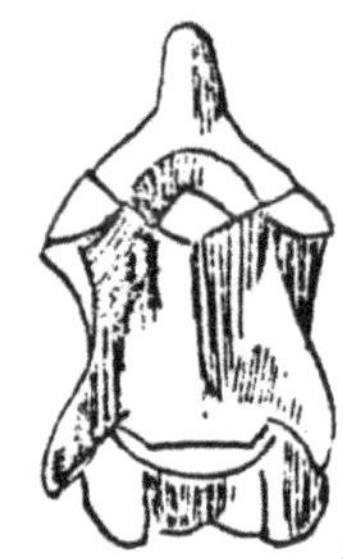

f

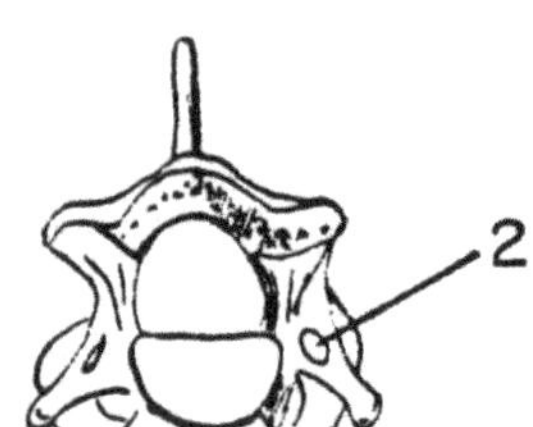

a

b

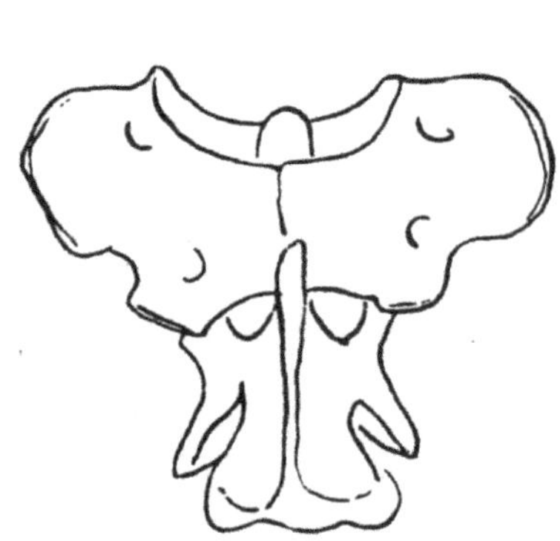

1:1

Felis catus L.

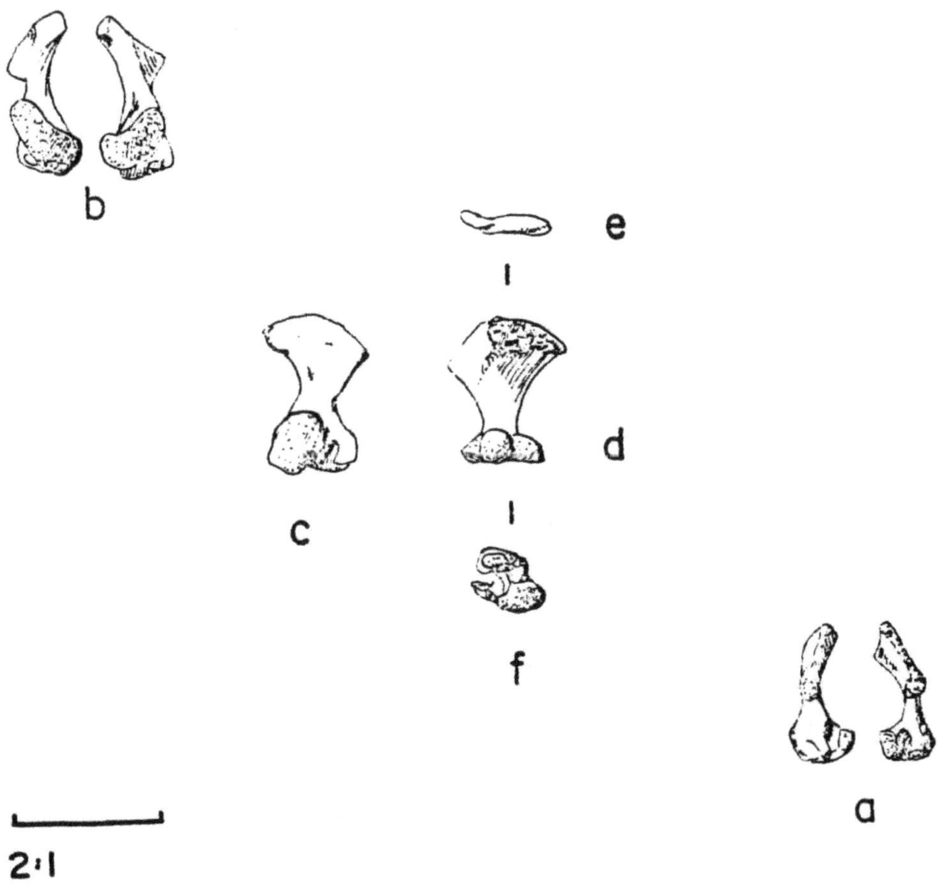

Rattus norvegicus

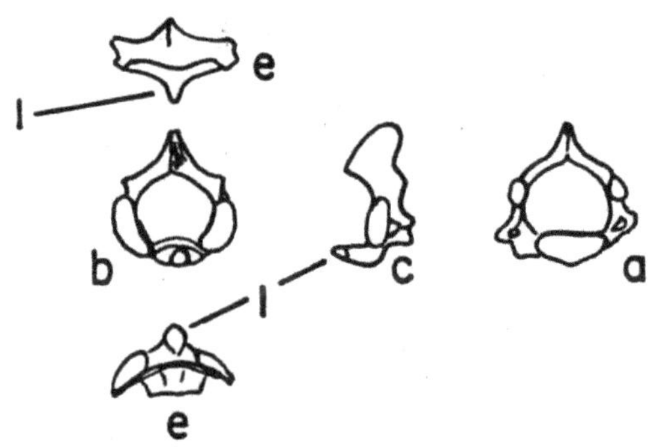

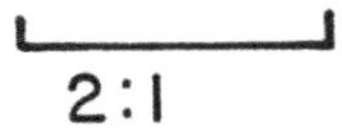

Plates V
Cervical Vertebrae

With the exception of the atlas and axis, the remaining cervical vertebrae are the transition into the thoracic series of vertebrae. The foramen transversarium (see figure 4-1, vertebrarterial canal) becomes progressively shorter posteriorly. The transverse process at first is directed downwards and backwards. Later the processes direction become more upturned and lateral. Cervical ribs are usually pointed forwards with a downward direction. The 7th vertebra can bear a demi-facet for the first rib. The zygapophyses (the articular surface for the next vertebra) are large, flattened; sometimes continuous with the root of the neural arch. The length of the centrum will diminish towards the rear of the series (Carter 1976:114-116). The illustrations of the cervical vertebrae in this text are representative of the middle of the series. This is not only demonstrate the transition between the cervical and thoracic vertebrae, but also the characteristics of the cervical vertebrae.

Views:
a - cranial
b - caudal
c - ventral
d - dorsal
e - lateral

Landmarks:
1 - Foramen transversarium

Equus caballus

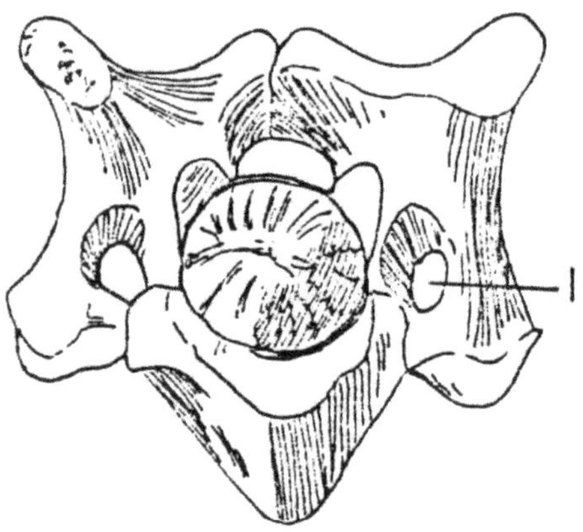

a

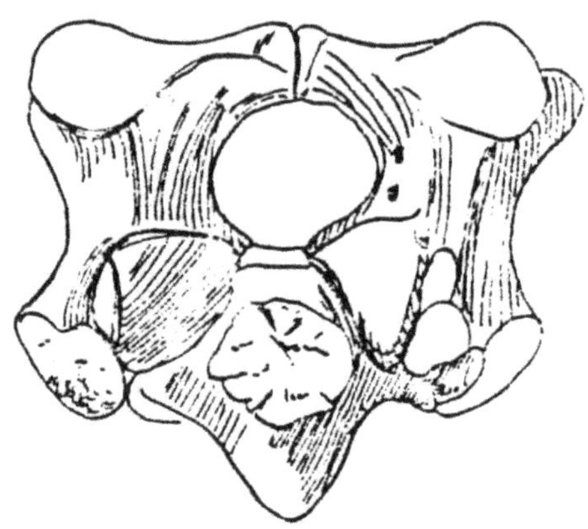

b

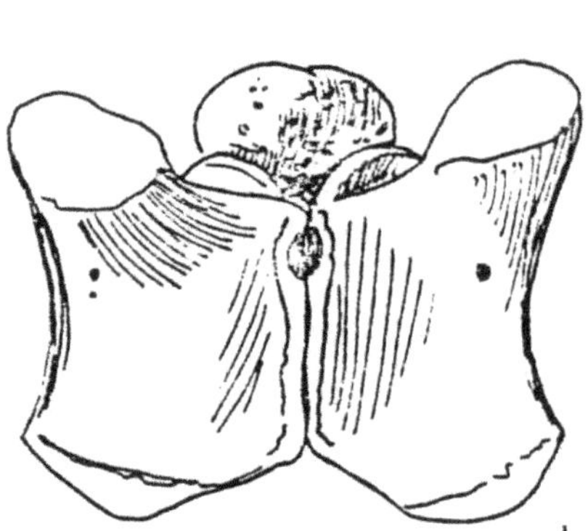

d

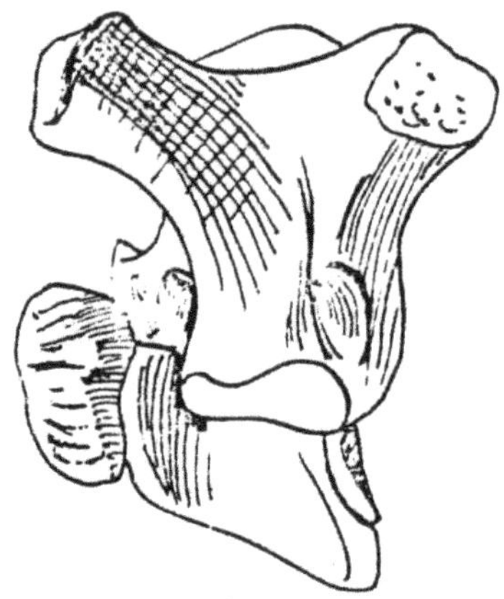

e

1:1

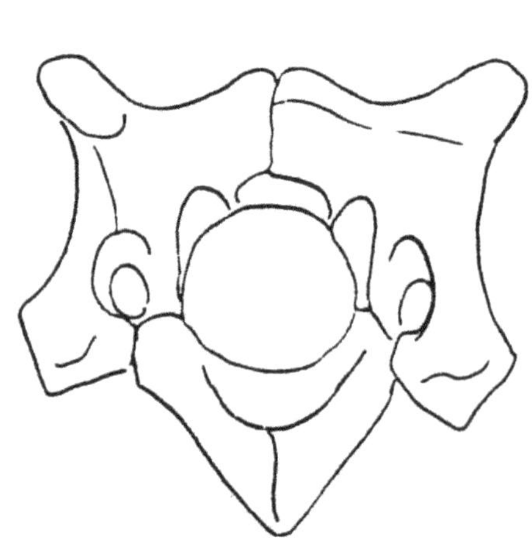

Bos sp.

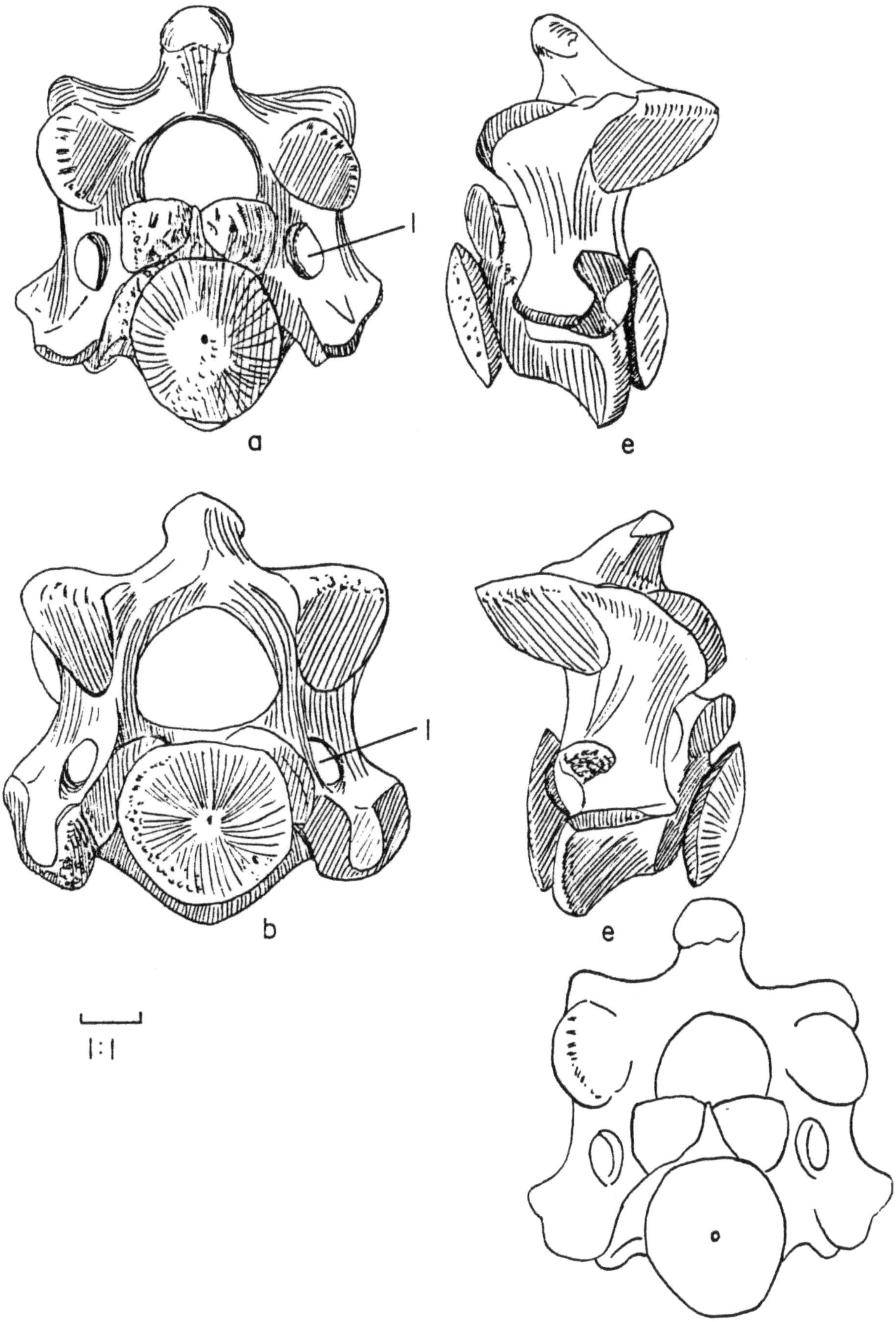

Ovis aries

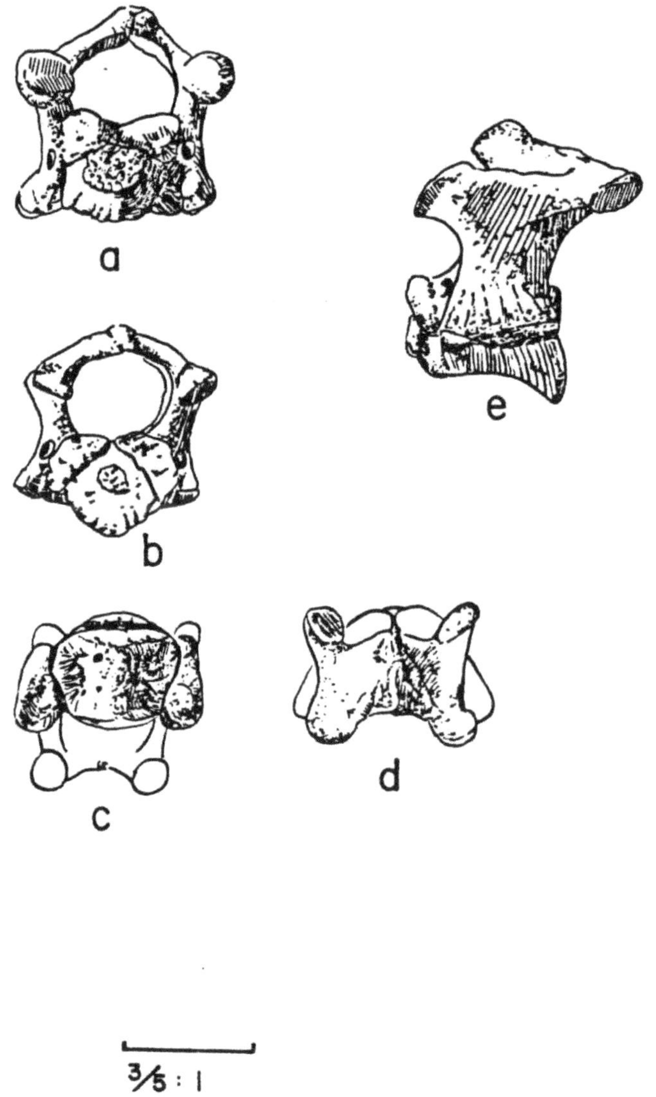

Sus scrofa

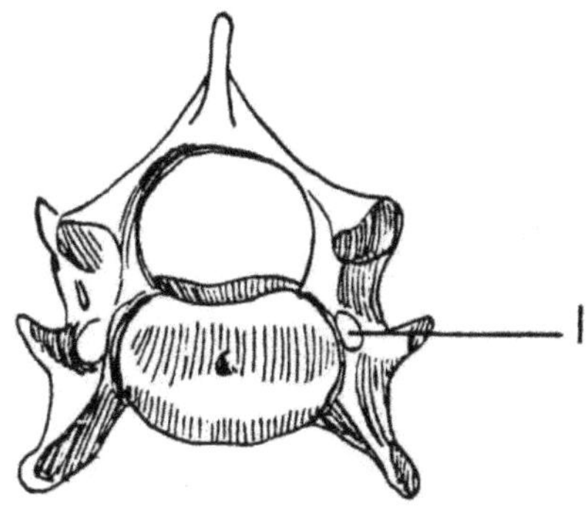

a

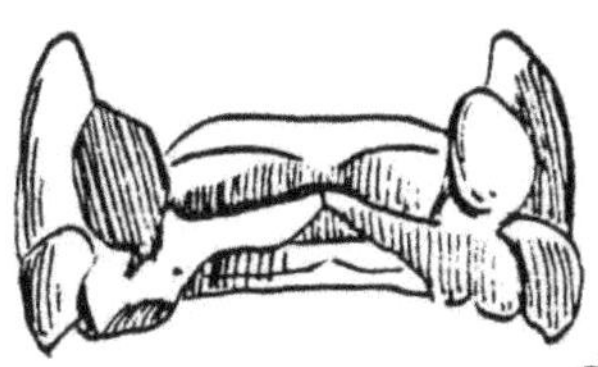

d

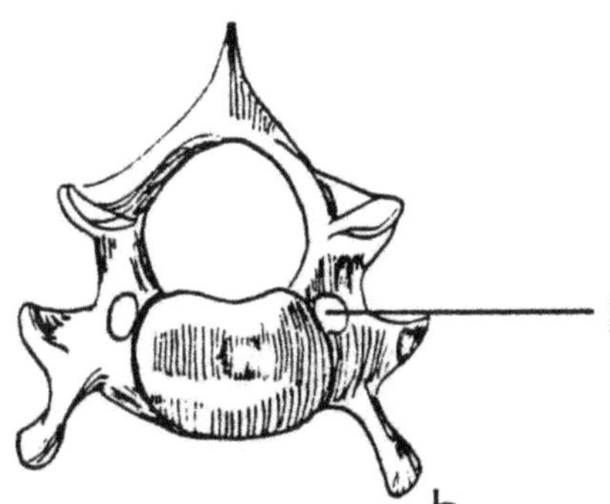

b

c

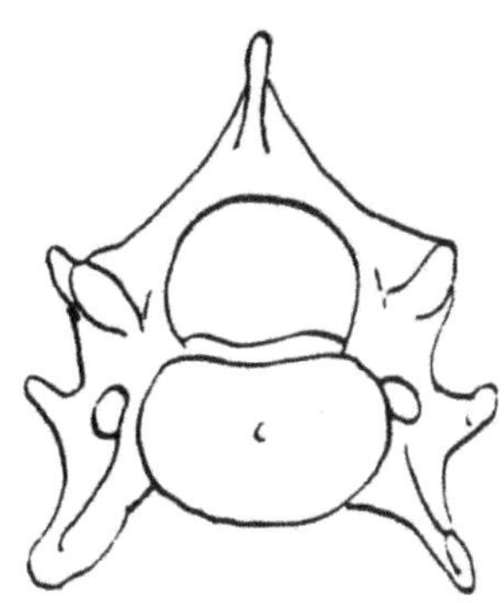

1:1

Canis familiaris

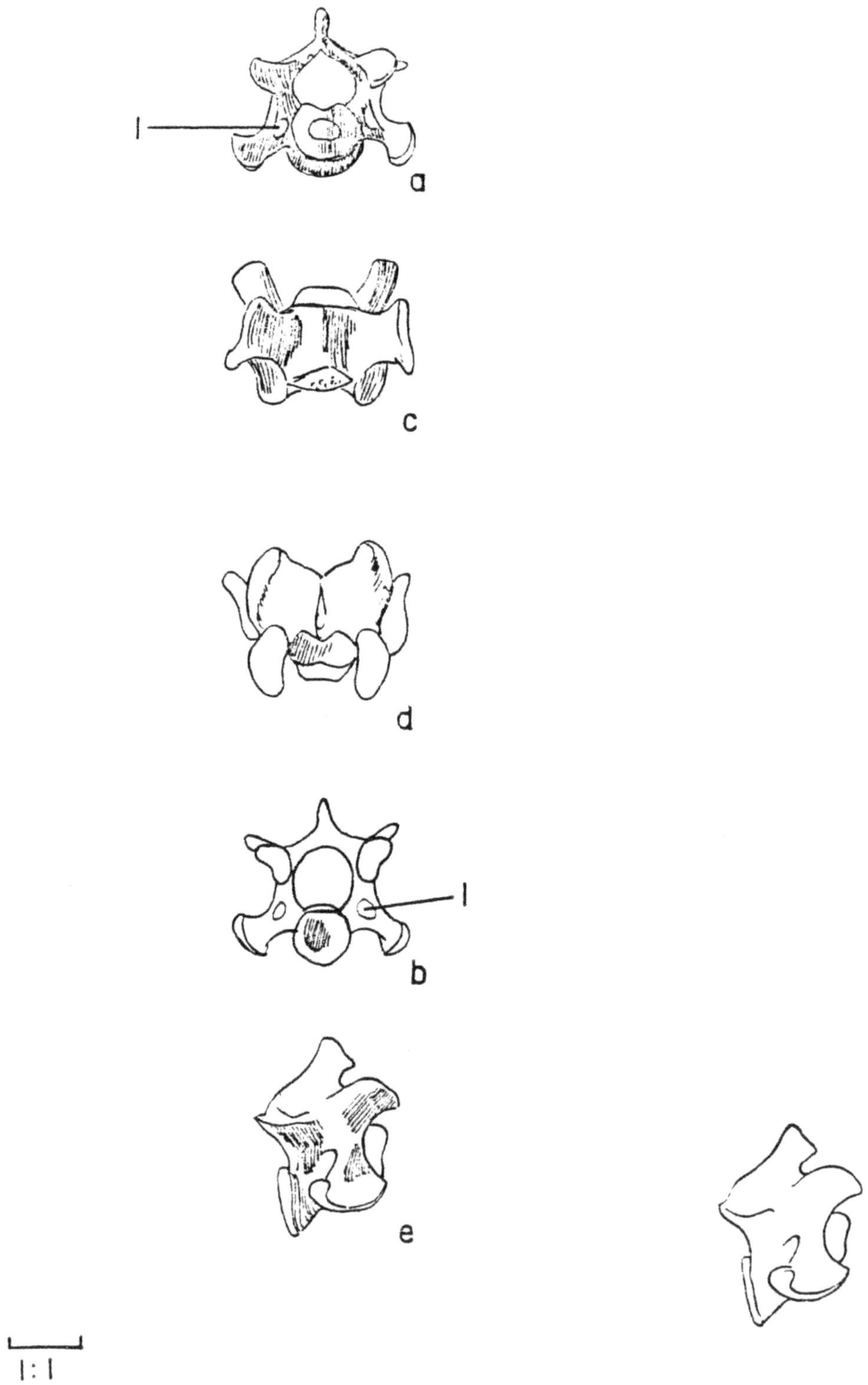

Plates VI
Thoracic Vertebrae

The anterior thoracic vertebrae differs from the posterior part of the thoracic series in that the posterior ones are more transitional to the lumbar vertebrae. Characteristic of the thoracic vertebrae are the demi facets for the articulation of the ribs. However, the size of the demi-facets varies, between animals and within the same animal. The size of the demi facets diminish posteriorly along the series. The smaller demi-facets can be seen on the last few vertebrae, and possibly lacking on the last thoracic. There is a general backwards slope, with the slope increasing from the 1st to the 8th, and decreasing after the 11th (sometimes known as the anticlinical vertebra). Beyond the 11th or 12th vertebrae, the thoracic series starts to resemble the lumbar vertebrae in general form. The transverse process around the 11th vertebra, change from a short thick process to a more longer, thinner process, as seen in the lumbar series (Cornwall 1956; Sisson and Grossman 1966; Carter 1976).

The zygapophyses on the 1st thoracic resemble those of the posterior cervical series. From the 2nd to the anticlinal vertebrae, the zygapophyses are small, oval in shape, and close together, nearing the neural arch. After the 11th vertebrae, the zygapophyses resemble those of the lumbar series.

Views:

a - cranial
b - caudal
c - ventral
d - dorsal
e - lateral

Equus caballus

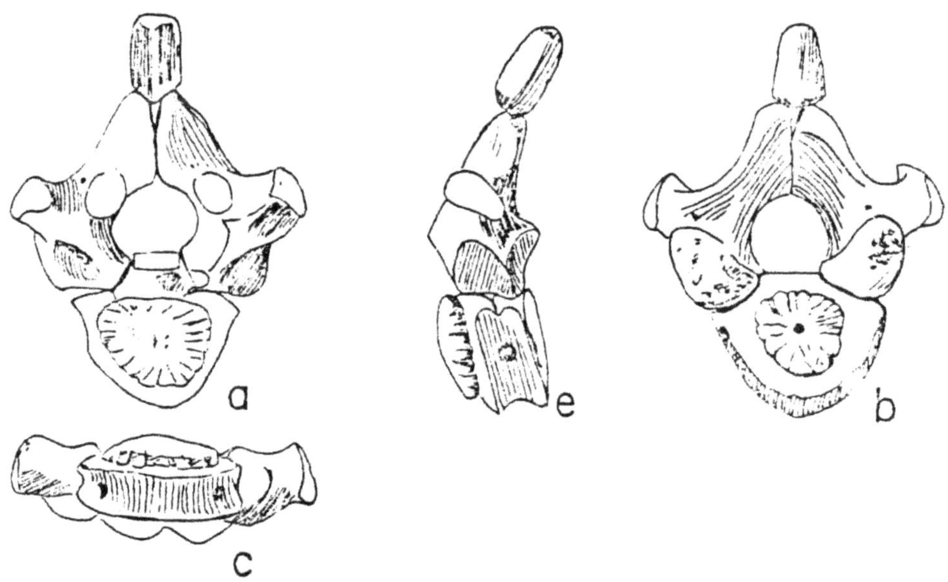

1:1

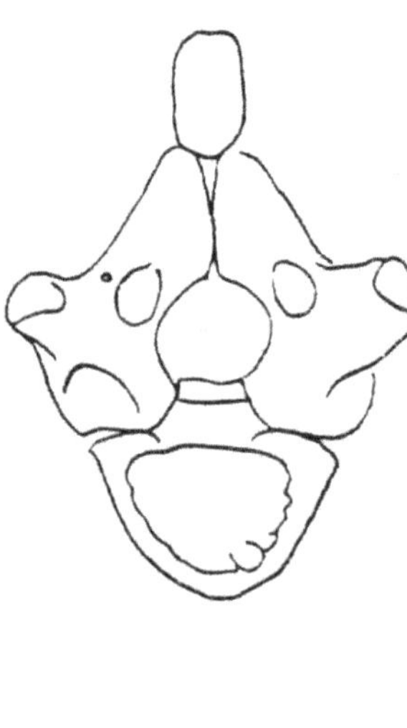

Bos sp.

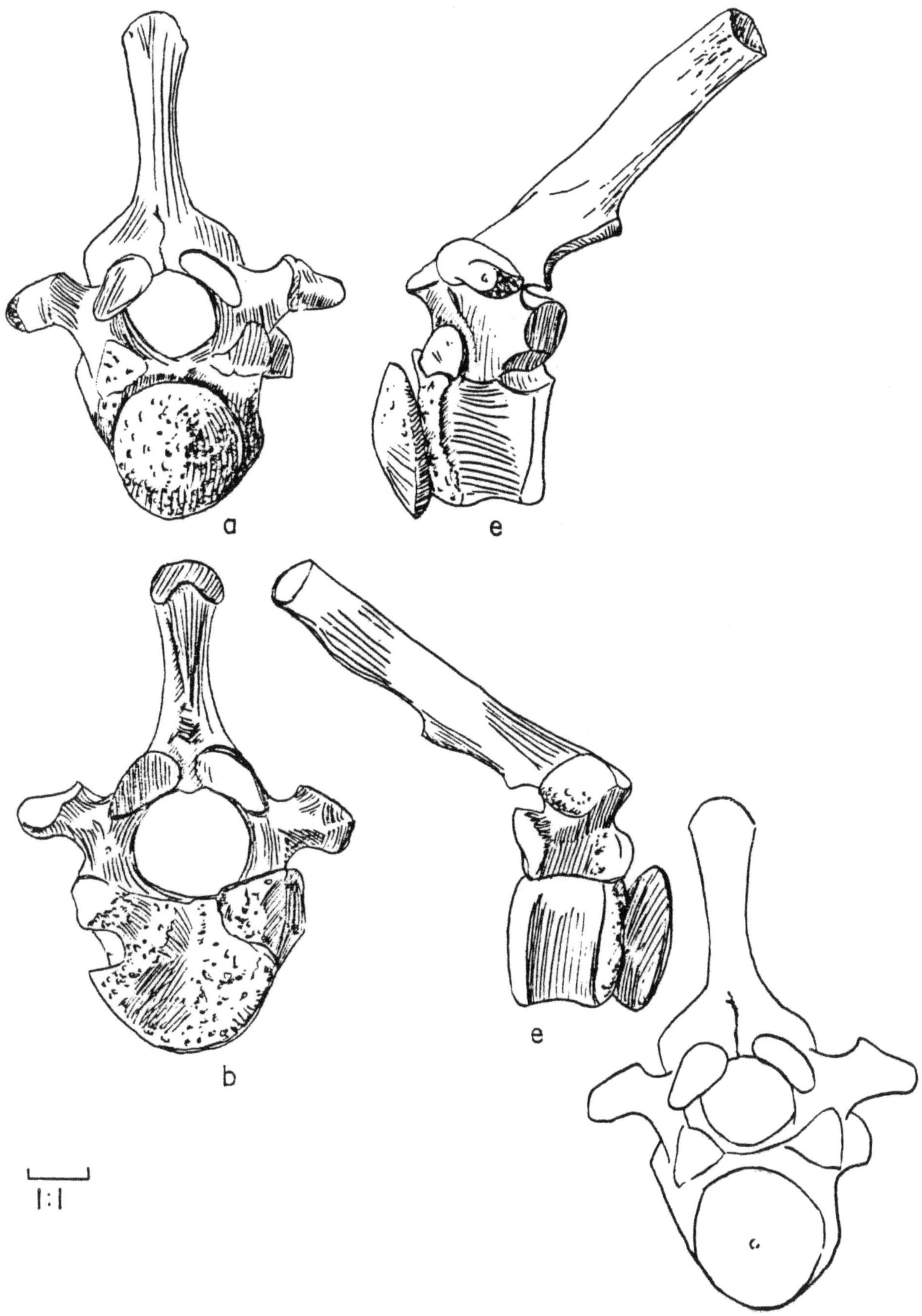

Ovis aries

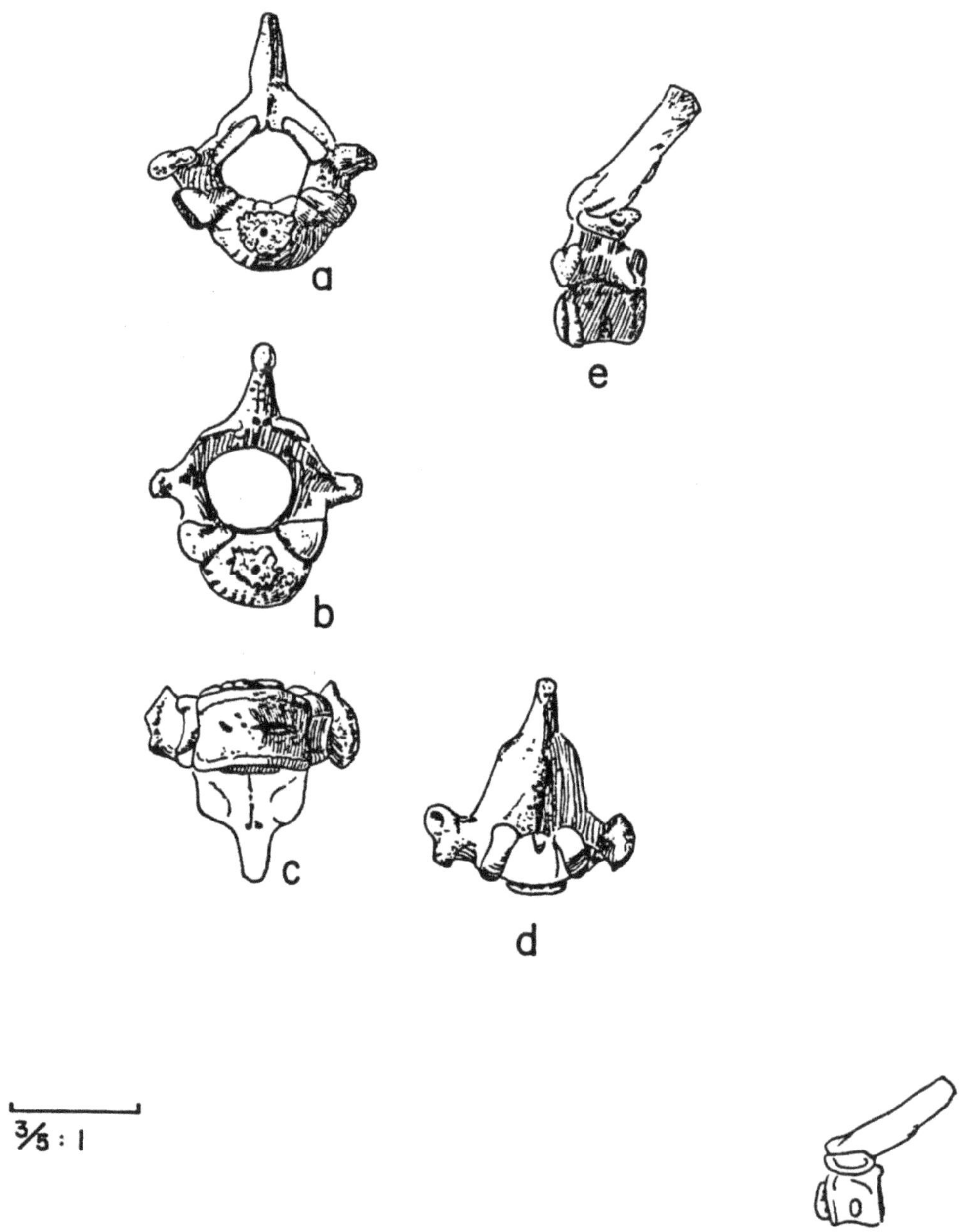

Sus scrofa

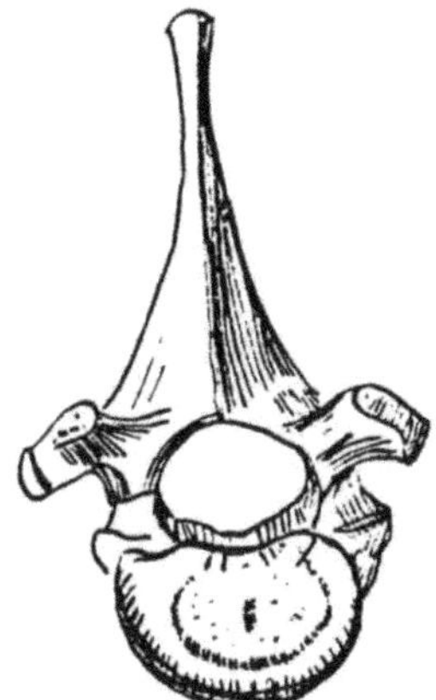

a

d

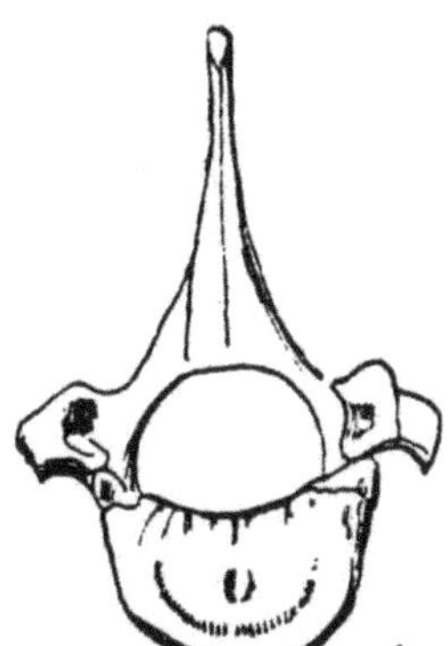

b

c

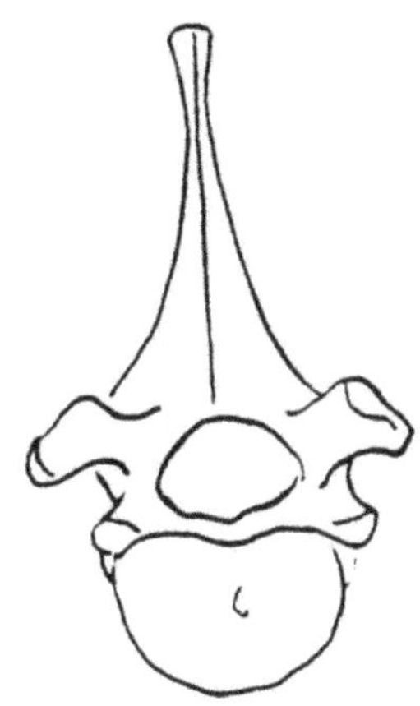

1:1

Canis familiaris

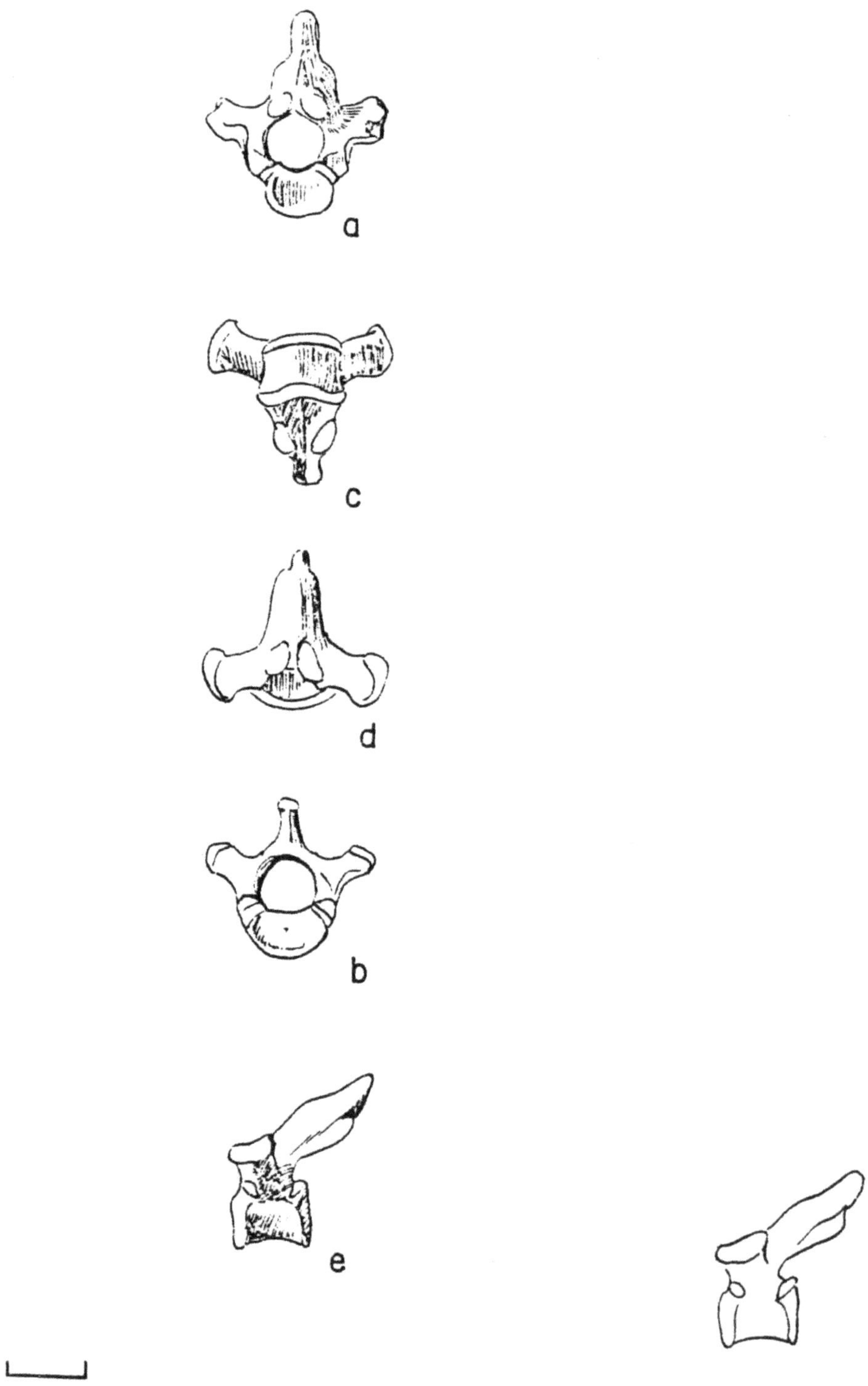

Plates VII
Lumbar Vertebrae

The zygapophyses are much more prominent in the lumbar series than in any other vertebrae series. The facets of the zygapophyses tend to be curved and positioned on a vertical axis. The neural spine becomes low, and less robust than the thoracic series. Often the neural spine is vertical of facing forward. The transverse processes are long and spreading, usually directed in a forward-downward direction. The centrum becomes more elliptical in cross-section, especially towards the posterior part of the series (Carter 1976:117 119).

Views:

a - cranial
b - caudal
c - ventral
d - dorsal
e - lateral

Equus caballus

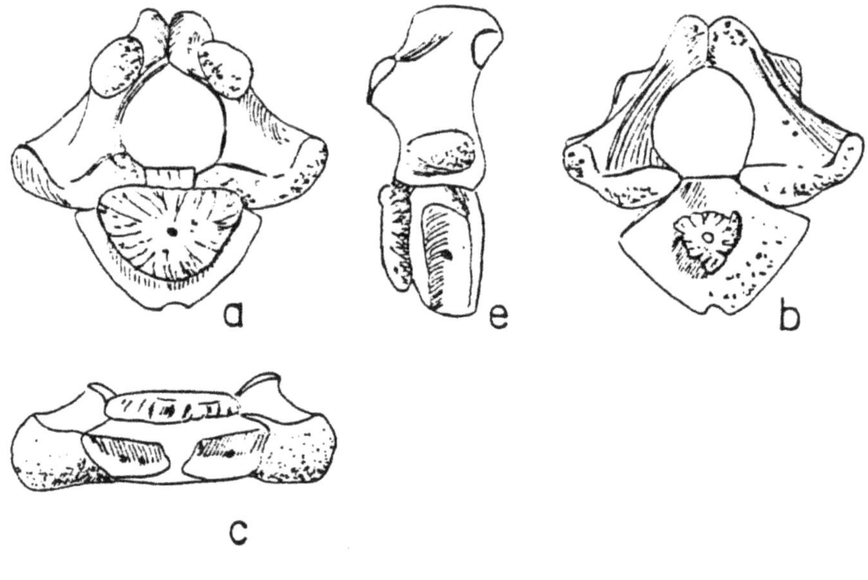

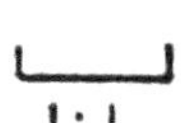

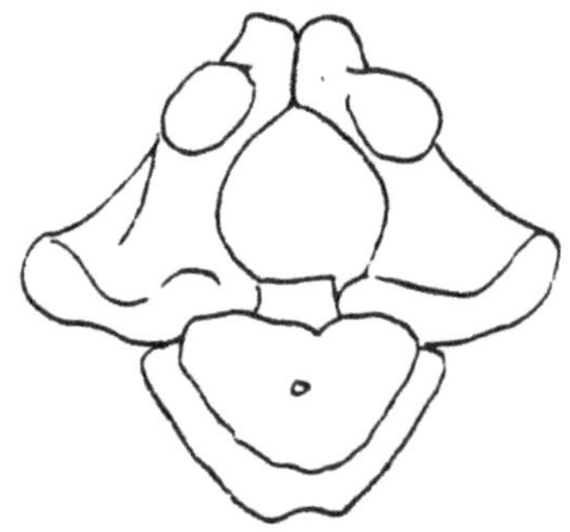

Bos sp.

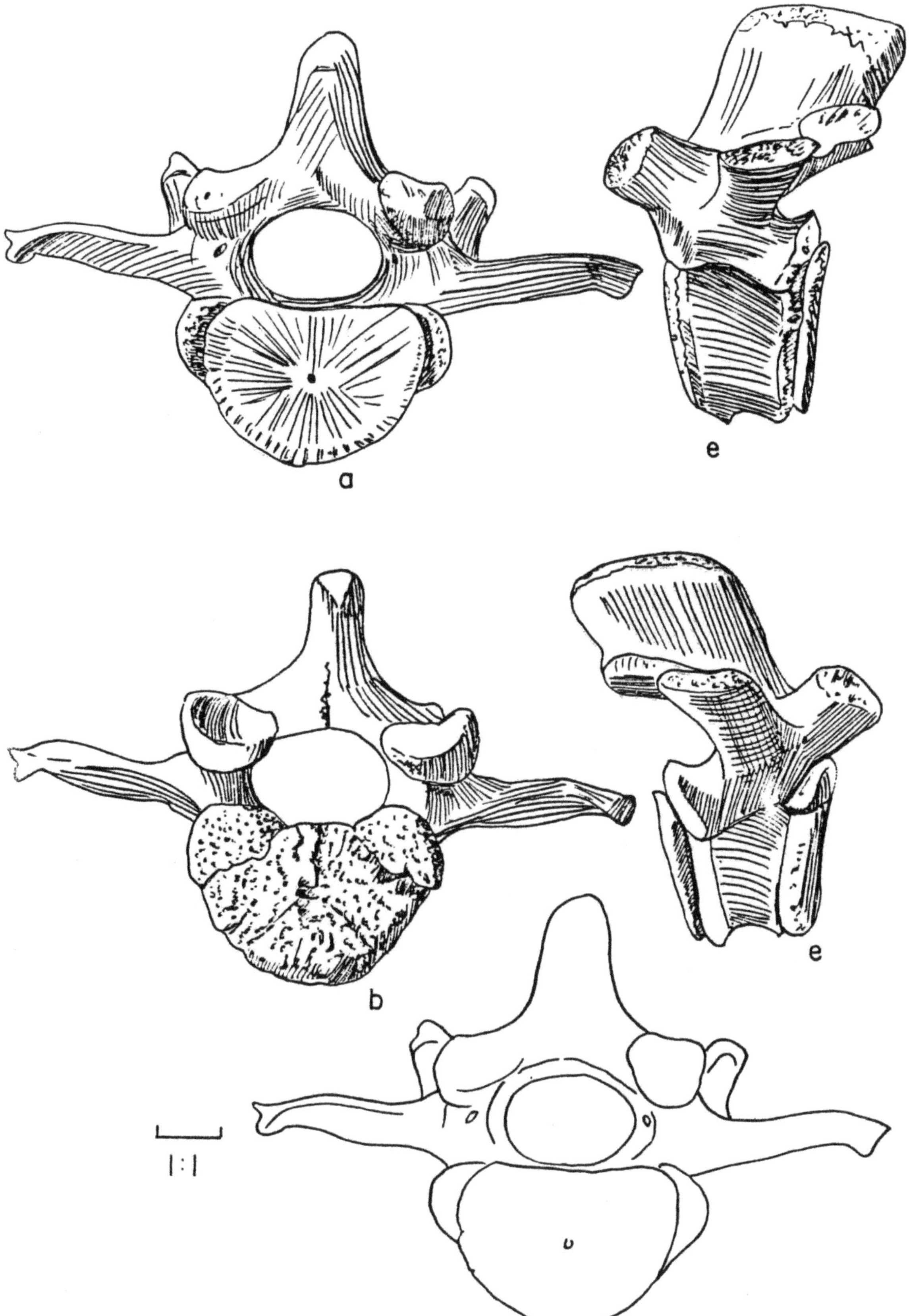

Ovis aries

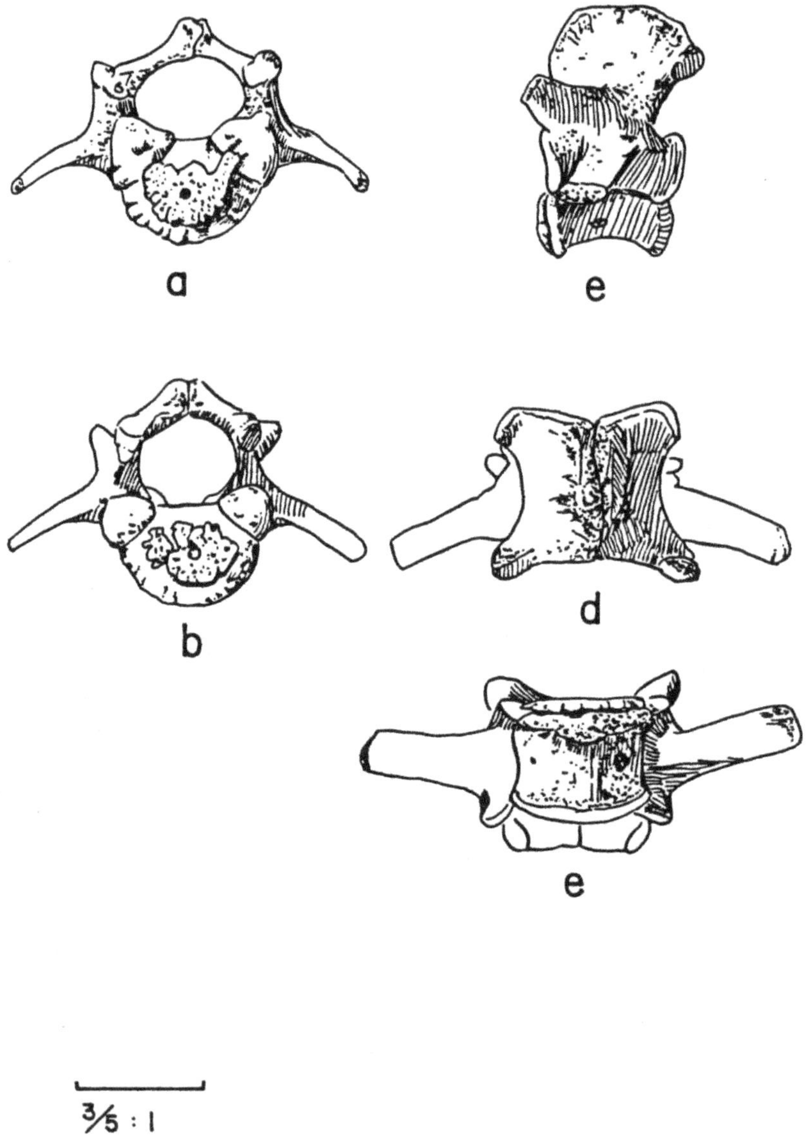

Sus scrofa

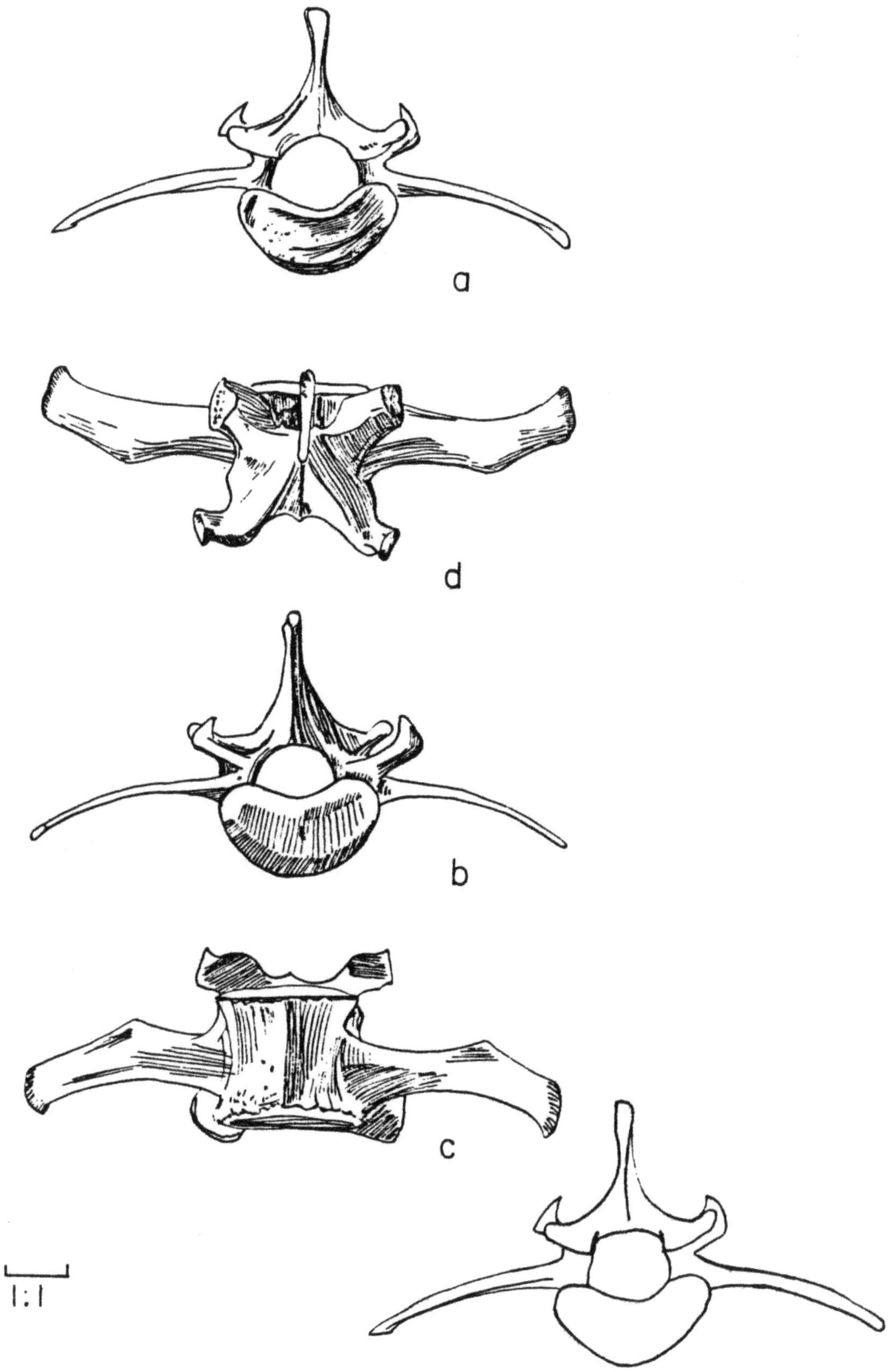

Canis familiaris

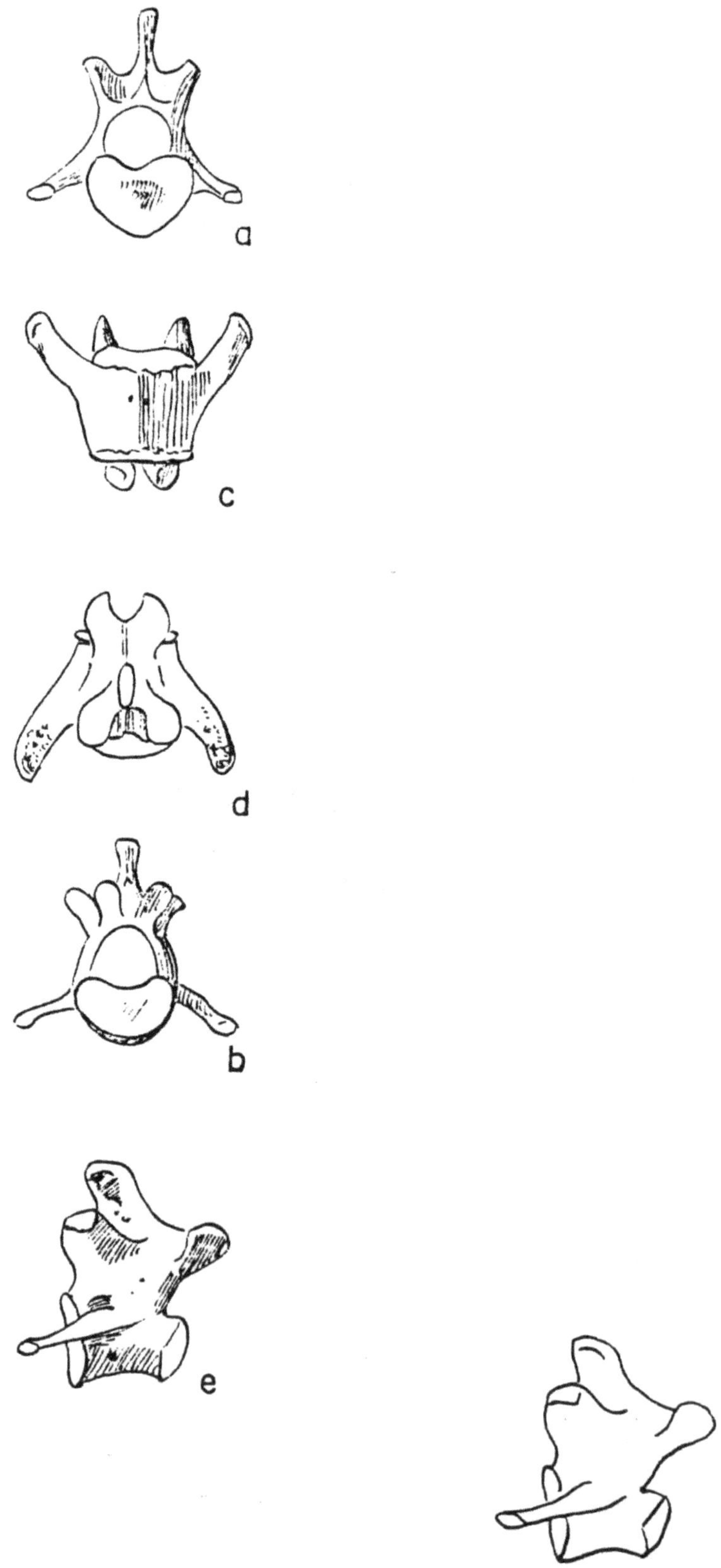

1:1

Plates VIII
Sacrum

The sacrum is composed of a set of fused vertebrae. The number of vertebrae can vary depending on the class of mammal. The neural spines of the sacral series can fuse together forming a continuous ridge of bone. The zygapophyses can also fuse together, forming bony ridges parallel to the neural spine. The transverse processes of the first sacral vertebrae are stout, and articulate with the ilium. The centra are spread more laterally than the other vertebrae series.

Carter (1976:119) states that sex can readily be determined from the sacrum for ungulate species. In males, the centra are shallow and triangular in cross section. Ventrally, there is a distinct blunt keel. Females possess a thinner centra, that is dorso-ventrally flattened without a keel.

Views:

a - dorsal
b - ventral
c - cranial
d - caudal
e - lateral

Bos sp.

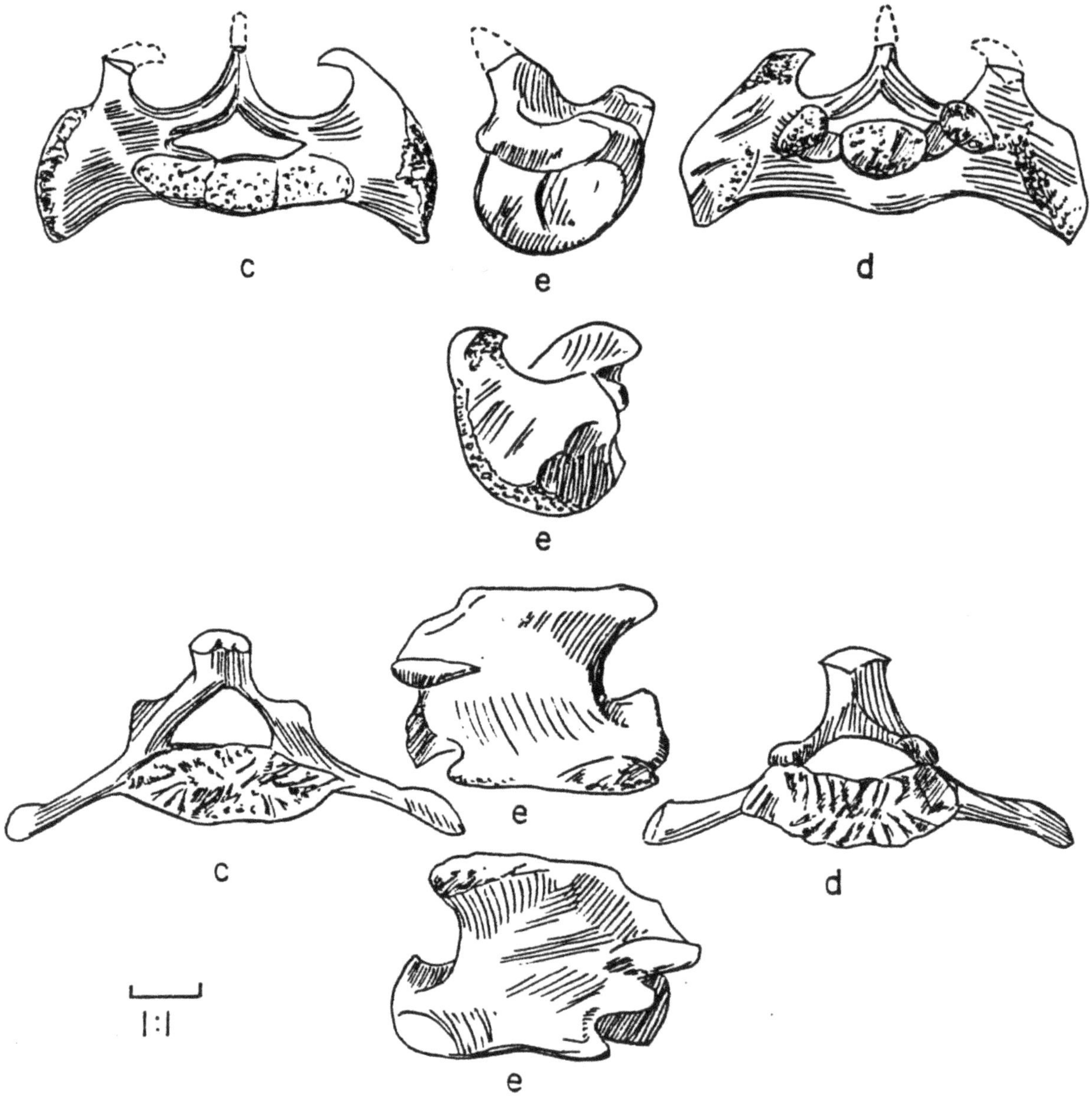

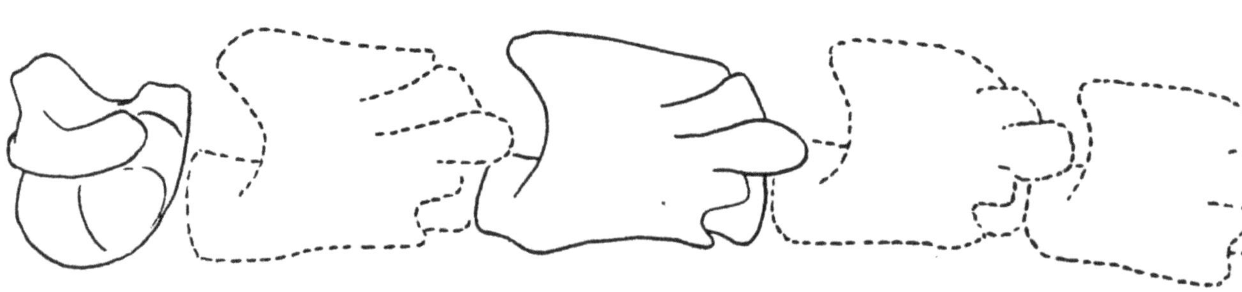

Ovis aries

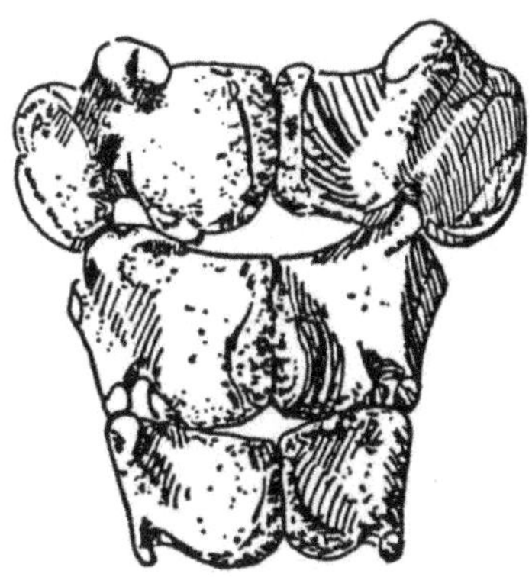

a

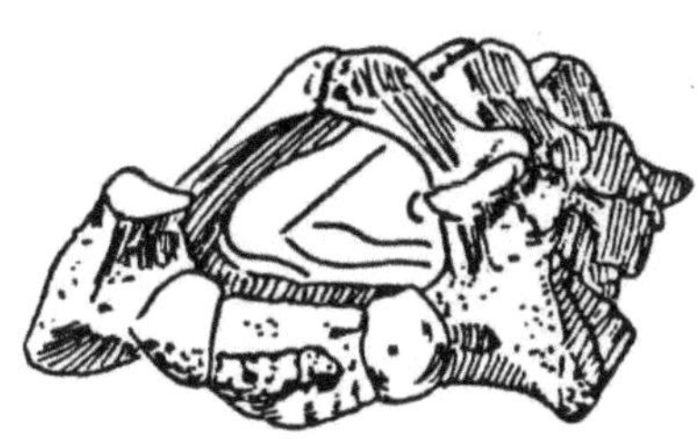

b

$^{3}/_{5}$: 1

Sus scrofa

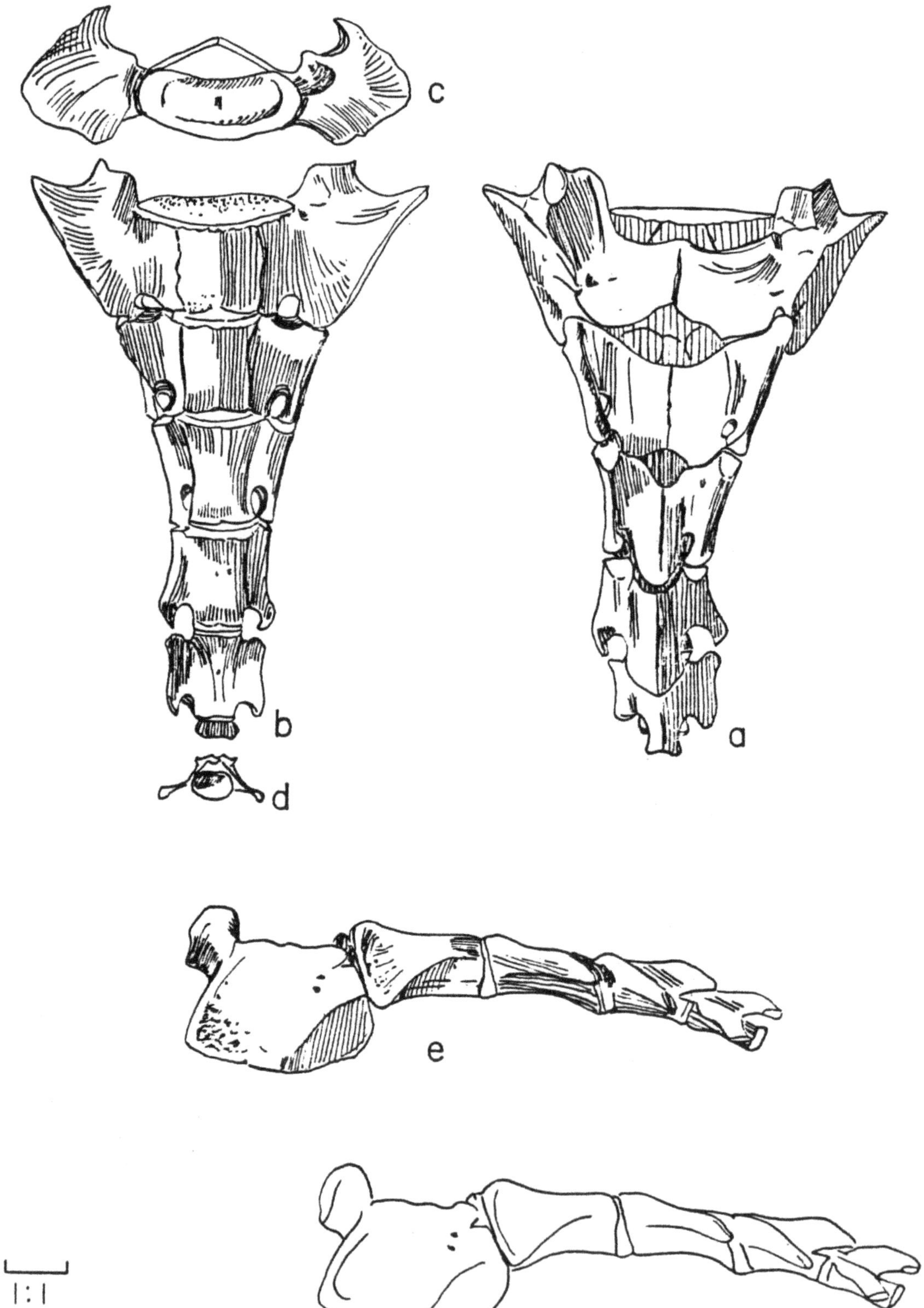

Canis familiaris

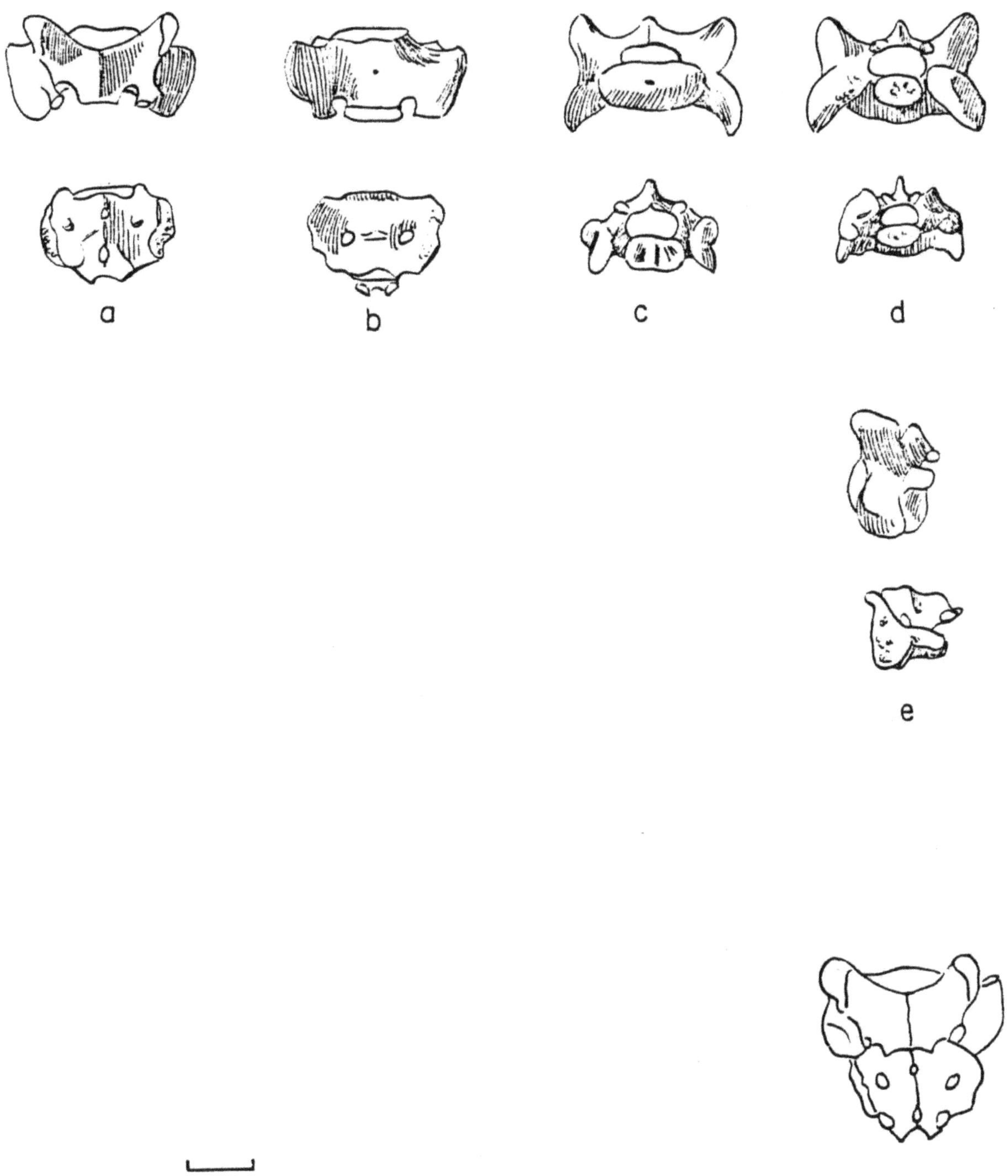

Appendicular Skeleton: Thoracic Limb
Plates IX
Scapula

The function of the scapula is to provide a steady, but movable base for the attachment of muscles in the fore limb. One set of muscles can move the scapula, and fix it in any position within its natural limits of travel. Another set arising on the scapula itself, holds the head of the humerus in constant contact with the glenoid cavity of the scapula. Thus, the scapula can move while it remains fixed by the opposed tension of the muscles whenever the limb is under load (Cornwall 1956).

The scapula can be described as occupying a dorso-lateral position in relation to the cranial part of the thorax. There is a prominent spine or ridge that divides the dorsal surface into two fossae, in quadrupeds, the pre- and post-spinous fossa (see figure 4-2). The thorax surface, (costal surface), is somewhat dish shaped and is known as the sub-scapular fossa. The muscle tendons arising from and lying within each fossae converge towards the periphery of the glenoid cavity, and insert at the tuberosities surrounding the neck of the humerus (Cornwall 1956). The distal part of the spine of the scapula can extend into a prominent, somewhat overhanging process, known as the acromion. The acromion, if strongly developed, can serve as protection for the shoulder-joint and adds to the articulation of the clavicle. The acromion can extend in a right angle to the spine, in a caudal direction and is known as the metacromial process. The coracoid, another process projects from the cranial margin of the glenoid cavity (Cornwall 1956).

Views:
a - dorsal
b - costal
c - glenoid fossa

Landmarks:
1 - Tuber scapulae
2 - Cranial end of the articulation
3 - Lateral edge of the articulation
4 - Tuber in <u>Ovis</u> beside lateral edge, small; in <u>Sus</u> along the cranial edge, larger
5 - Medial edge in <u>Ovis</u> round, in <u>Sus</u> somewhat incurved

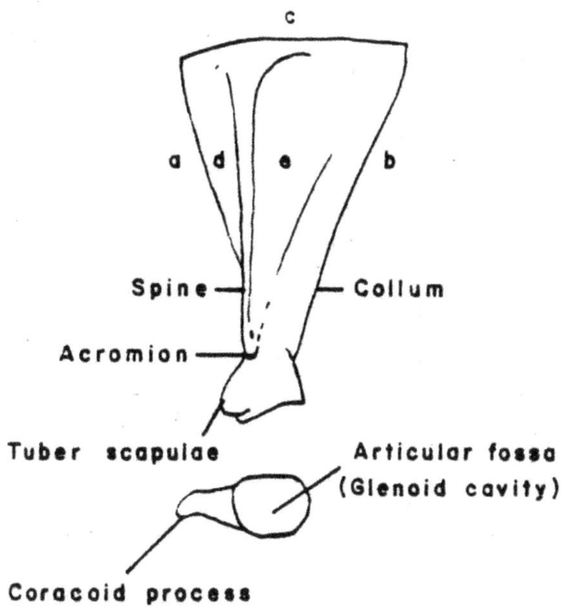

Figure 4-2, the basic architecture of the scapula of Ovis. a - Margo cericals, b - thoracius, c - Margo vertebralis, d - Pre-spinous fossa, e - Post-spinous fossa.

The general form of the scapula is indicative of the class of mammal (see figure 4-3). While the spine of the scapula can be indicative of the class or species of mammal (see figure 4 4).

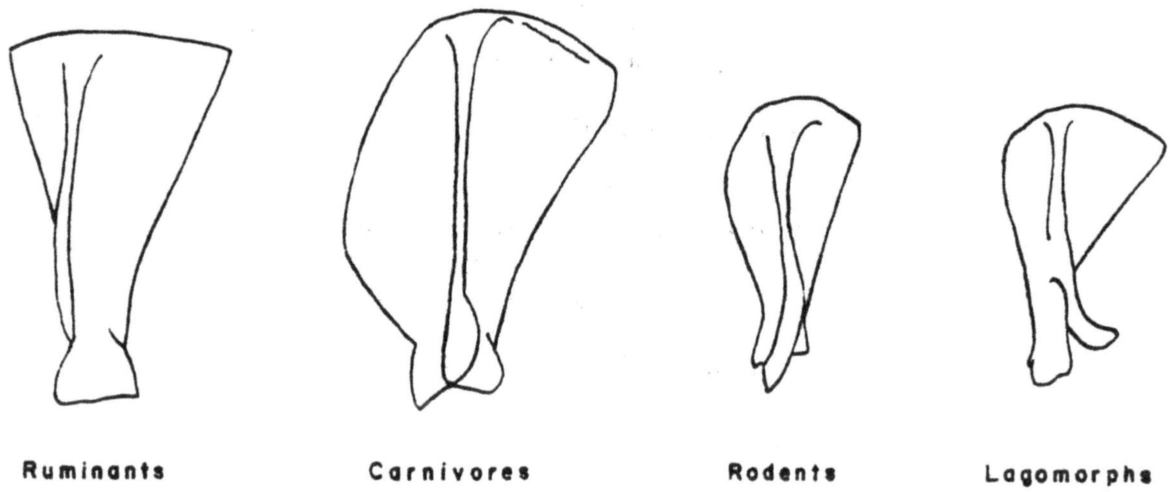

Figure 4-3, the general form of the scapula is indicative of the class of mammal. In rodents the acromion extends over the glenoid cavity. Lagomorph have the acromion extending over the neck of the scapula (collum).

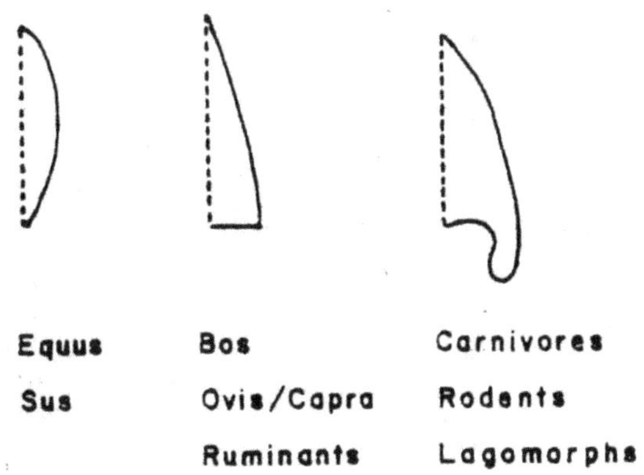

Figure 4-4, the spine as seen in cross section, can be indicative of the class or species of mammal (redrawn after Schmid 1972: figure 29).

Equus caballus

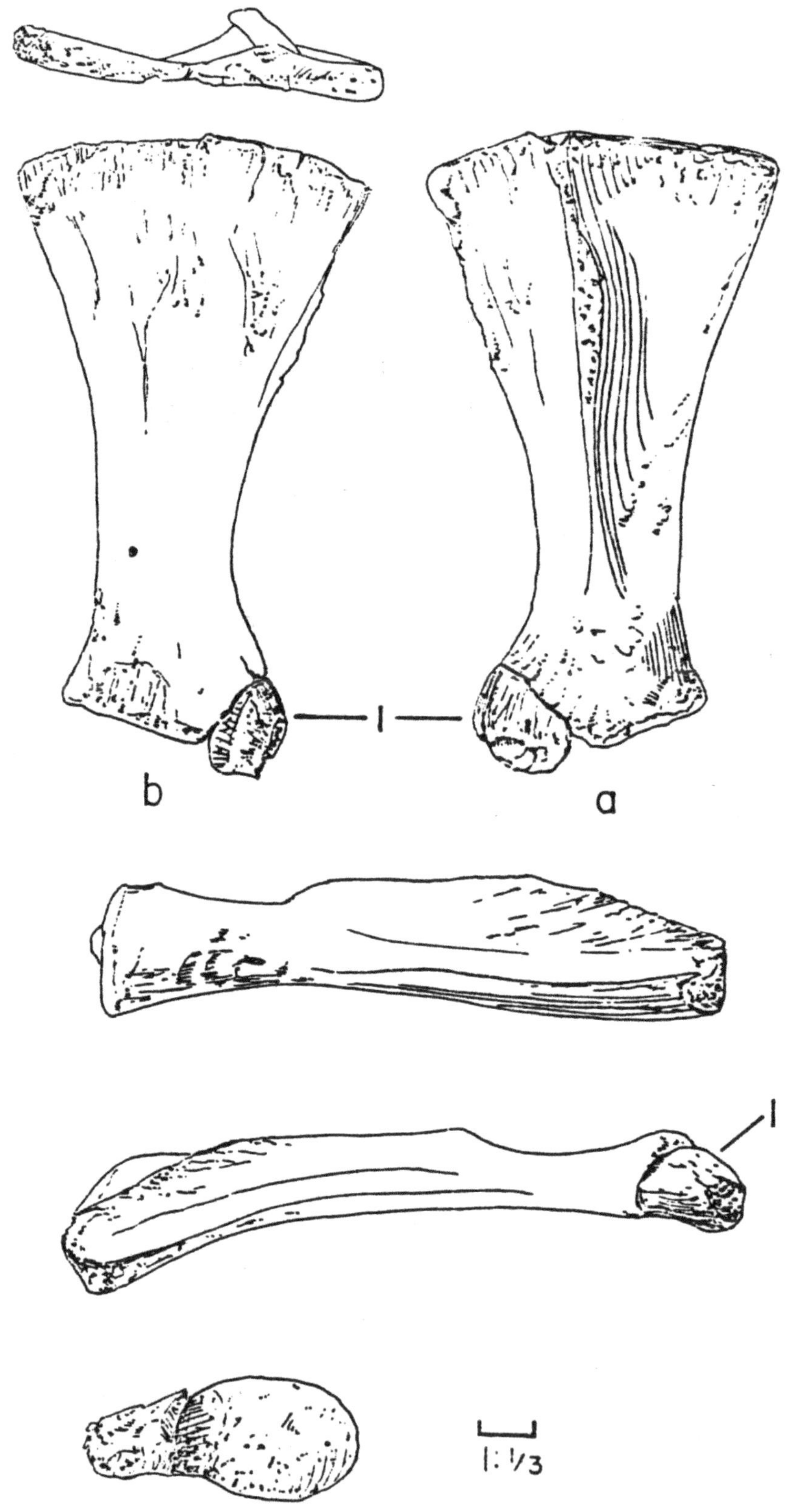

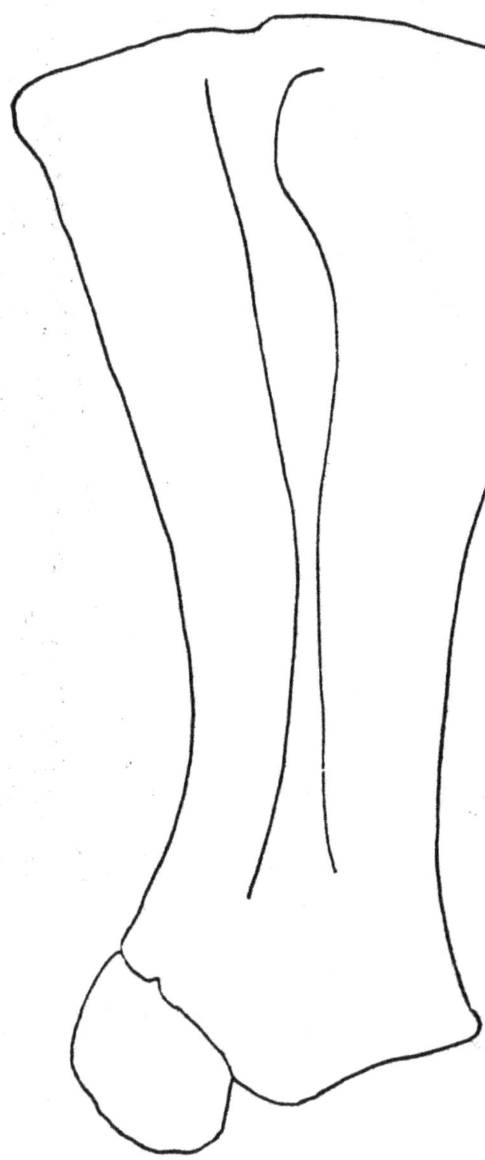

Bos sp.

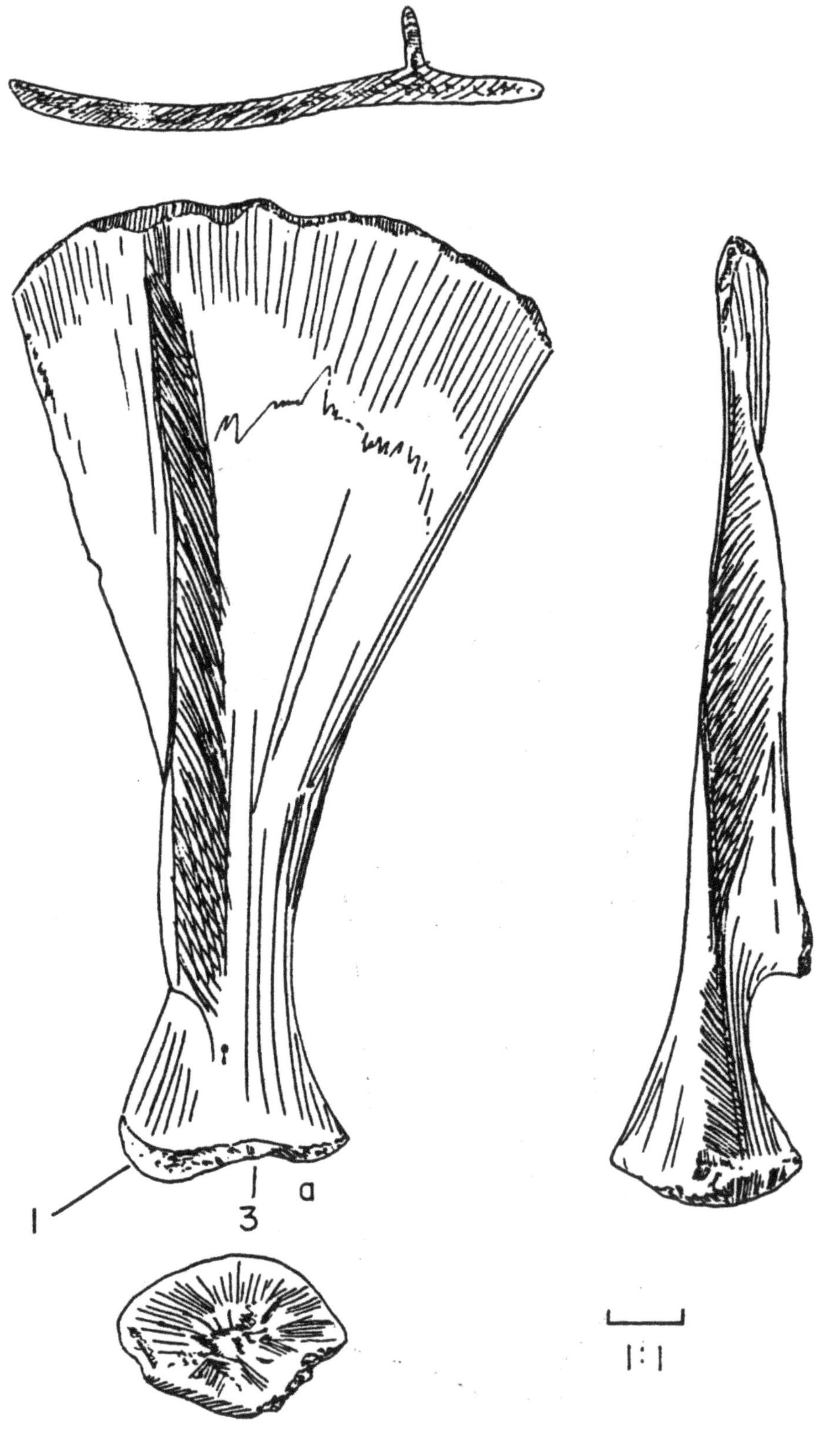

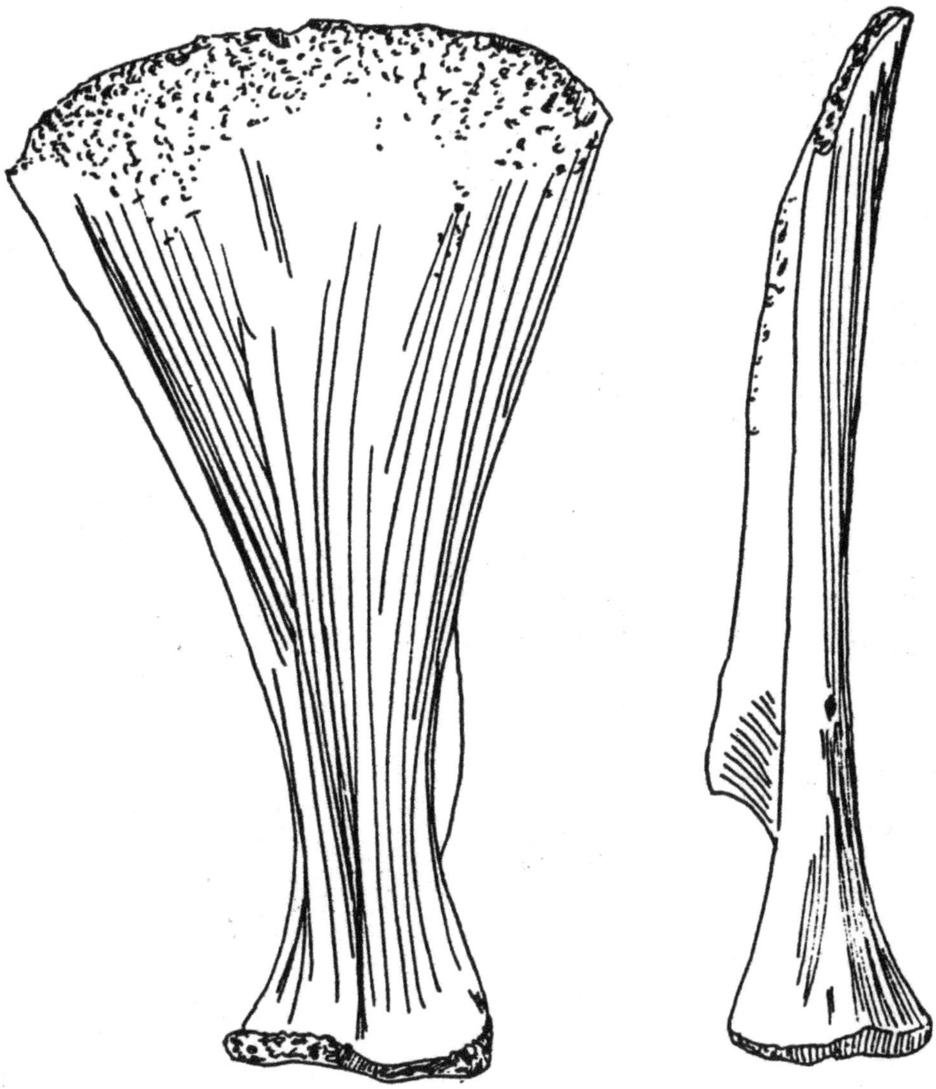

b

Ovis aries

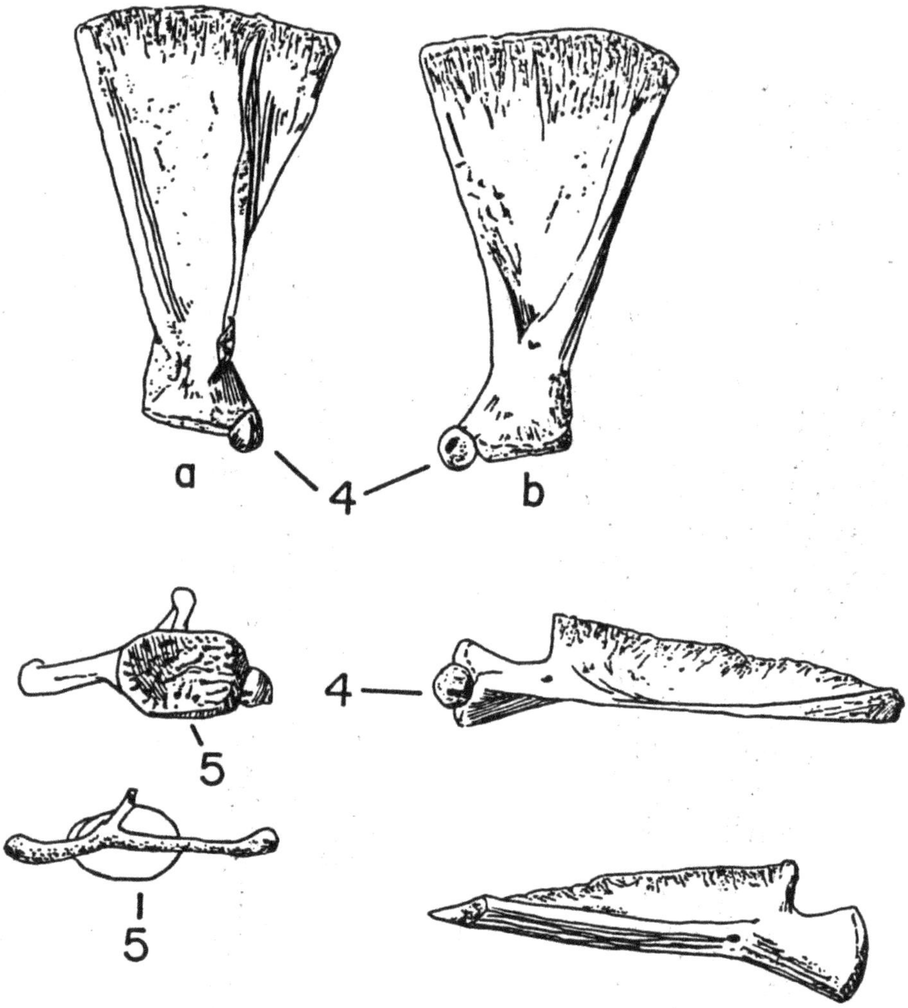

1:1

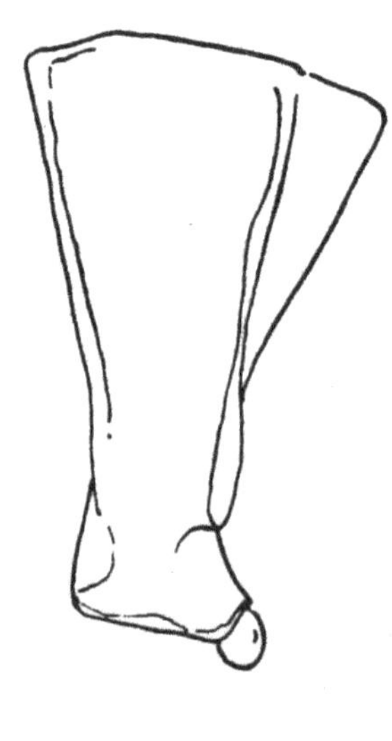

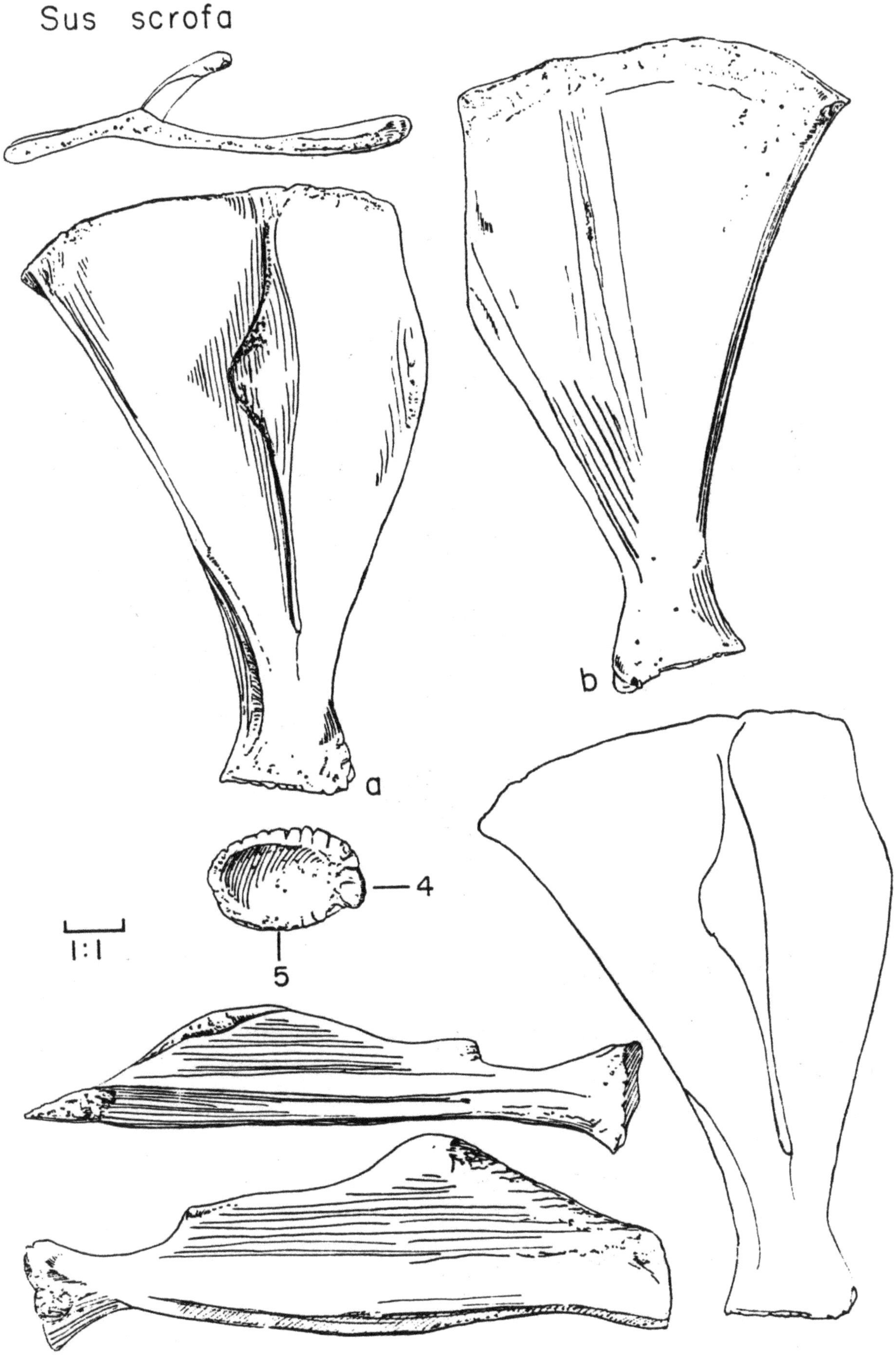
Sus scrofa
a
b
4
5
1:1

Canis familiaris

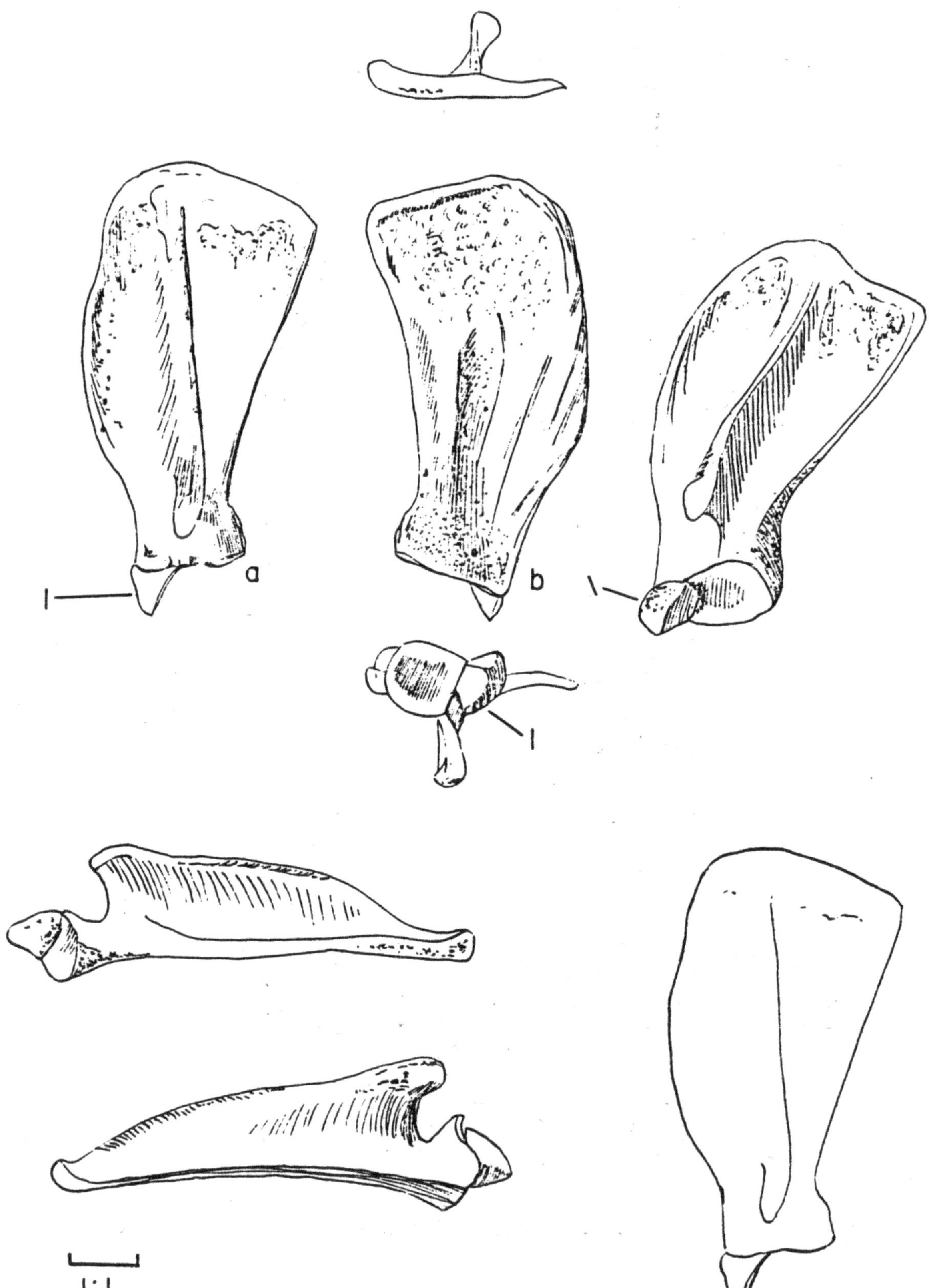

Felis catus L.

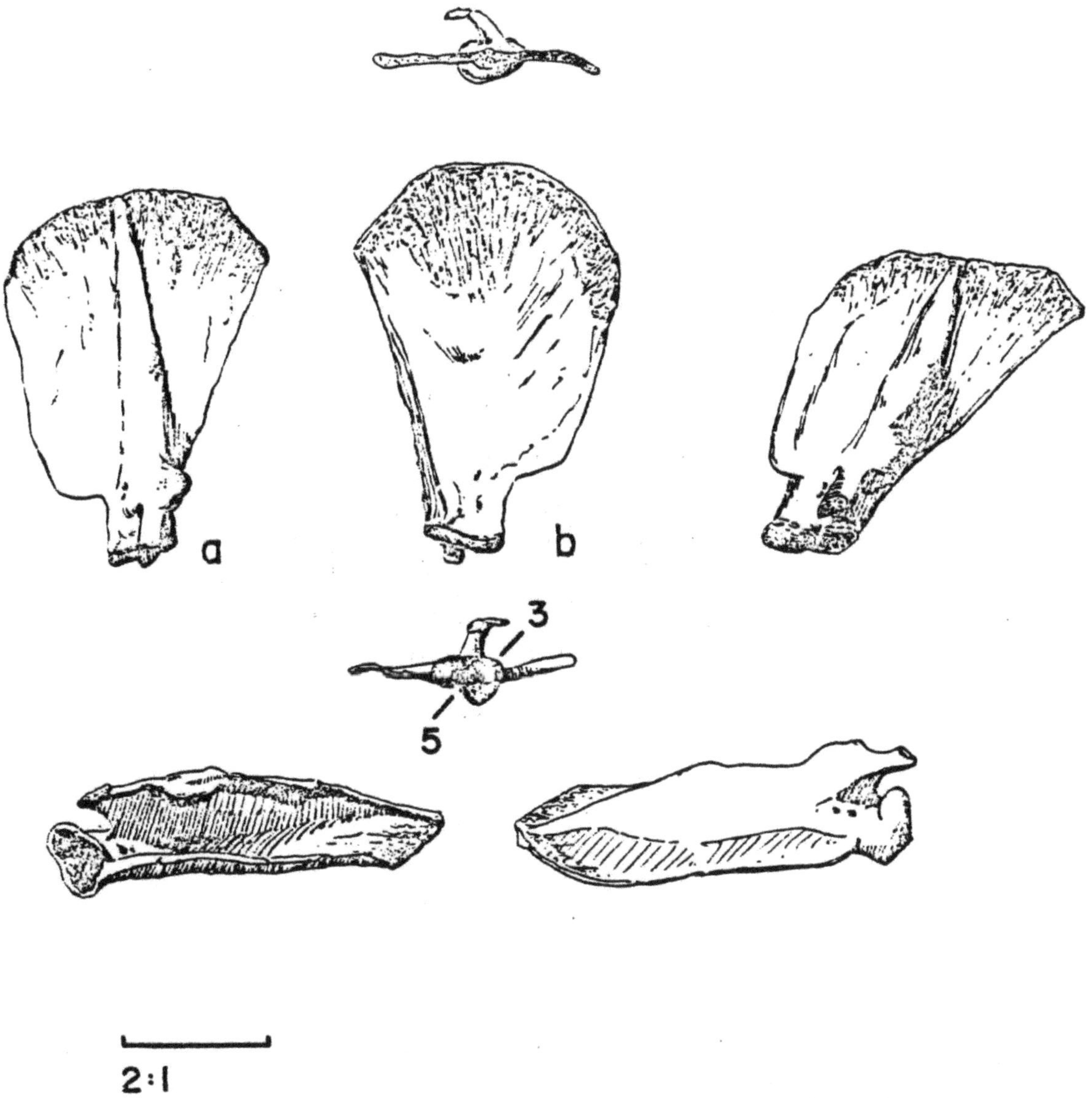

Rattus norvegicus

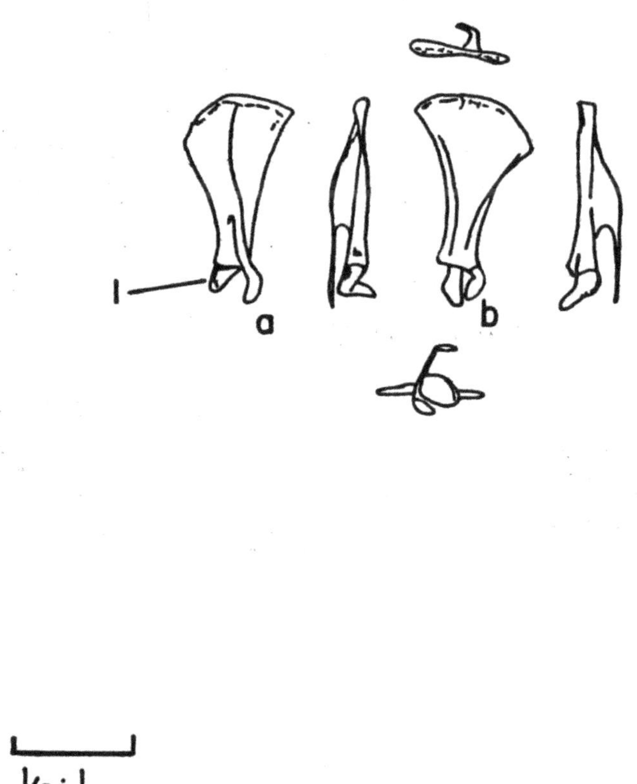

Plates X
Humerus

The humeral head articulates with the glenoid cavity of the scapula forming a ball-and-socket joint. The head of the humerus is more or less spherical in shape, and projects somewhat dorsally and caudally from the main axis of the bone (figure 4-5). The humerus has a neck, but it is usually obscure. There are two tuberosities, the greater (lateral and lesser (medial) that are located at the base of the neck, on the anterior aspect of the bone. The tuberosities afford the attachment for the scapula muscles which keep the head of the humerus and the glenoid cavity in articulation. The tuberosities are separated by a groove (biciptal groove), in which the tendon of the long head of the bicep muscle is positioned (Cornwall 1956; Sisson and Grossman 1966; Bass 1971).

The shaft of the humerus is usually cylindrical in cross section, except for those mammals adapted to aquatic environments such as seals. The shaft might have a prominent ridge crossing its anterior aspect, diagonally, running medially and distally. This ridge is often referred to as the deltoid ridge.

The distal end of the humerus is often transversely elongated into the lateral and medial epicondyles. The medial epicondyle is usually the more prominent of the two. Just below the epicondyles is the trochlea, which is a cylindrical pulley-shaped articular surface. Lateral to the trochlea is a small rounded knob known as the capitulum. Immediately above the trochlea, anteriorly and posteriorly, are a pair of fossae which articulate with the coronoid process of the ulna (coronoid fossa - anteriorly), and the olecranon process of the ulna (olecranon fossa - posteriorly). The olecranon fossa is the deeper of the two fossae, and is separated from the coronoid fossa by a thin partition of bone (septum). In some species, such as Sus the septum can actually be perforated, and is termed as the supra-trochlear foramen or septal aperture (Cornwall 1956; Bass 1971).

The supracondylicum foramen (entepicondyle foramen) is characteristic of the families of Felidae and Mustelidae (see figure 4-6). This foramen occurs where the medial epicondyle is supported by a prominent ridge running up the shaft of the humerus. This ridge is often grooved, and perforated above the epicondyle, allowing the brachial artery and nerve to pass through (Cornwall 1956; Sisson and Grossman 1966; Schmid 1972).

Views:
a - posterior
b - anterior
c - medial
d - proximal
e - distal
f - lateral

Landmarks:
1a - Tuberculum majus
1b - Tuberculum minus
2 - Tuberositas deltoidea (deltoid ridge)
3 - Proximal edge of trochlea
4 - Foramen supratrochleare
5 - Foramen supracondylicum

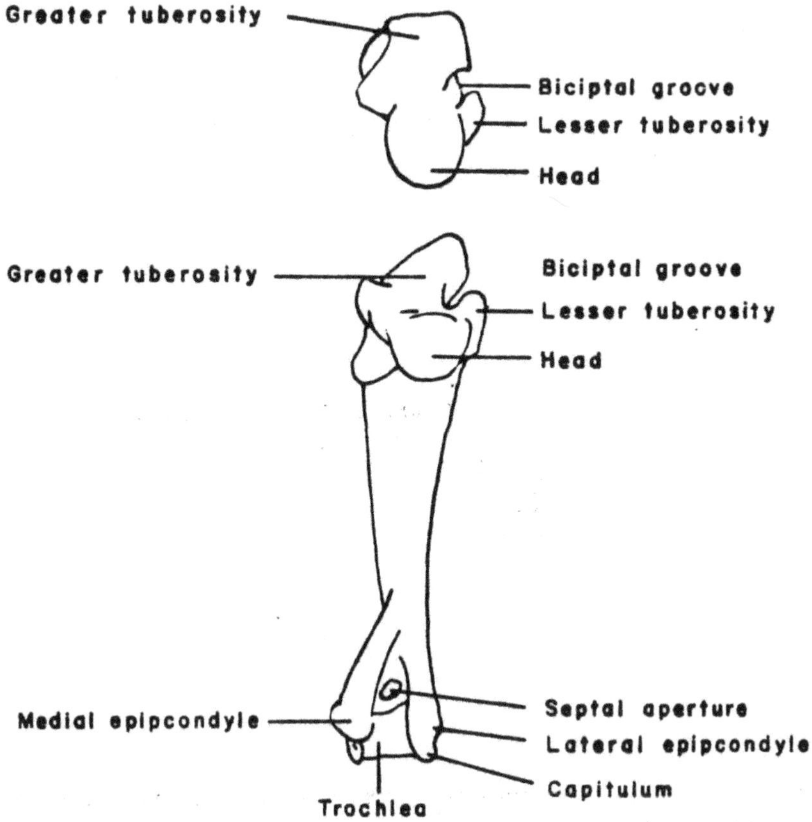

Figure 4-5, The basic architecture of the humerus of Sus.

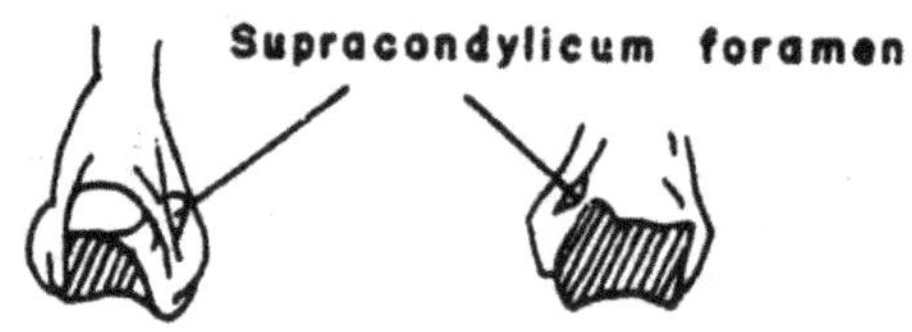

Figure 4-6, The supracondylicum foramen above the medial epicondyle (redrawn after Schmid 1972: figure 36).

Equus caballus

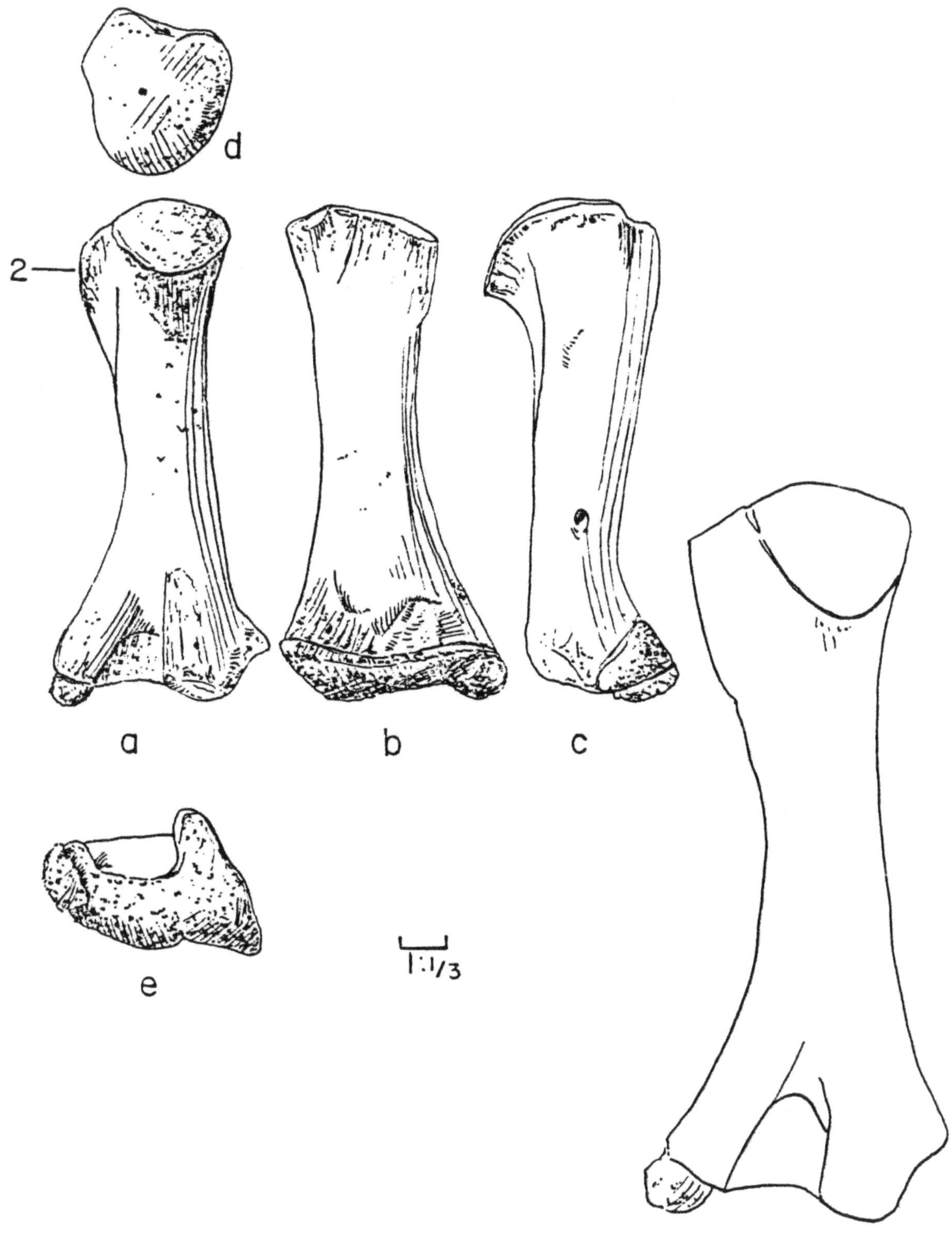

Bos sp.

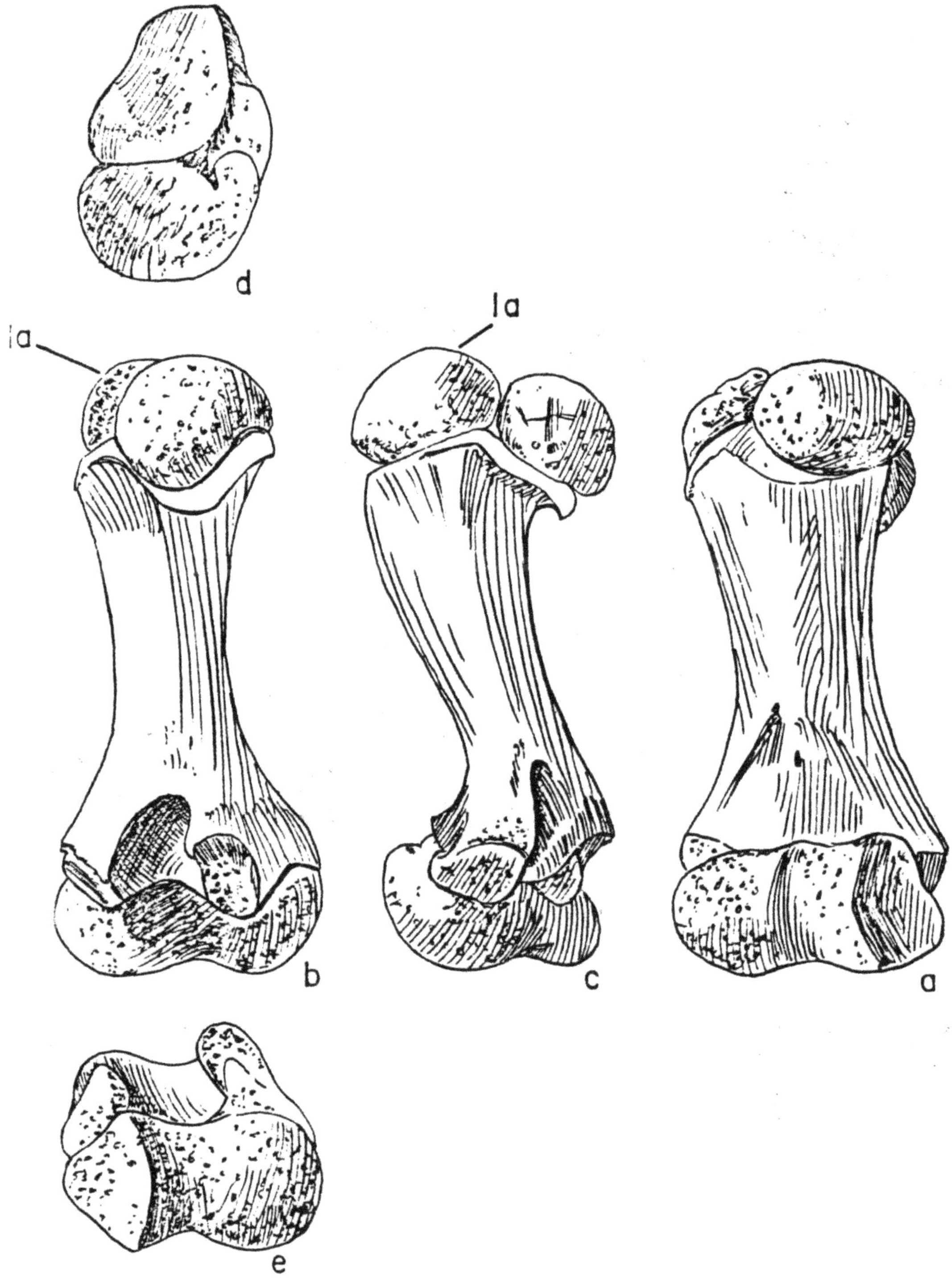

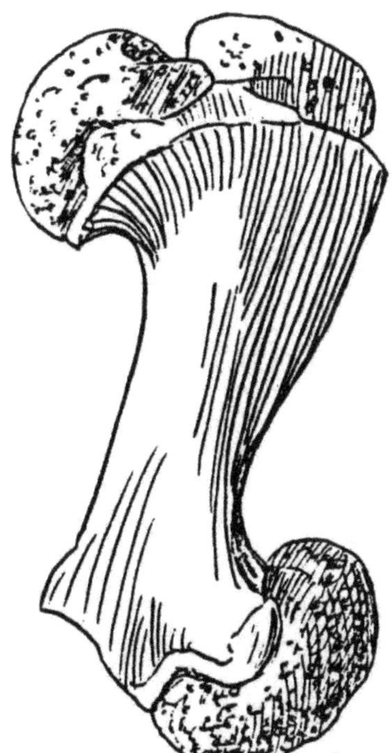

f

1:½

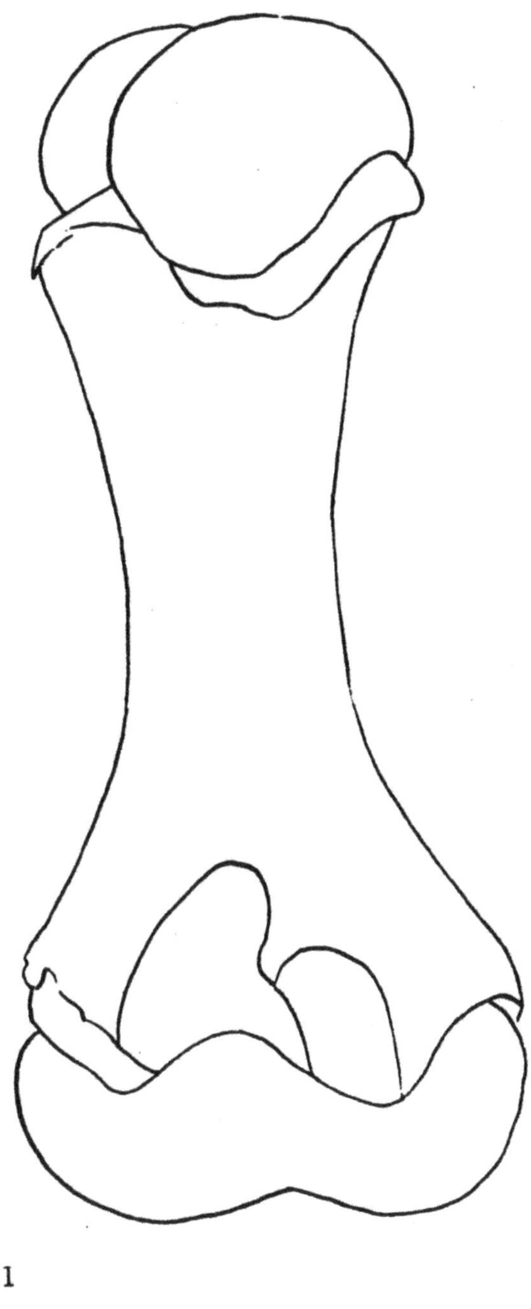

Ovis aries

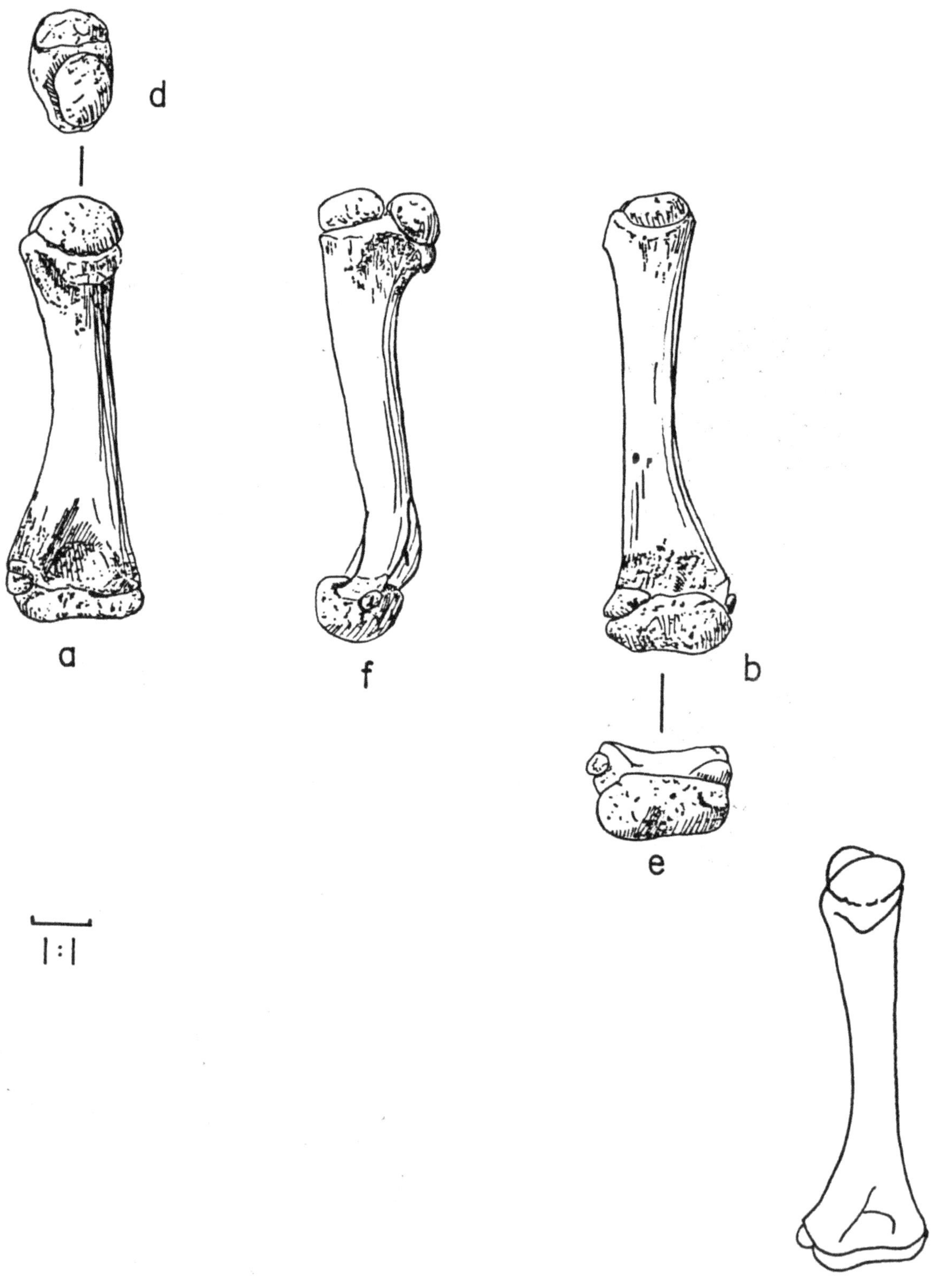

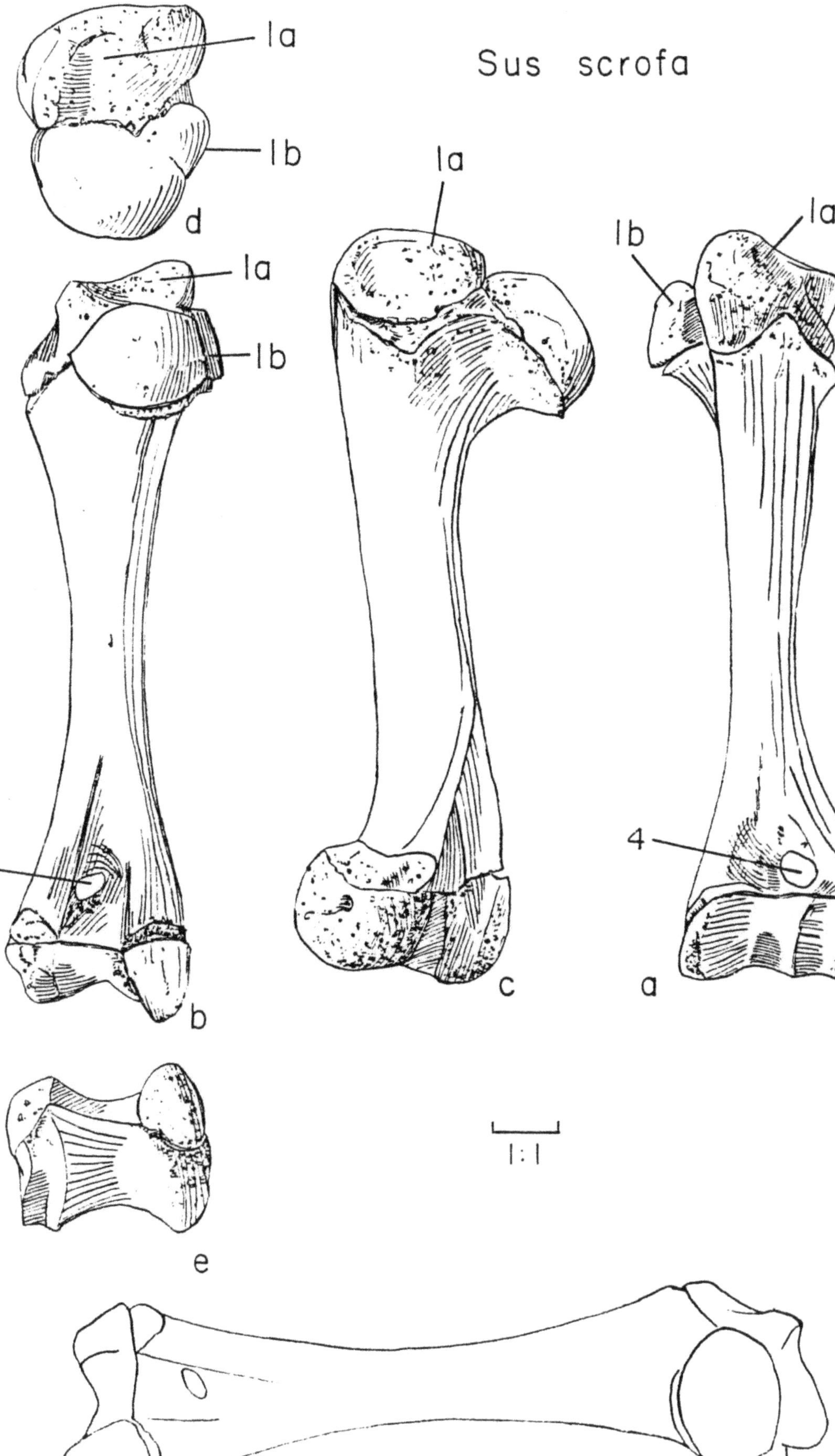
Sus scrofa
la
lb
d
la
lb
la
lb
la
4
4
b
c
a
e
1:1

Canis familiaris

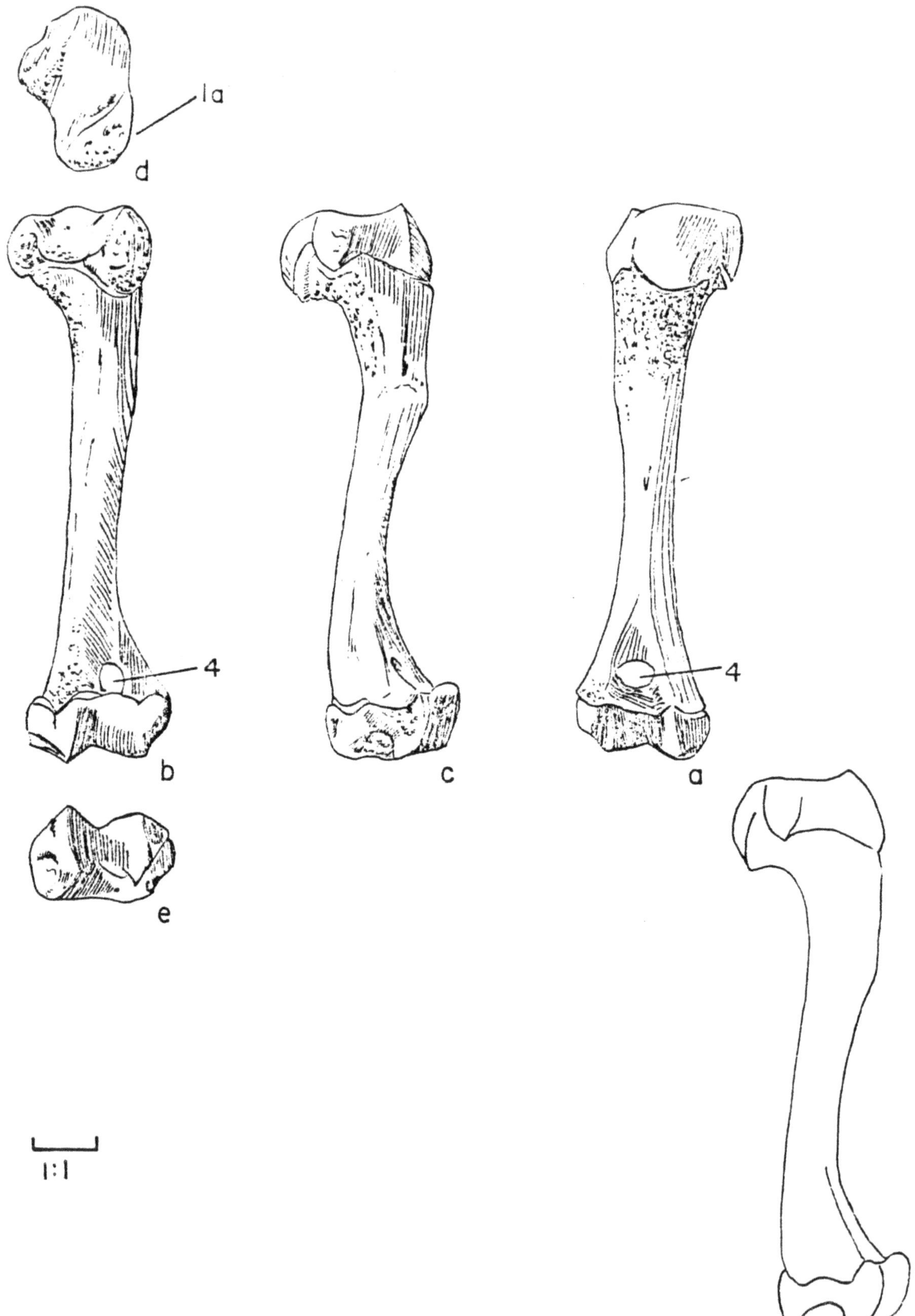

Felis catus L.

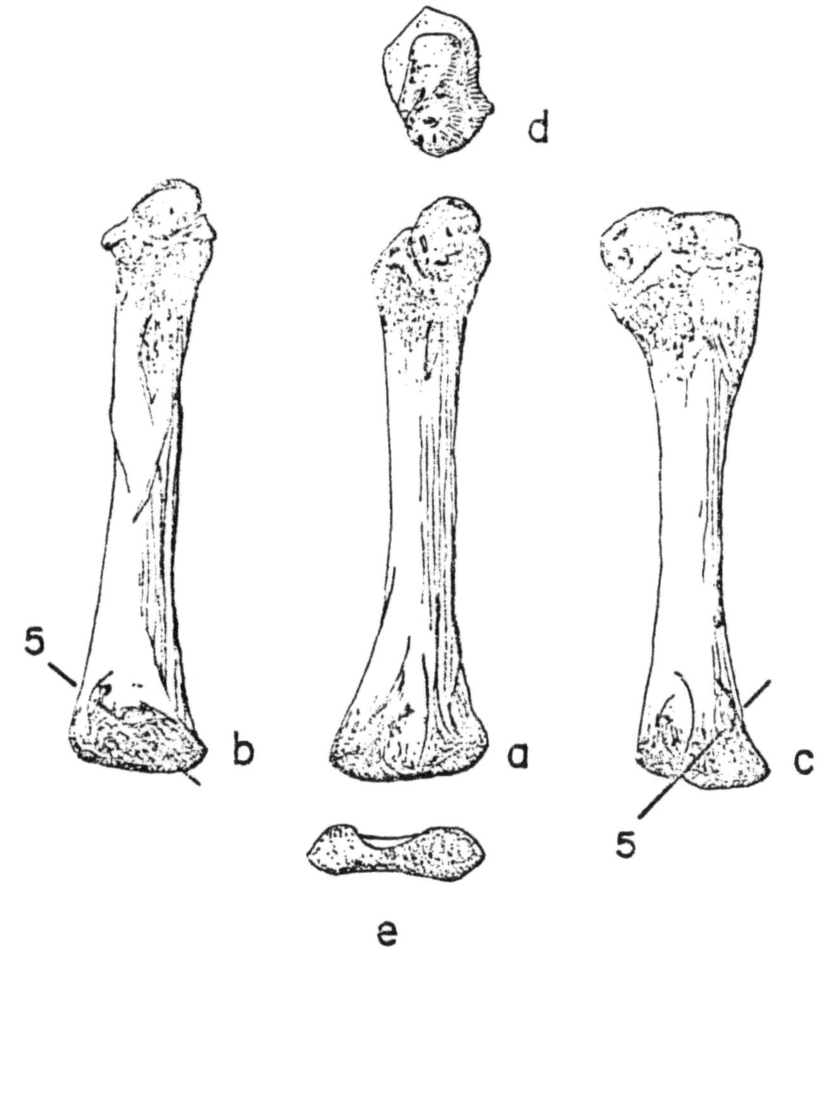

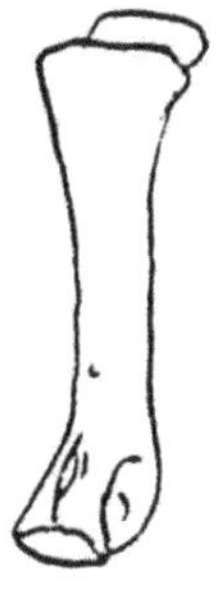

Rattus norvegicus

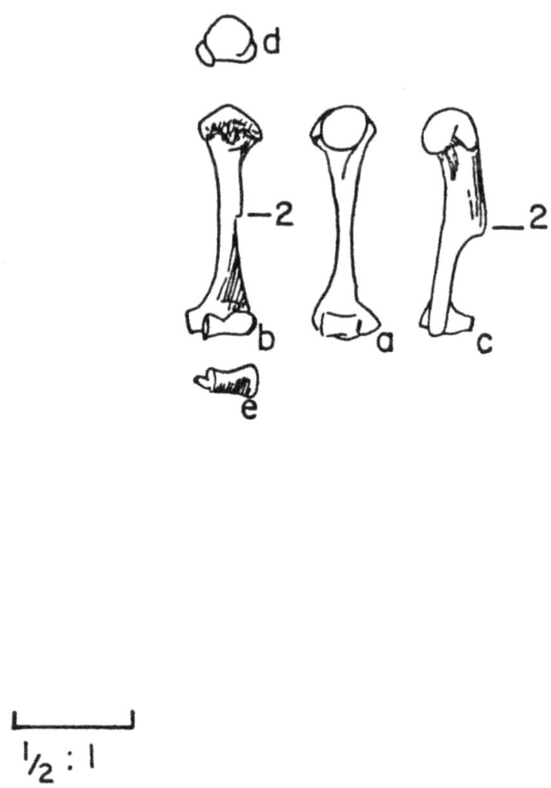

Plates XI
Radius

In carnivores, rodents, and lagomorphs, the radius is a separate movable bone that lies along the lateral side of the ulna. In these classes of mammals, the radial head is cylindrical in shape. The head articulates with the capitulum of the distal humerus, and can rotate about its own axis against the articulation of the ulna (radial notch).

However, in the ungulate species, the radius becomes the principal bone of the forearm. The radius is made immovable by either ligaments or by ankylosis to the ulna. By becoming immobile, the radius has come to be positioned in front of the ulna, supporting nearly the whole load borne by the upper arm. In doing so, the ulna's distal end is reduced, and in the extreme only the radius will articulate with the carpus (Cornwall 1956; Ellenberger and Dittrich 1959; Sisson and Grossman 1966).

Views:
a - anterior
b - posterior
c - proximal
d - distal
e - medial
f - lateral

Landmarks:
1 - Medial area, in Sus the whole articulation is broad and rectangular; in carnivores the articulation has a round surface
2 - Medial Tuberosity

Equus caballus

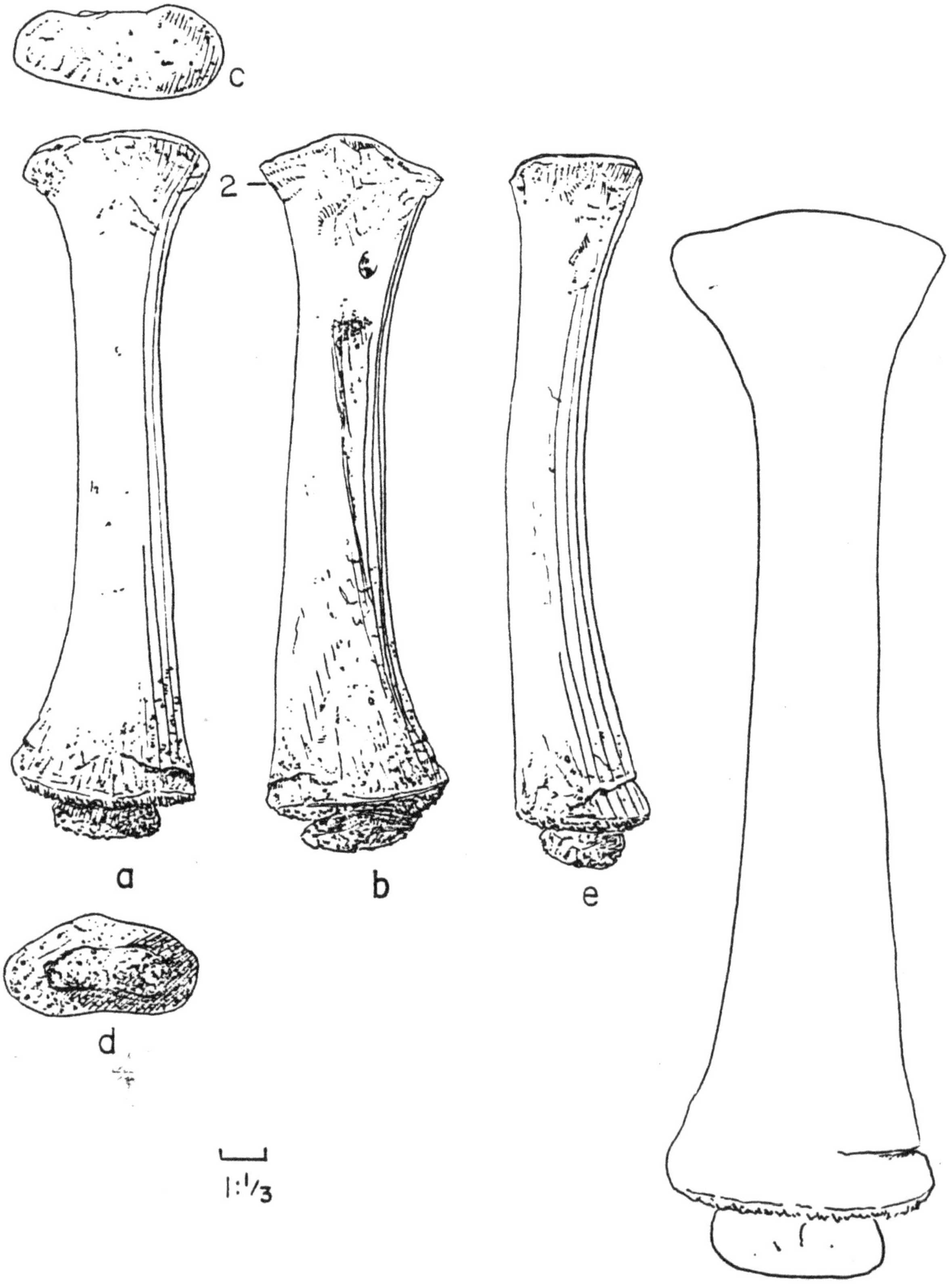

Bos sp.

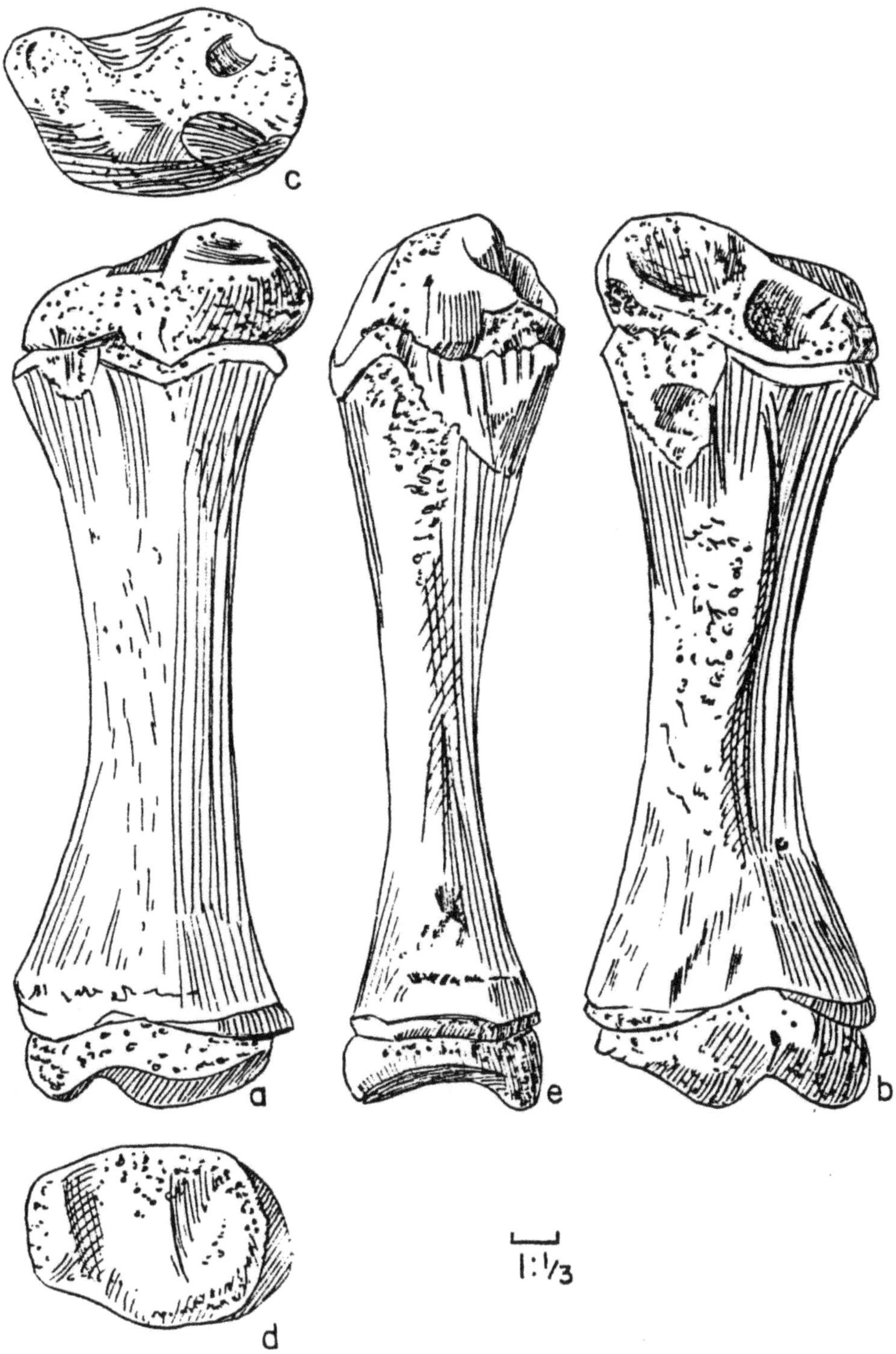

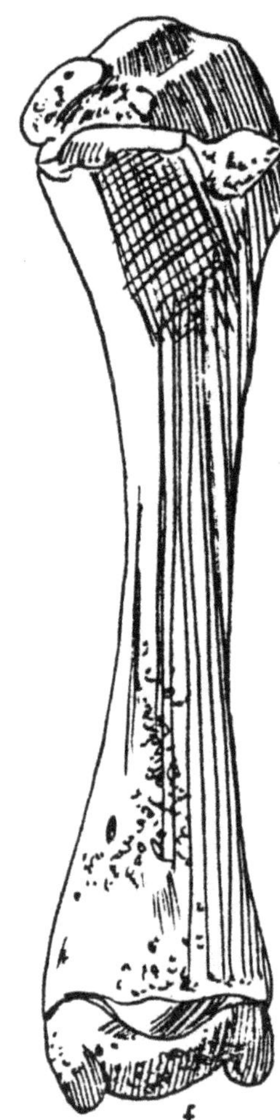

f

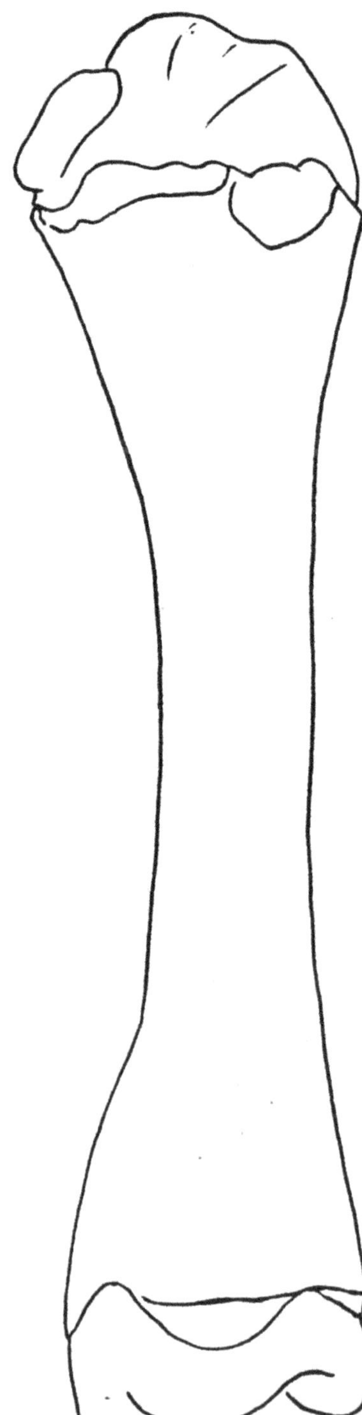

Ovis aries

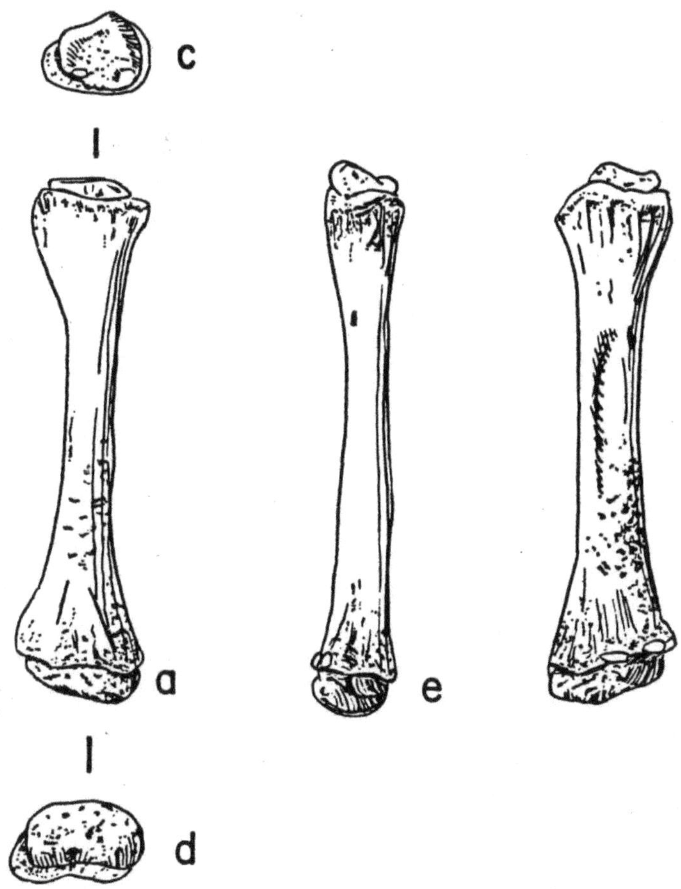

1:1

Sus scrofa

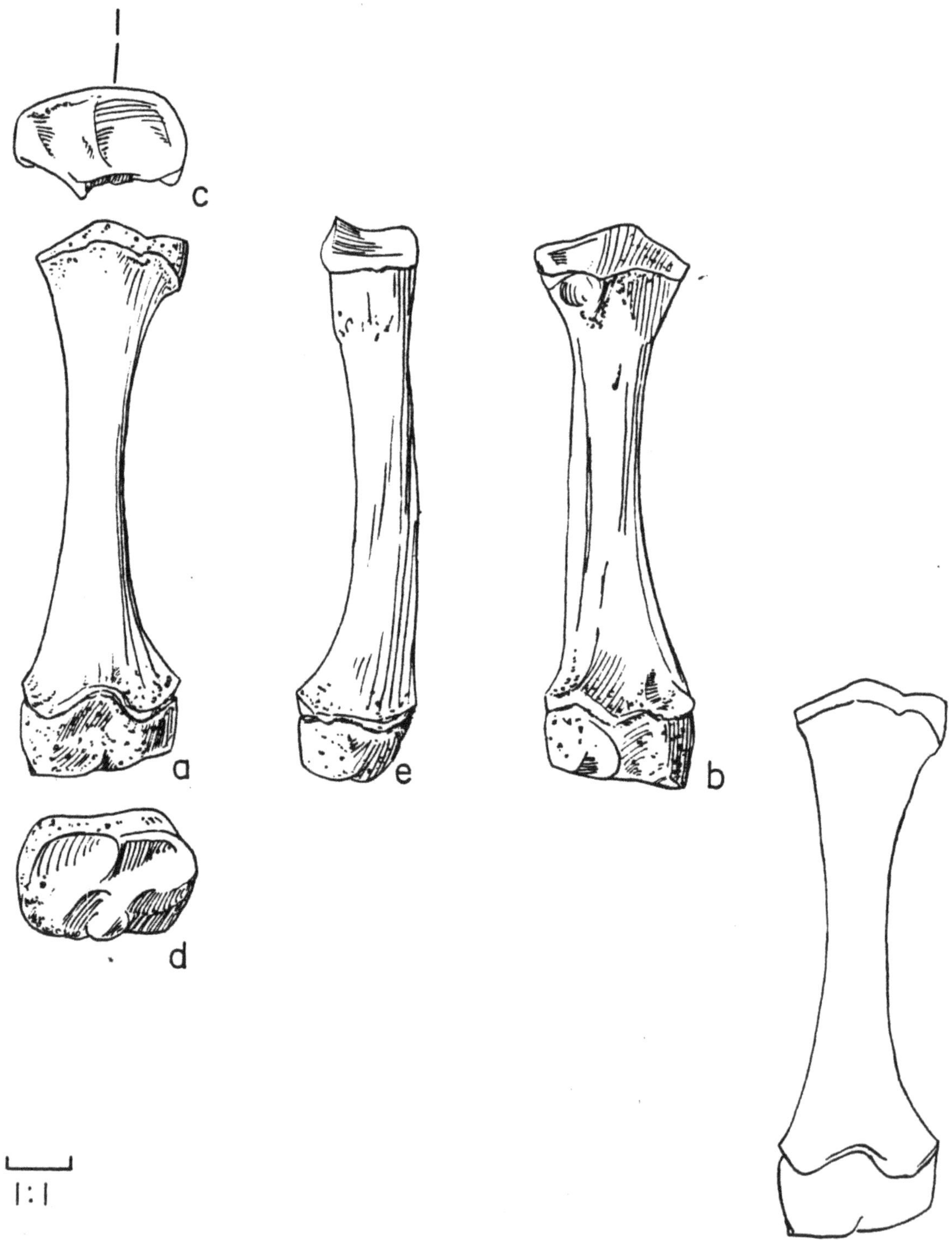

Canis familiaris

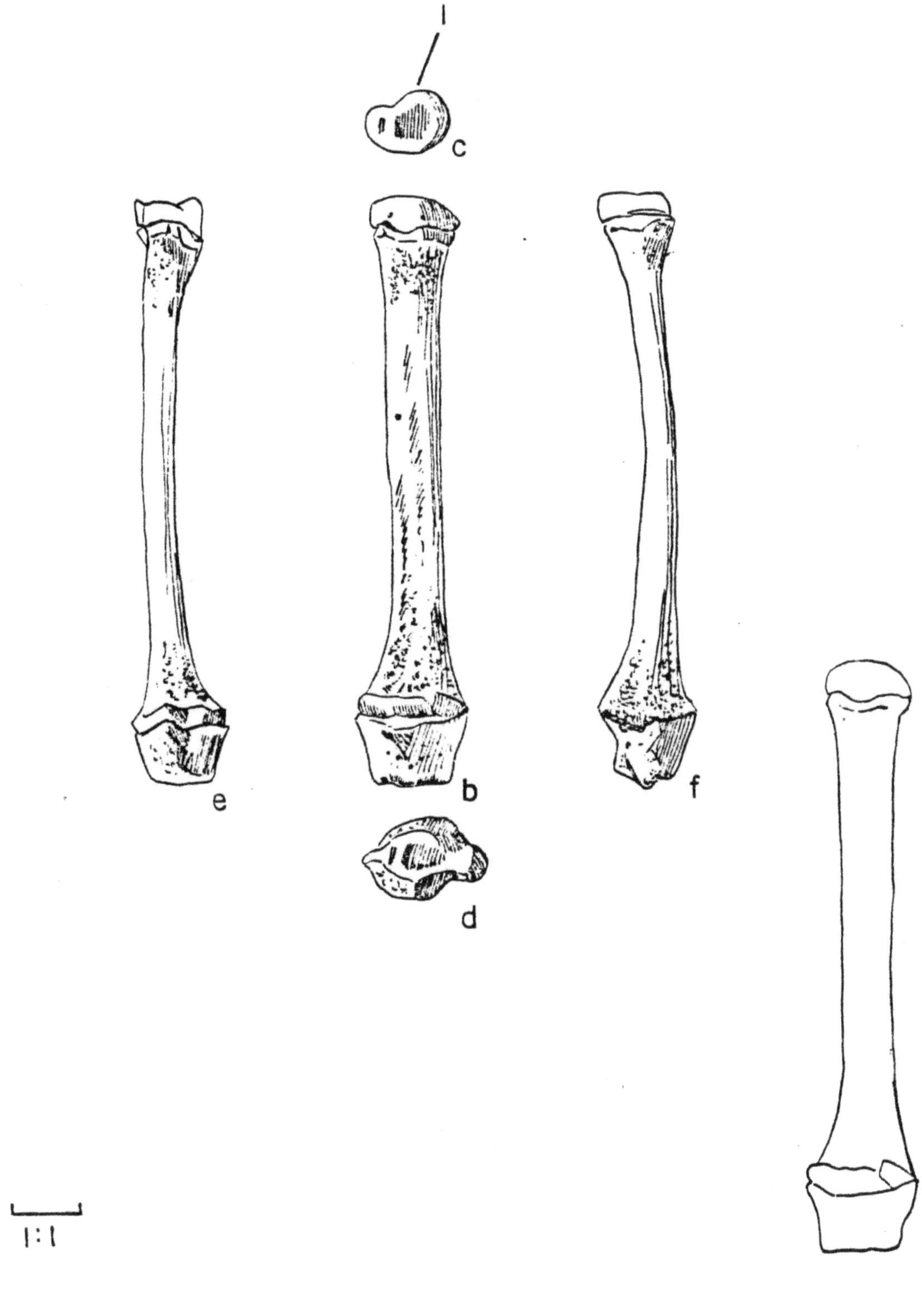

Felis catus L.

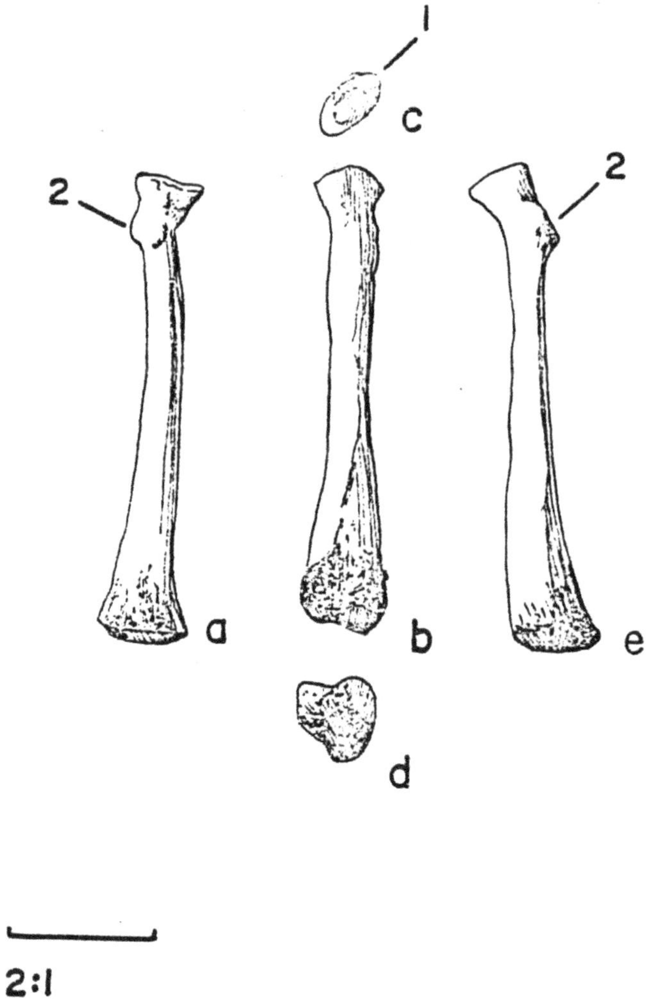

Rattus norvegicus

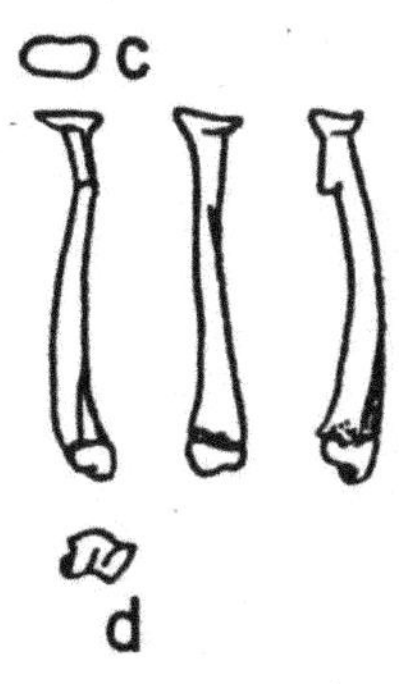

1/2 : 1

Plates XII
Ulna

The ulna articulation with the humerus by its trochlear notch (see figure 4-7). This notch is delimited on the anterior side by the coronoid process, and posteriorly by the olecranon process. The ulna's shaft and distal end are often reduced in size and independent in the specialized quadrupeds (Cornwall 1956; Sisson and Grossman 1966; Schmid 1972).

In most ungulate species, both the ulna and radius are fused, and known as the zygopodium. Equus is the only species among the ungulates represented in this volume that lacks a distal third of the ulna in the adult form. The juvenile equine ulna is almost the same length as the radius. The carnivores, rodents, and lagomorphs have a separate ulna and radius in comparison to the ungulates.

Views:
a - lateral
b - anterior
c - proximal
d - distal
e - medial
f - posterior

Landmarks:
1 - Processus olecrani
2 - Incisura semilunaris
3 - Processus coronations lateralis

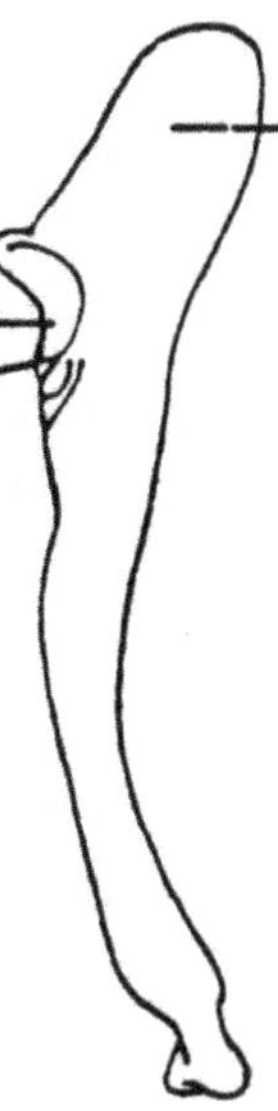

Figure 4-7, The basic architecture of the ulna of <u>Sus</u>.

Equus caballus

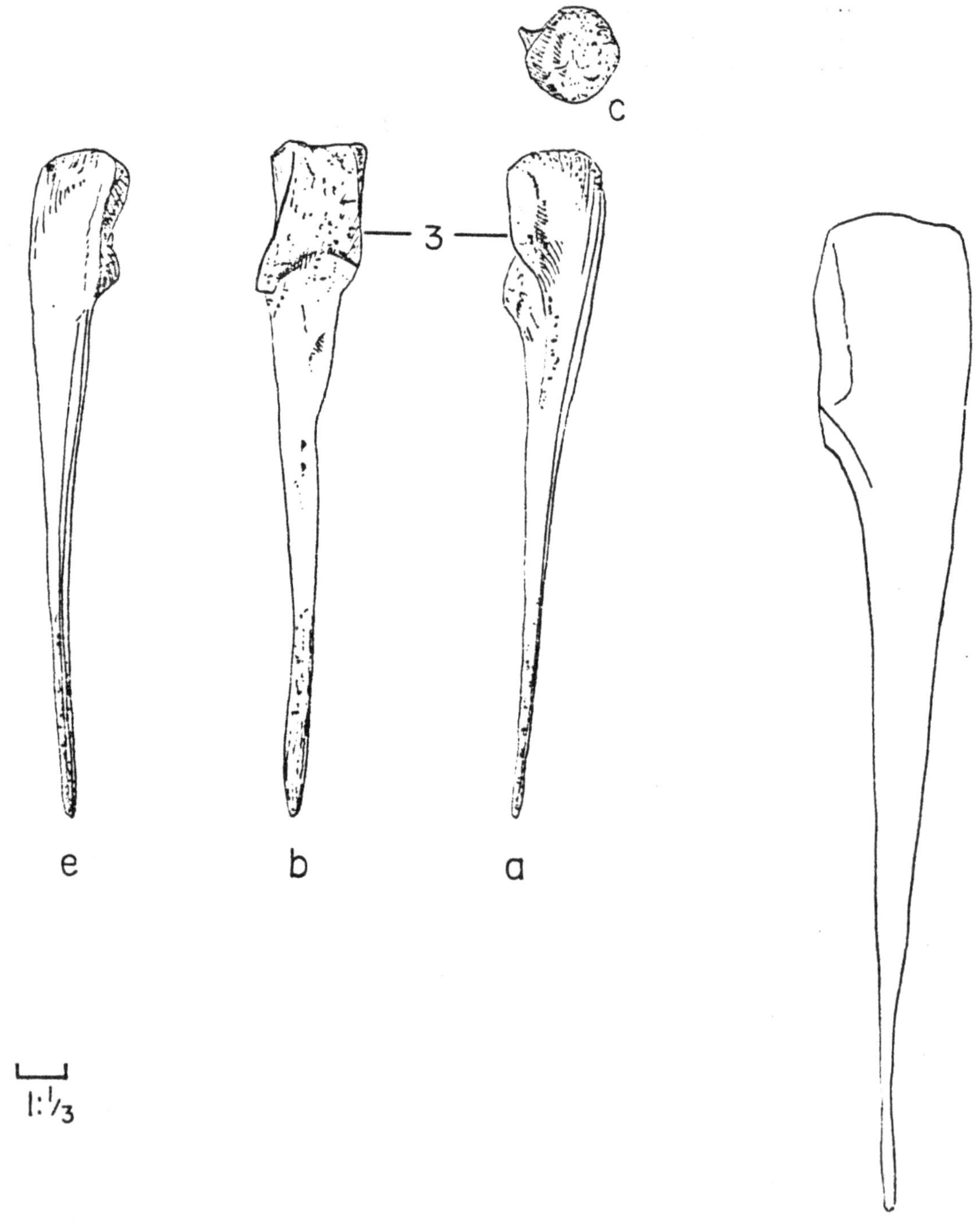

Bos sp.

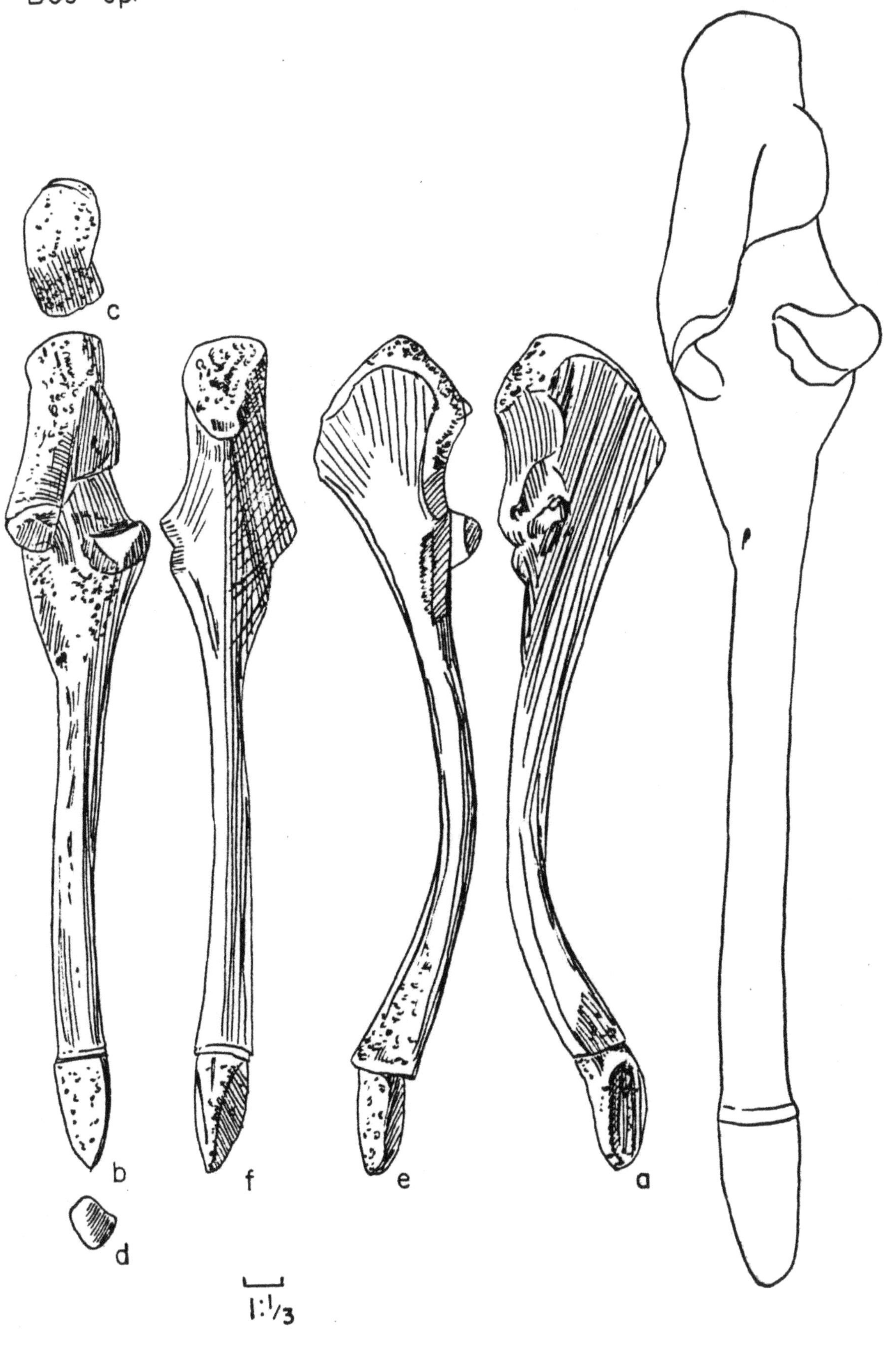

Ovis aries

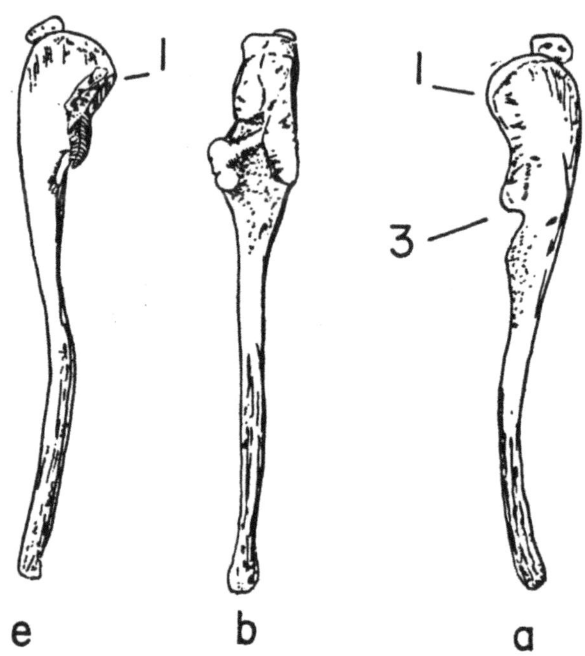

1:1

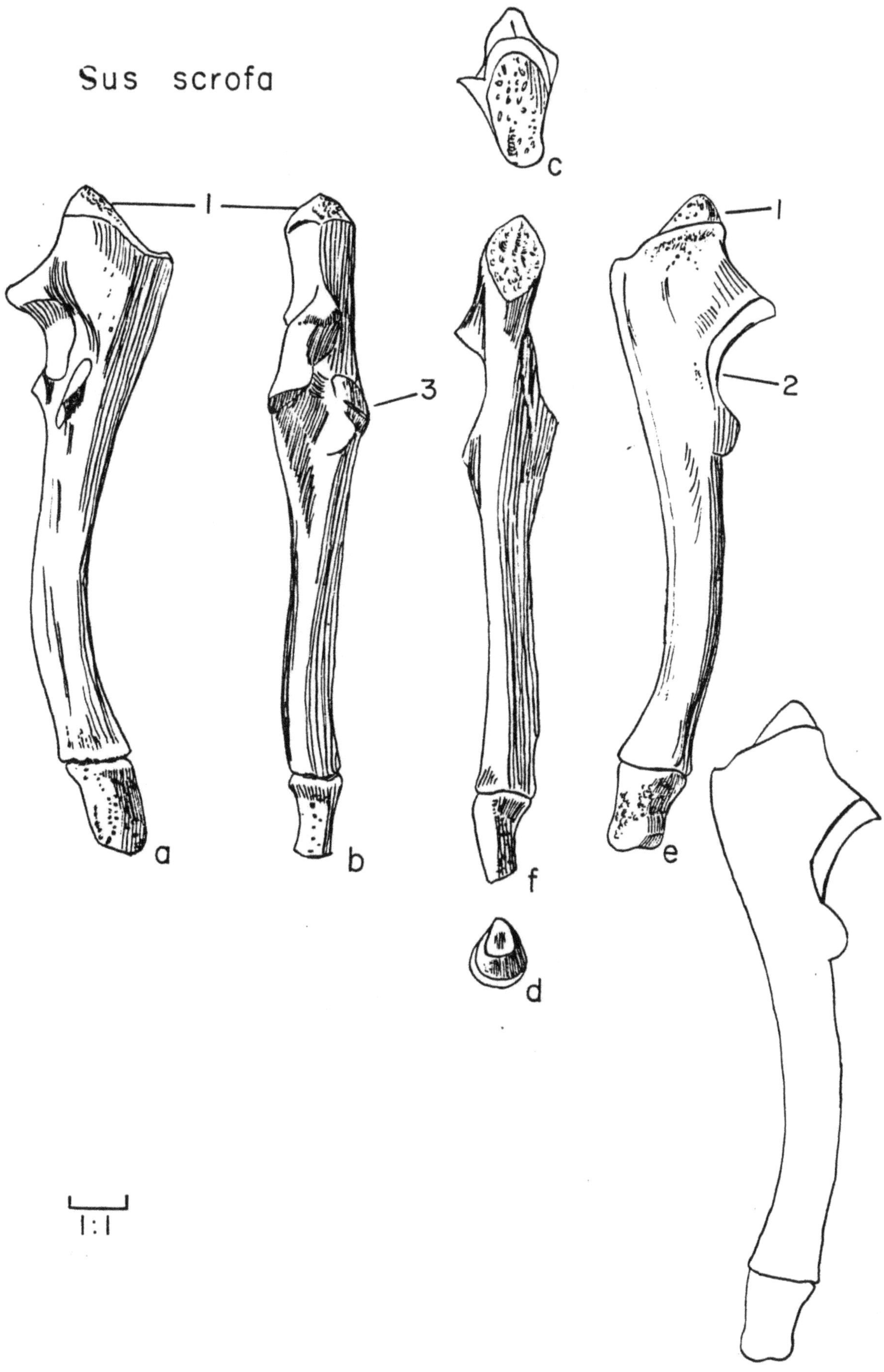
Sus scrofa
c
1
1
1
3
2
a
b
f
e
d
1:1

Canis familiaris

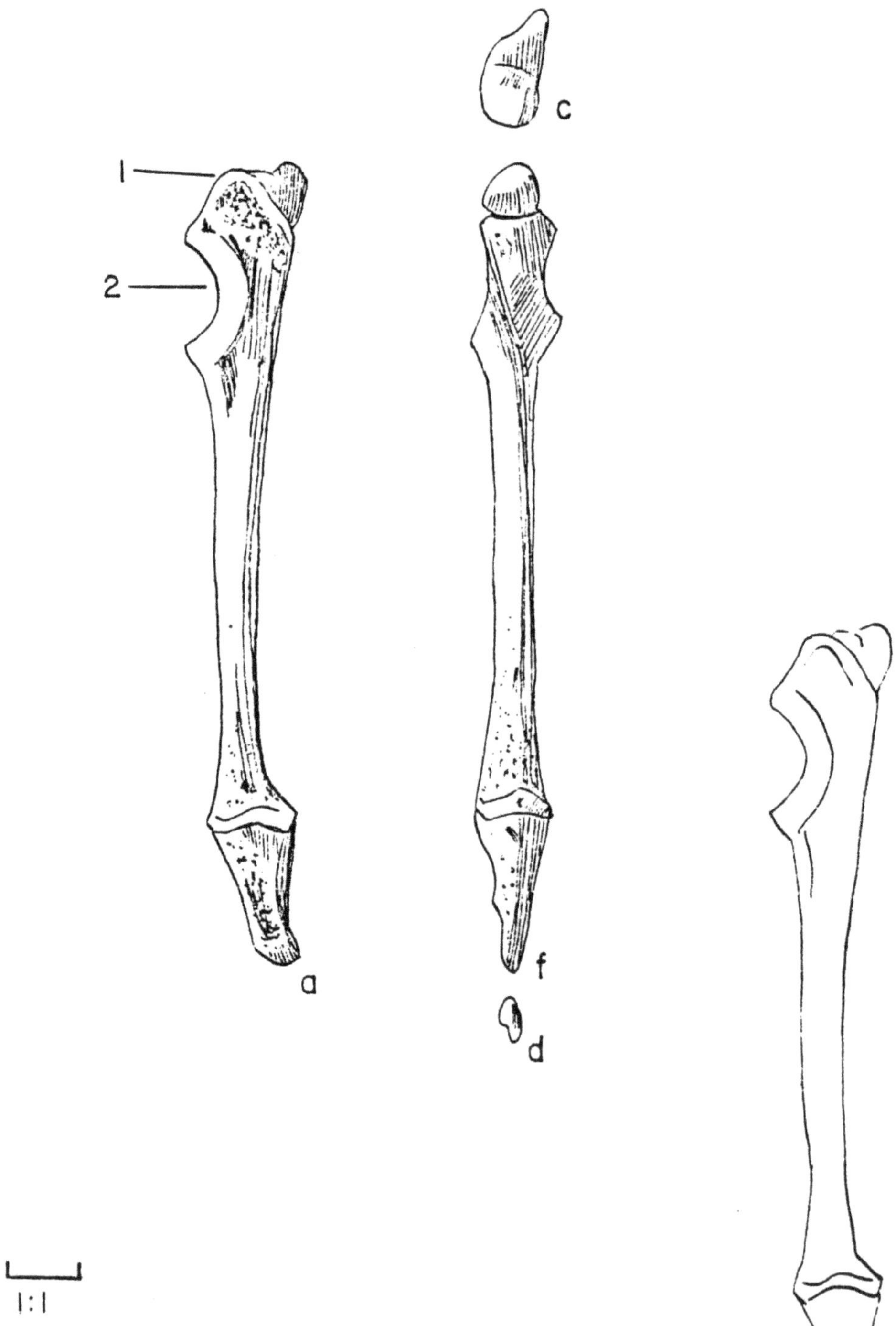

Felis catus L.

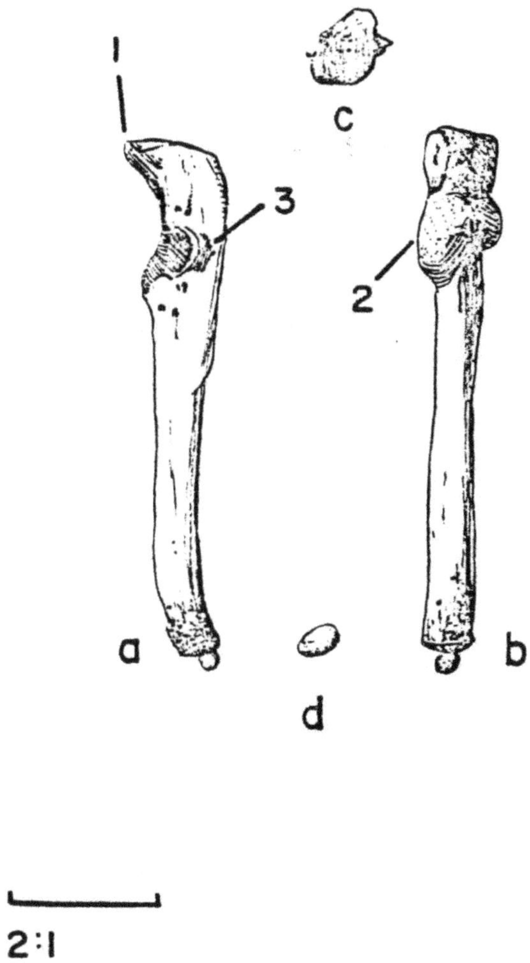

Rattus norvegicus

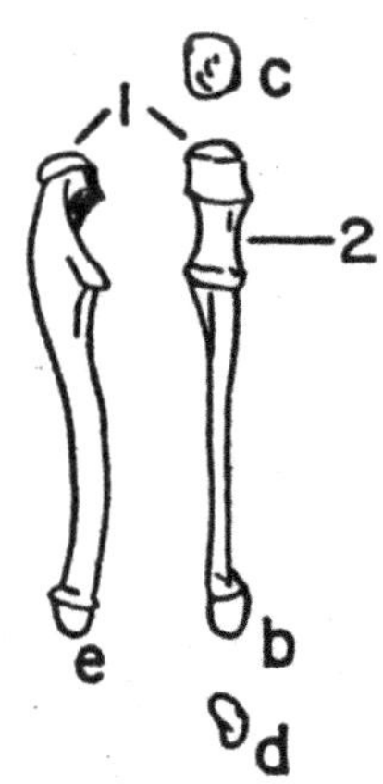

1/2 : 1

Plates XIII
Metacarpus

The metacarpus is a set of five bones, which articulate with the carpus forming joints, but having little latitude in movement. However, the metacarpus does form a very mobile hinge-joint for the proximal ends of the digits (Cornwall 1956; Sisson and Grossman 1966).

In many of the more specialized orders, the basic bony units of the metacarpus have degenerated or are missing. Such reductions can often characterize various mammalian groups (see figure 4-8). The metacarpal equipment of Canidae, Felidae, and Sus are complete, but degenerative in Canidae. In Sus the metacarpal III and IV are much more strongly developed than II and V, which are shorter and thinner at the proximal end. The genus Equus represents the most extreme reduction of the metacarpus. The "canon bone" (metacarpal III) of Equus is elongated and strongly developed. On the posterior side of the metacarpal III are the metacarpals II and IV, which taper off to a point near the middle of metacarpal III. Metacarpals II, III, IV articulate with the carpals, but only III articulate with the digits. The metacarpal complement of Bovidae (Bos and Ovis/Capra) is complete, but the lateral metacarpals are very degenerate. The "cannon bones" (metacarpals III and IV) of bovidae are much shorter and more heavily built than Equus.

A distinction of the metacarpus and metatarsus of Equus and Bovidae can be at the neo-natal to juvenile age range, by comparison of the shaft and to a lesser degree the articular surface of the proximal end. In general, the metacarpal tends to be shorter and of more stocky construction. In cross section (figure 4-9) the metacarpal is somewhat elliptical as compared to the metatarsal's oval-like shape.

Views:
a - anterior (dorsal)
b - proximal
c - lateral
d - posterior (volar)
e - distal
f - medial

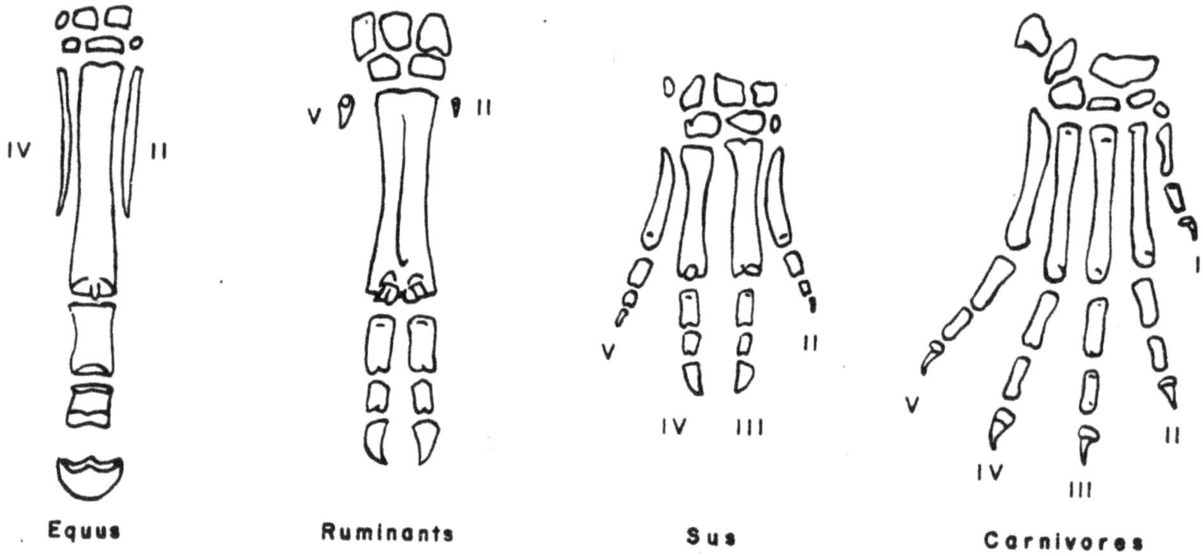

Figure 4-8, Basic pattern of the manus in different mammalian groups.

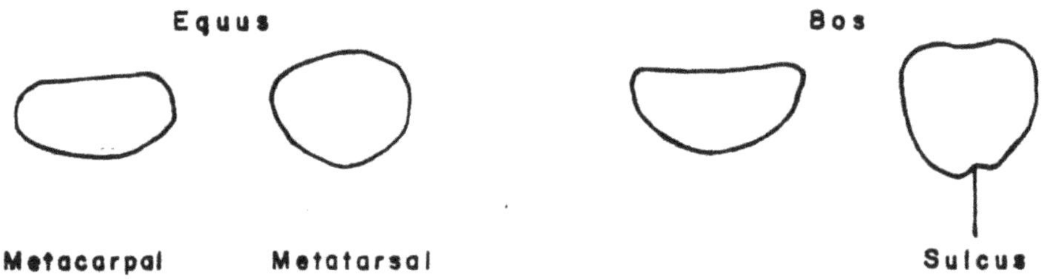

Figure 4-9, Distinction between the metacarpal and metatarsal by cross section (redrawn after Schmid 1972). The neo-nates still have their proximal articular surfaces in a stage of development, making this distinction difficult.

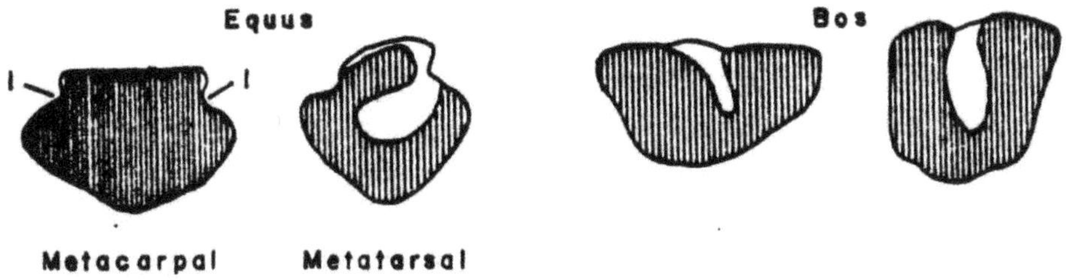

Figure 4-10, The distinction between the metacarpal and metatarsal. The vertically shaded area is the articular surface. 1 - the indentation for the insertion of metacarpals II and IV.

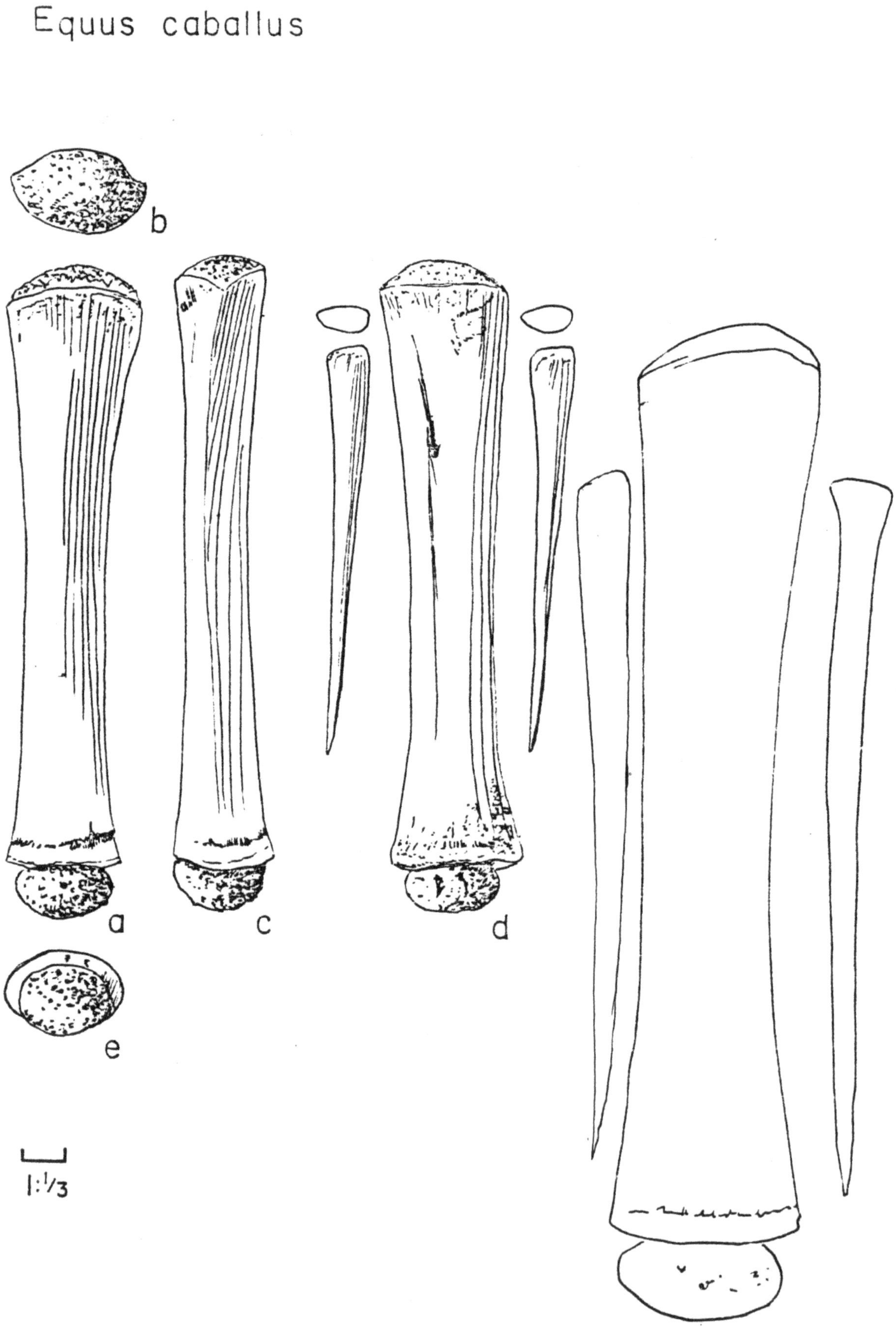
Equus caballus
b
a
c
d
e
1:1/3

Bos sp.

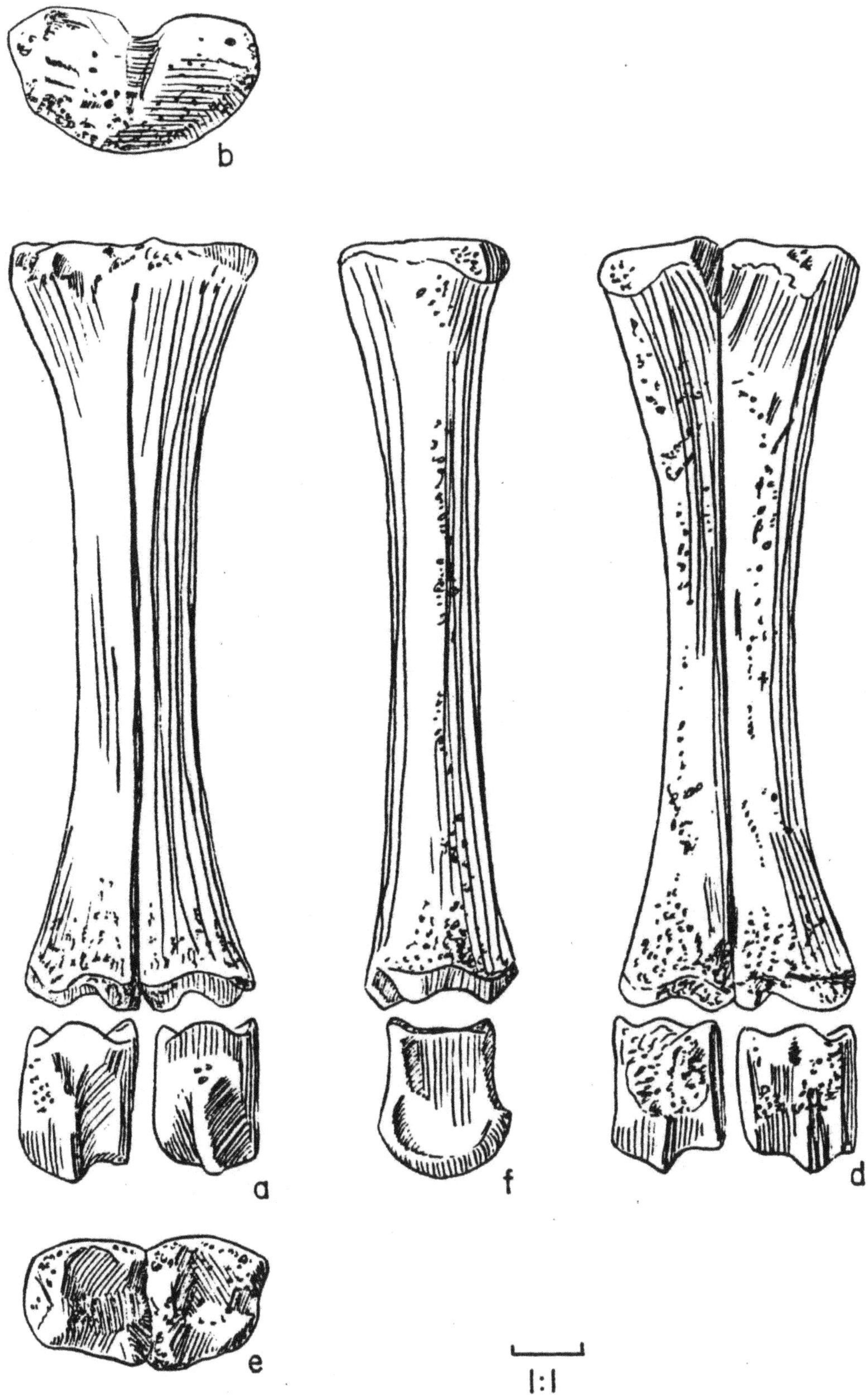

c

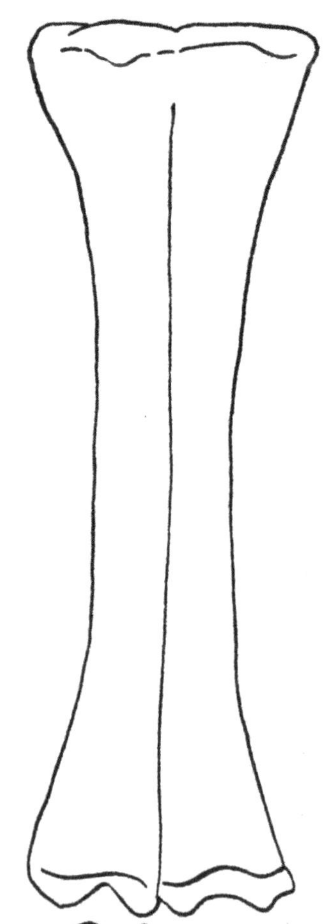

Ovis aries

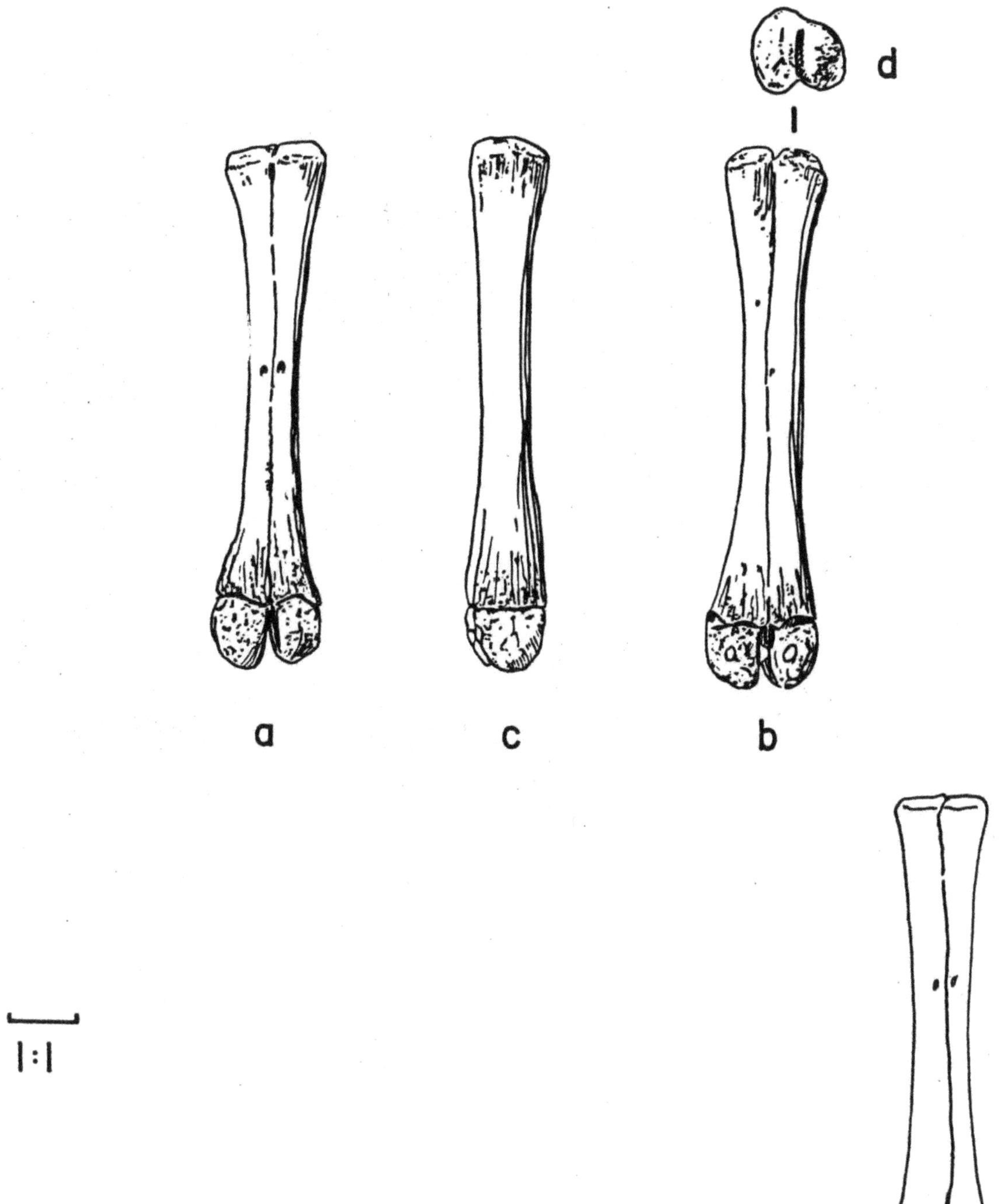

Sus scrofa

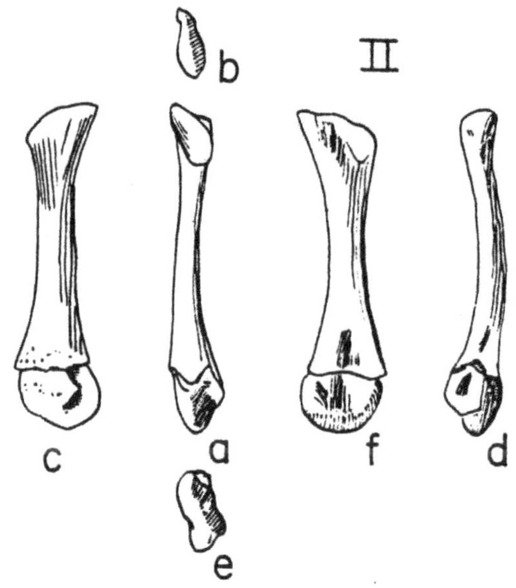

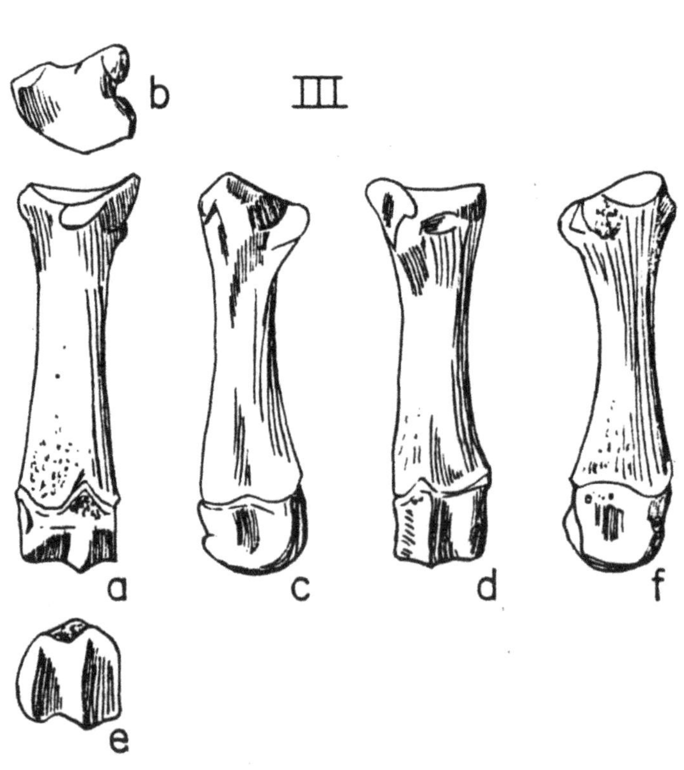

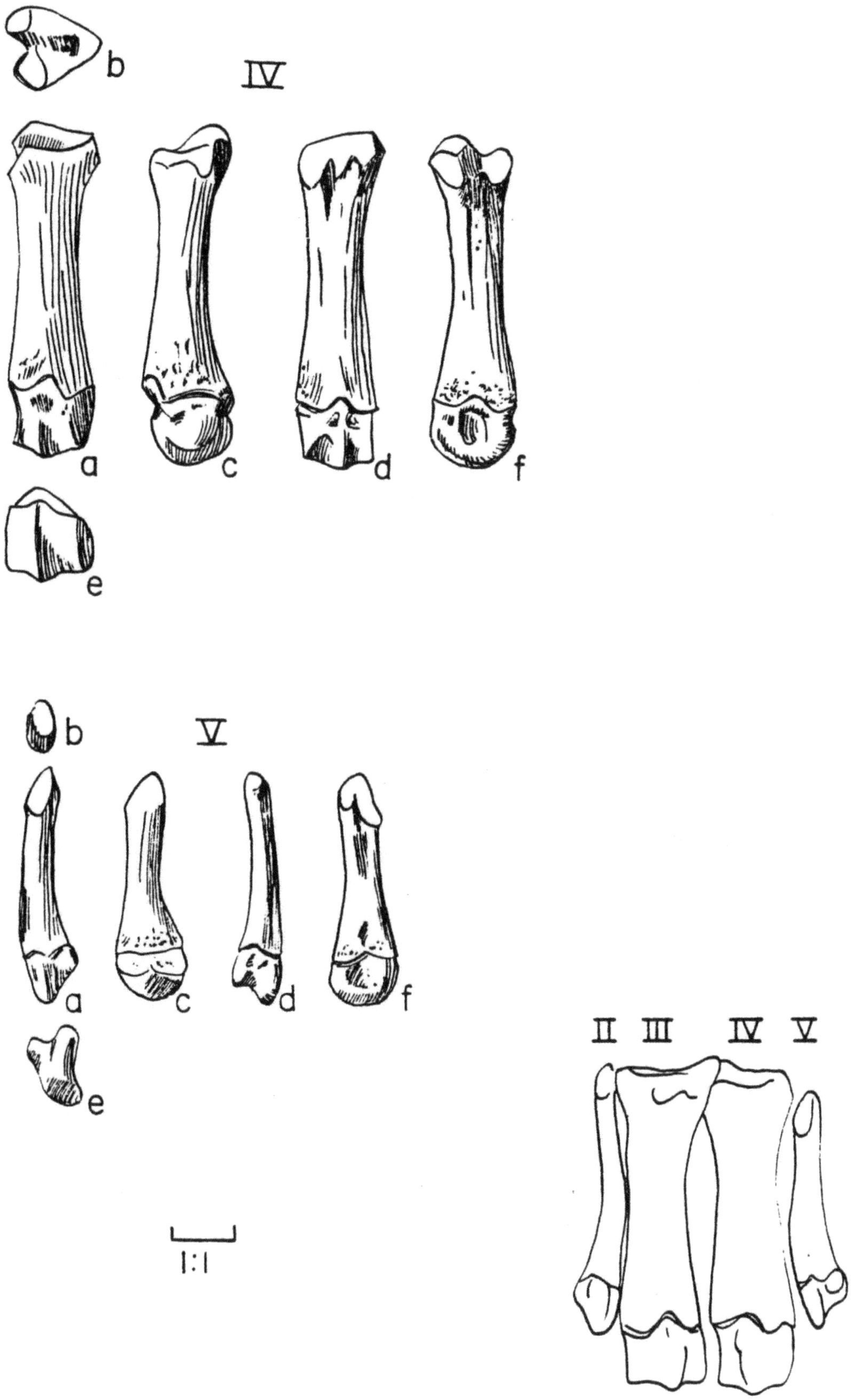

IV
a
b
c
d
e
f
V
a
b
c
d
e
f
1:1
II III IV V

Canis familiaris

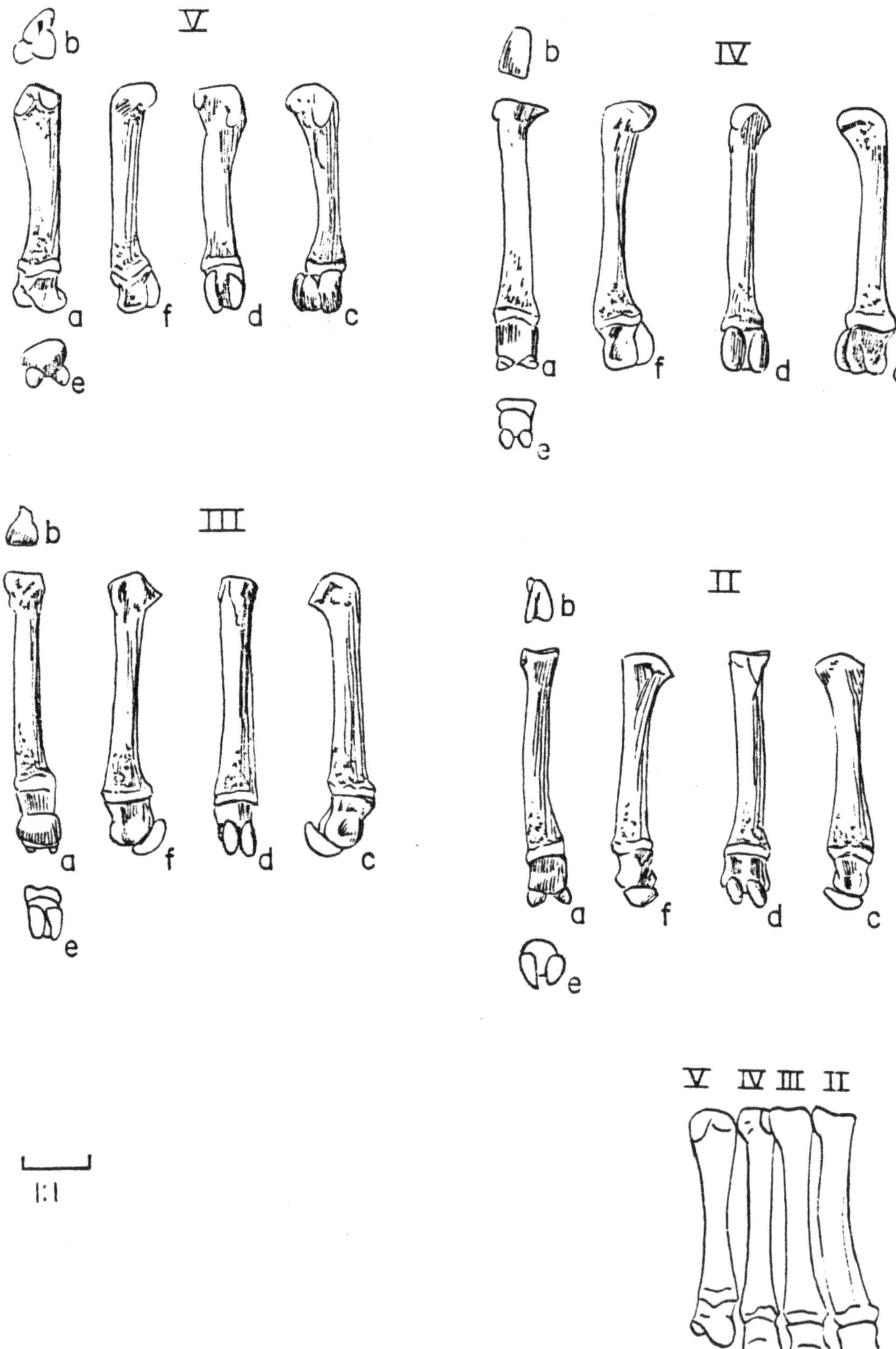

Appendicular Skeleton: Pelvic Limb
Plates XIV
Innominate (Pelvis)

The hip-gridle consists of a bony structure of three distinctive portions, the two innominates (ossa innominata) meeting each other ventrally and connected dorsally with the sacrum (see figure 4-11). The innominate's main function is to link the hind limb to the axial skeleton. However, unlike the muscle attachment of the shoulder-gridle, the pelvic-gridle is not itself movable in relation to the axial skeleton. The other important functions that the pelvis serves, are to assist in the support of the internal organs, and to serve as an area of attachment for the muscles of the hind limb (Cornwall 1956; Sisson and Grossman 1966; Bass 1971).

The innominate (os innominatum) consists of three bones, the ilium, ischium and pubis. In youth, these bones are separate, but become fused together as adult stature is reached. These three join together at a junction to form a hemispherical-like cup known as the acetabulum. The acetabulum articulates with the head of the femur, forming one of the major ball-and-socket joints of the body. The ilium articulates on its dorsal side with one side of the first sacral vertebra's transverse process. The ischium projects caudally, and bending ventrally meets the pubis, forming what is generally referred to as a ramus. The right and left pubes join ventrally about the midline of the body, at the pubic symphysis. The ischium and pubis enclose a large oval-like opening known as the obturator foramen (Cornwall 1956; Sisson and Grossman 1966; Bass 1971).

Similar to the scapula, the general form of the innominate can be characteristic of mammalian groups. In rodents the bones are light and slender. The ilium scarely expands dorsally, while the pubes and ischia are large and diverge caudally. In carnivores, the pelvis is narrow and long. The ilium is slender in shape, and somewhat parallel-sided to its other half. The ischium is broad, with laterally directed tuberosities. The pubic symphysis is long, formed in part by the ischium, as are the pubes. The ilia of Equus are very long, dorsally broad, but contracted above the acetabulum, making a slender neck. The ischium and pubis are short, creating an almost oval obturator foramen. Concerning the artiodactyls (Bovidae and Sus), the whole pelvis is elongated. The ilium is relatively shorter than Equus, and expands dorsally, but forming no neck about the acetabulum. The ischium tends to be long, and wide caudally. The pubic symphysis is long as compared to Equus, which is short. The obturator foramen is oval in shape. Pigs have an acetabular notch (incissura acetabuli), which is open more widely and ventrally, but in ruminants (cow, sheep/goats) it is closed along the ventral margin of the acetabulum's articular surface. Ruminants also possess a depression on the lateral-medial surface of the ilium, just above the acetabulum, known as the supra-acetabular fossa. This same fossa can be seen in Sus, but on the medial side.

Views:
a - lateral
b - medial
c - proximal
d - distal
e - anterior
f - posterior

Landmarks:
1 - Incissura acetabuli
2 - Supra-acetabular fossa

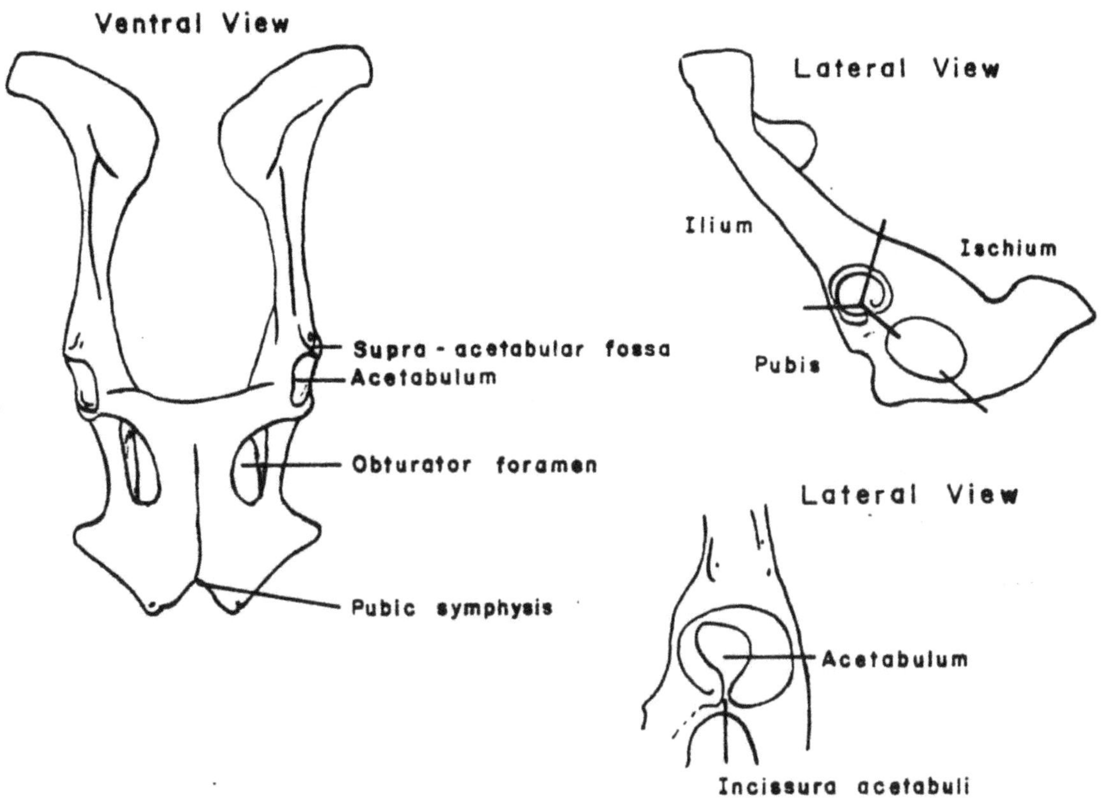

Figure 4-11, The basic architecture of the innominate of <u>Ovis</u>.

Equus caballus

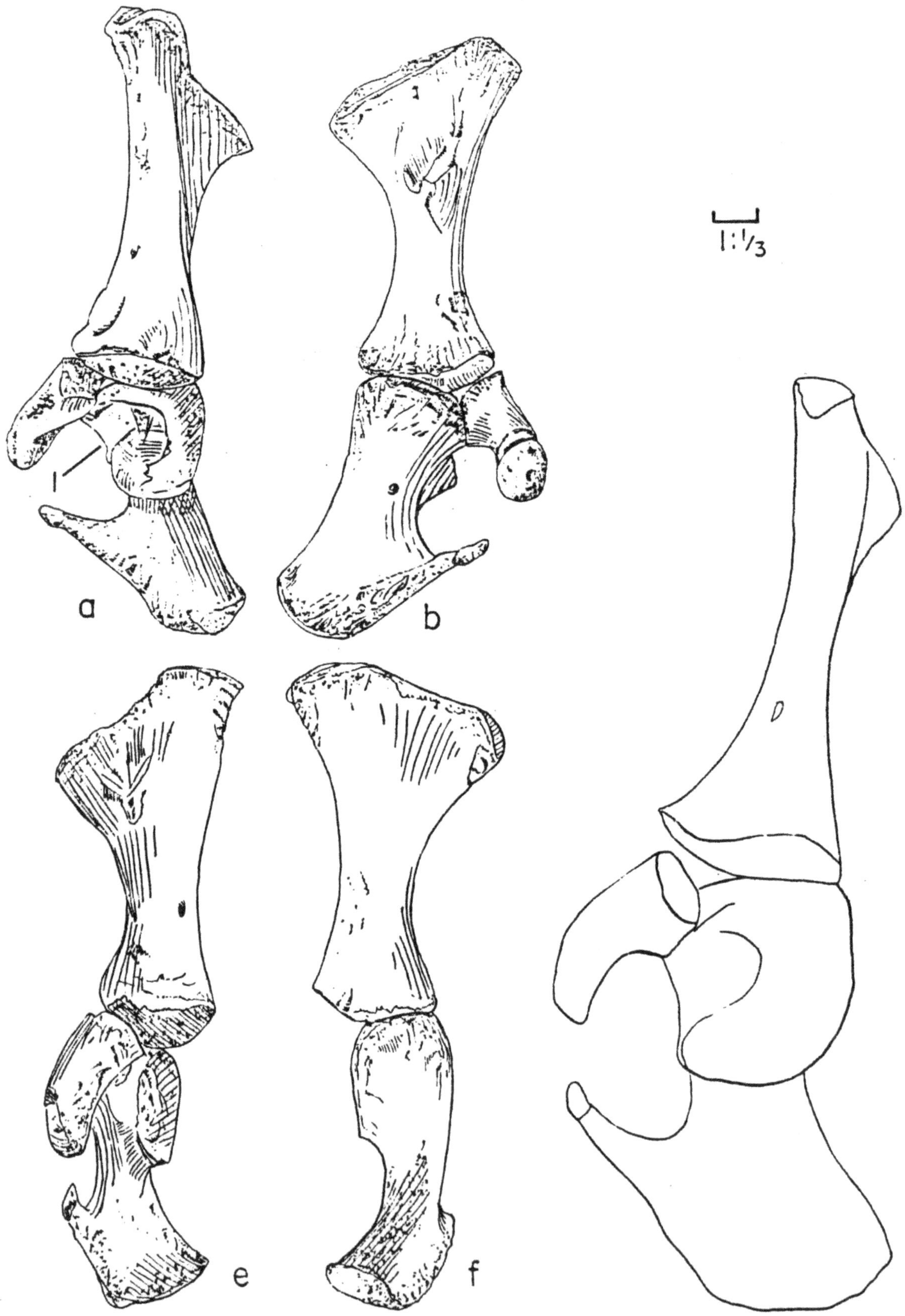

Ovis aries

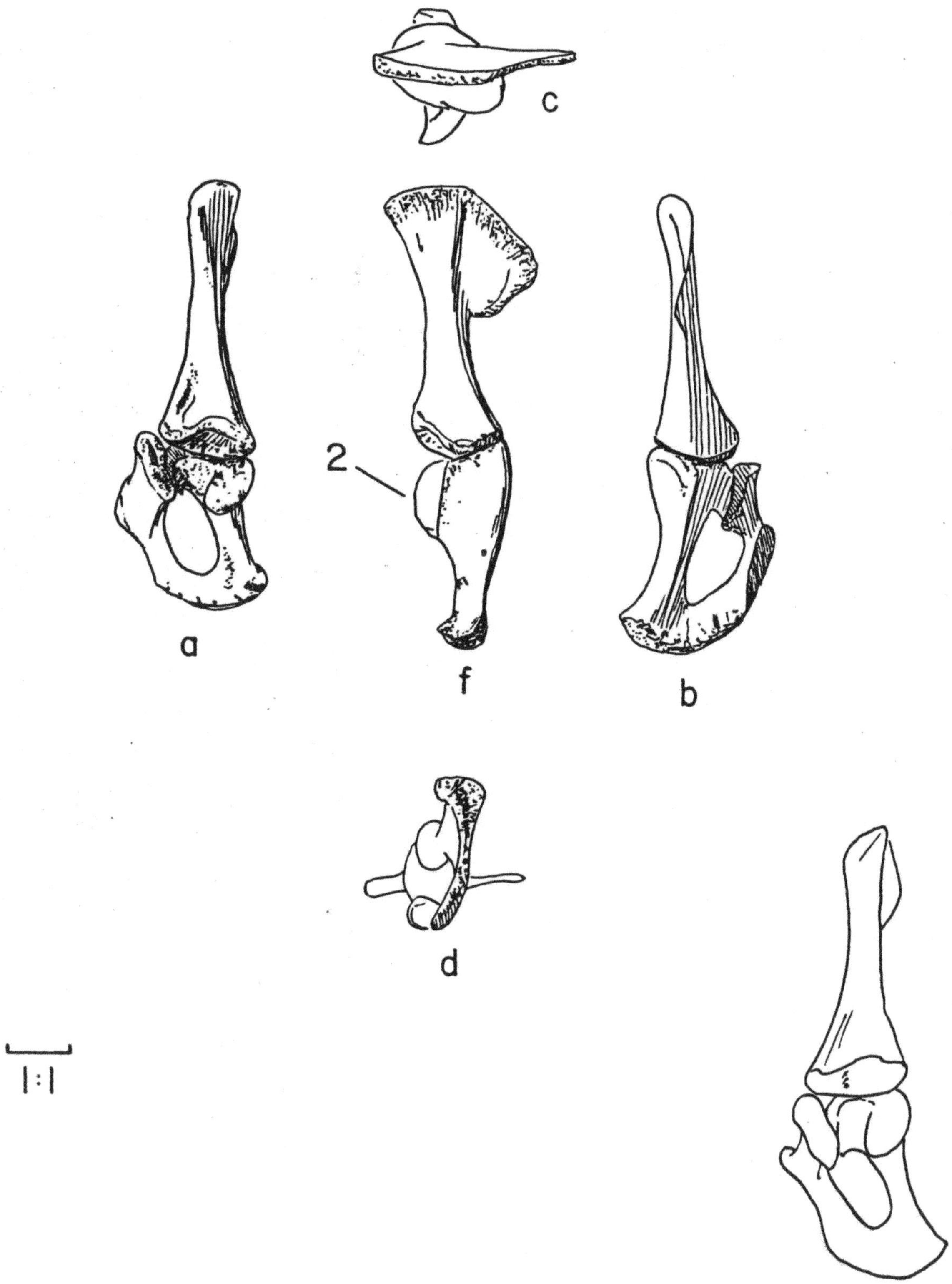

Bos sp.

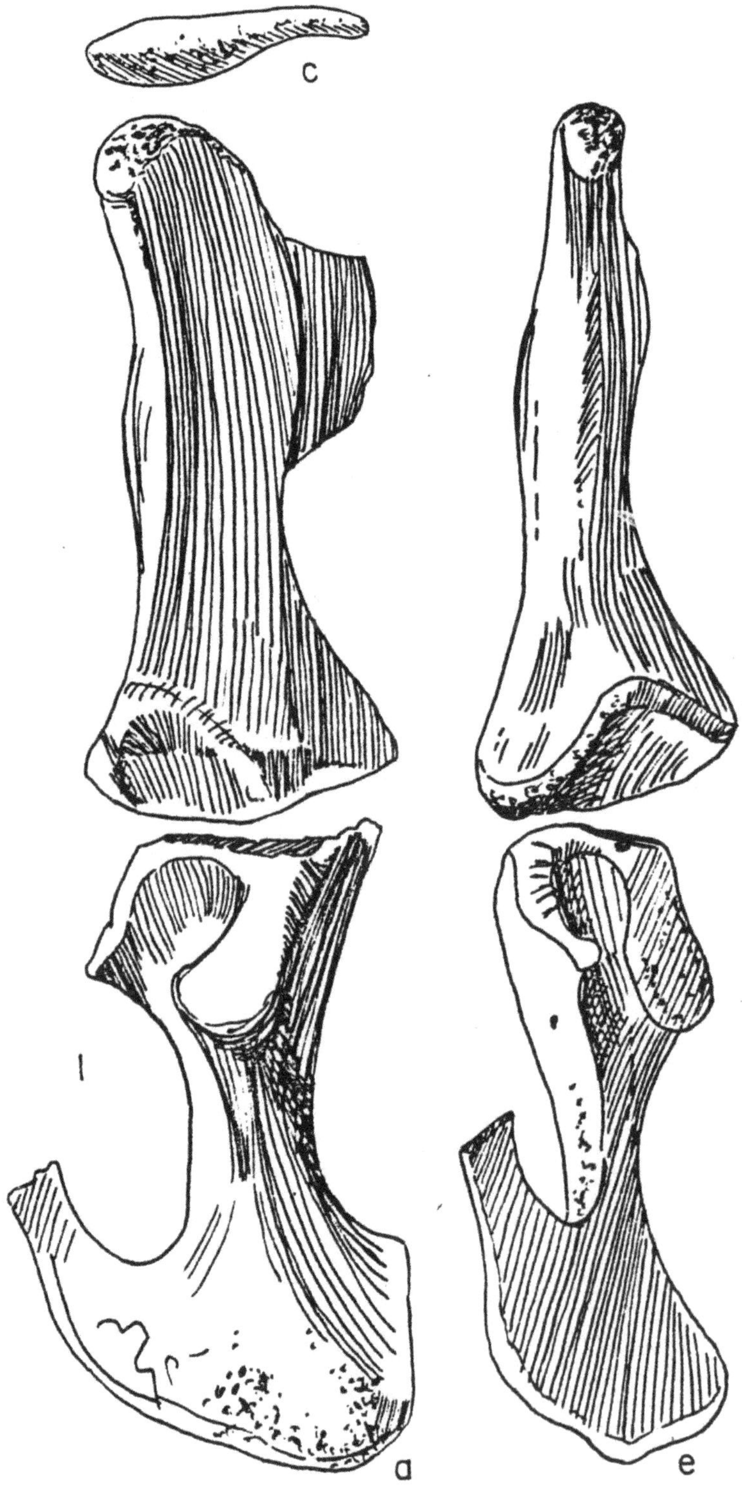

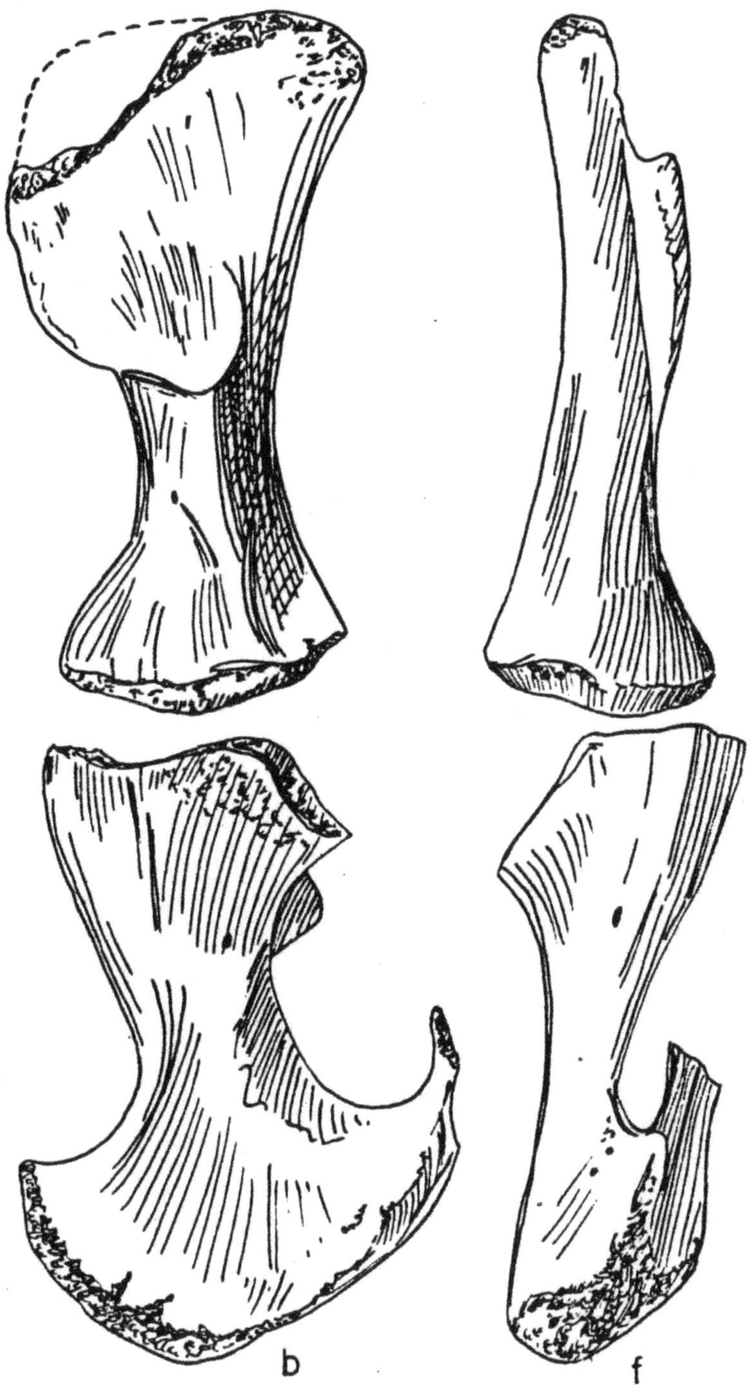

See comparison charts for
scaled contour drawing.

Sus scrofa

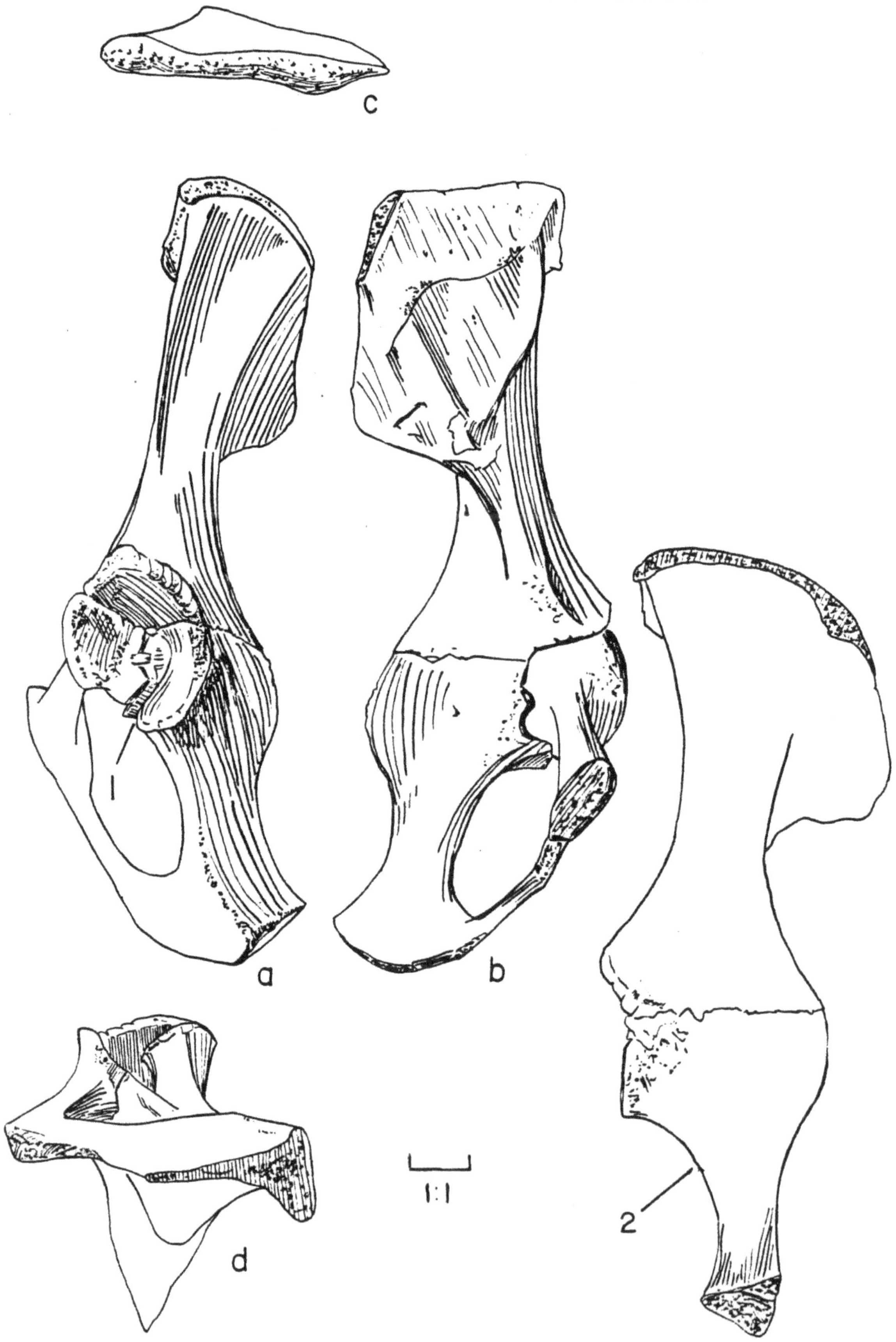

Canis familiaris

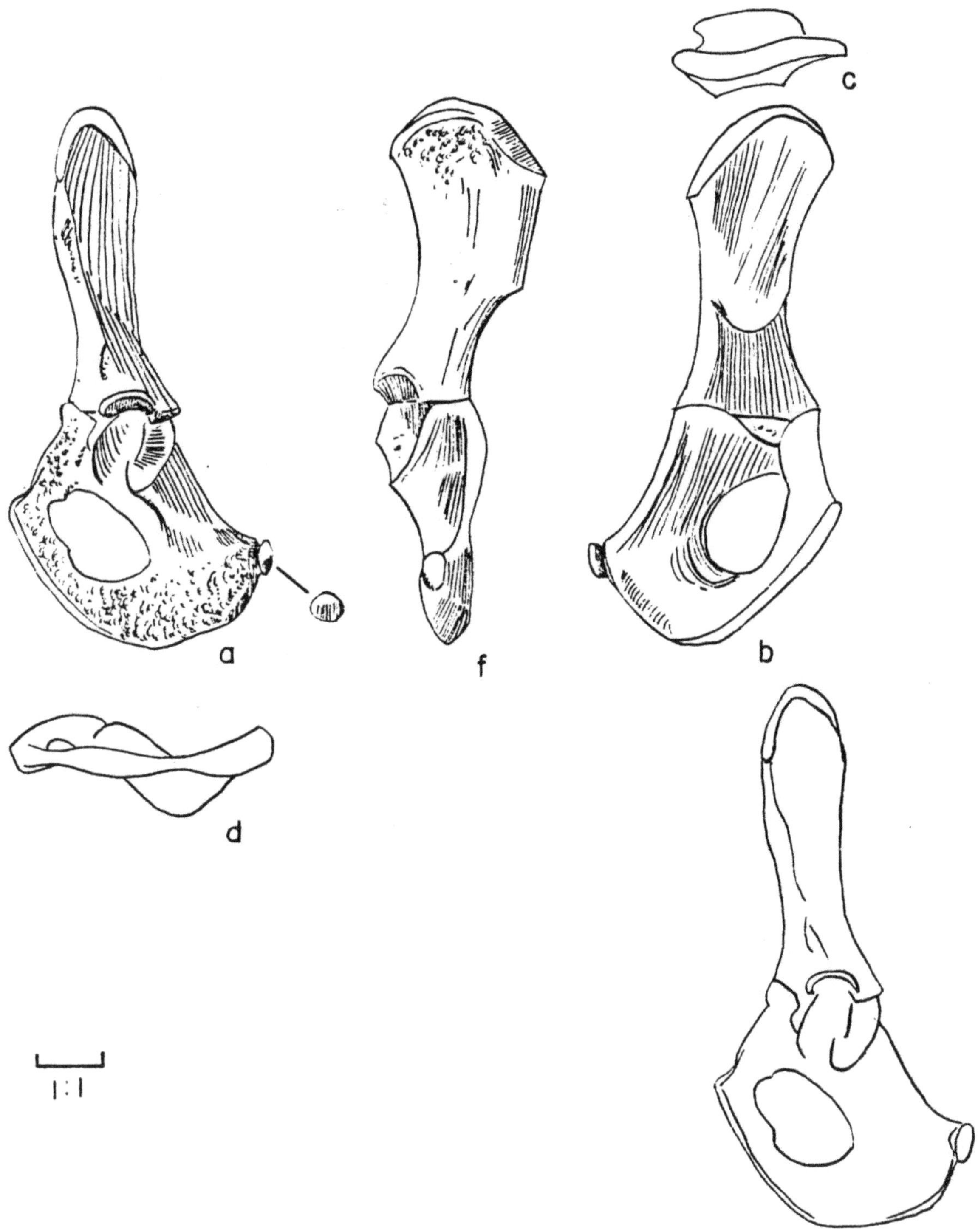

Felis catus L.

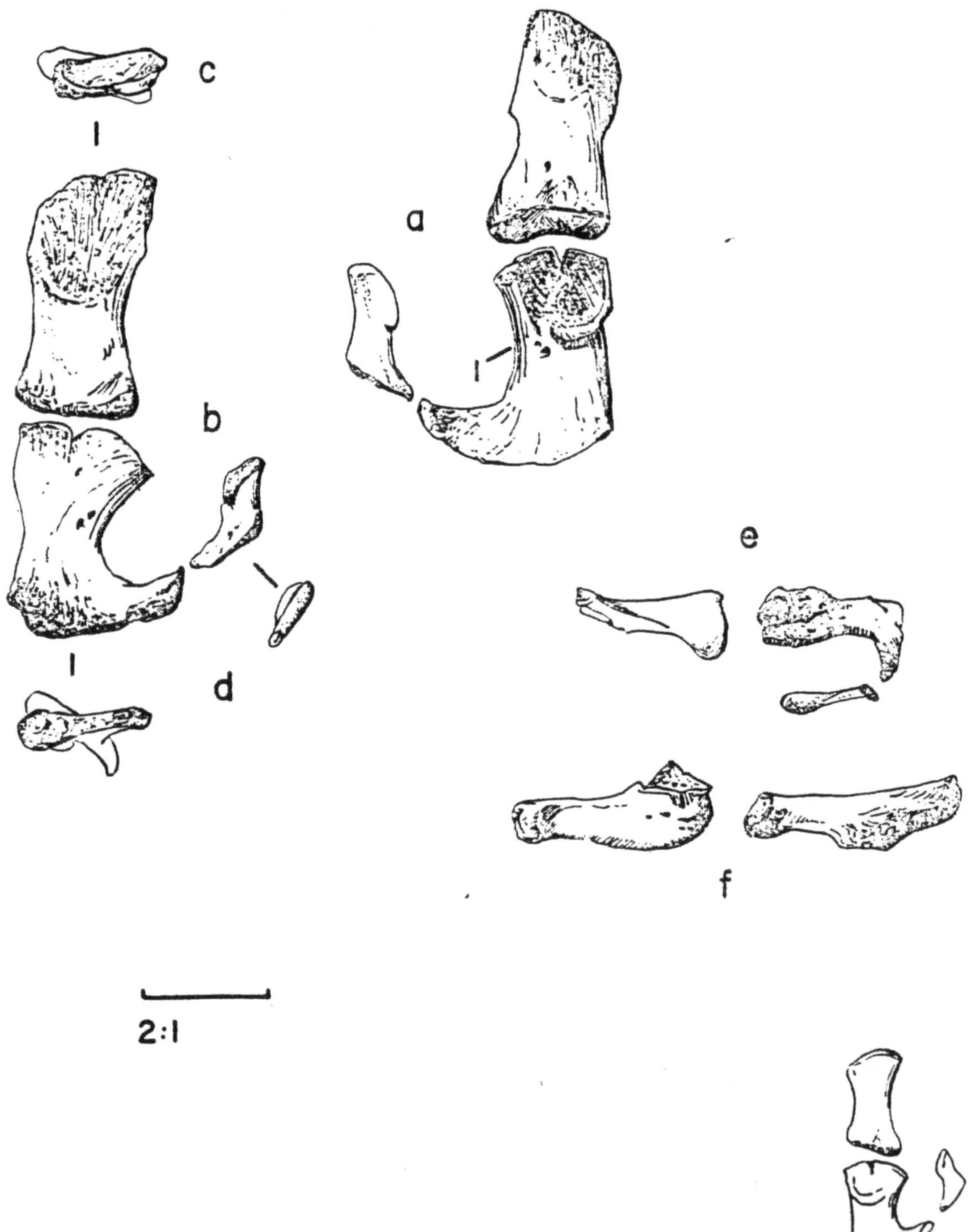

Rattus norvegicus

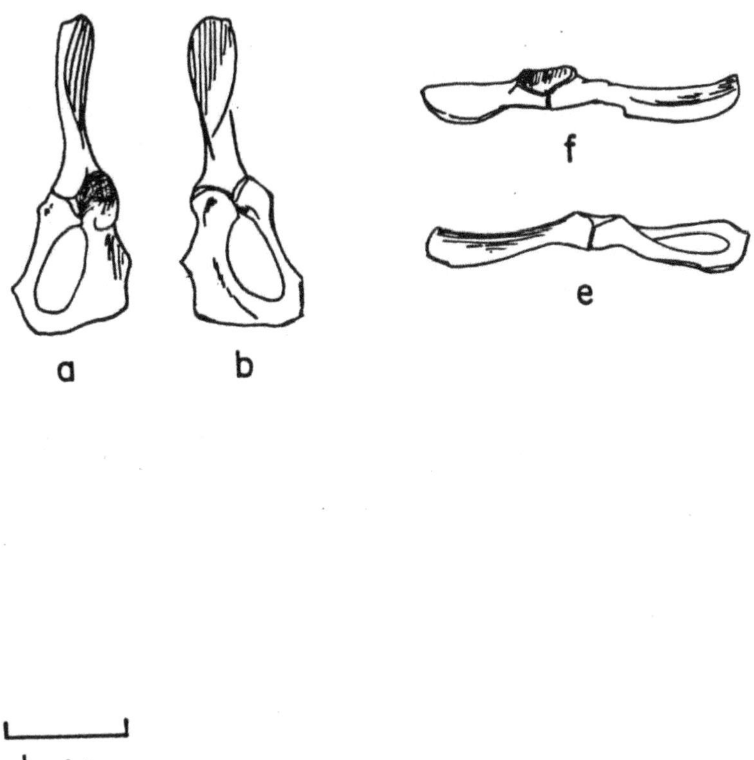

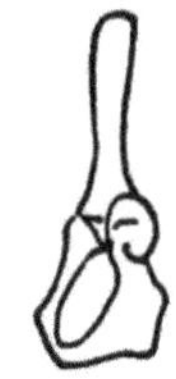

Plate XV
Femur

The femur is one of the largest long bones in the body. If examined in cross section, the shaft of the femur is roughly cylindrical. The femur is curved longitudinally along a sagittal plane, with the convex surface on the anterior aspect. The femur's head is roughly hemispherical in shape, and articulates with the innominate, forming a ball-and-socket joint at the acetabulum. Unlike the humerus, the head of the femur is set on a long neck (see figure 4-12). The angle taken from the axis of the head and neck is obtuse, for most mammalian species. Characteristic of the femoral head, is a small pit in the center of the articular surface, known as the fovea capitis. This pit serves as the attachment of a ligament on the femoral head, that fastens it to the acetabular floor (Cornwall 1956; Sisson and Grossman 1966).

Towards the base of the neck is a large process for muscle attachment known as the greater trochanter. In many mammalian species, the height of the greater trochanter is greater than the height of the neck and head. Separating the greater trochanter from the lesser trochanter is an oval-depression, or groove known as the trochanter fossa (digit fossa or fossa trochanteria). On the medial side of the shaft, below the neck, and somewhat posterior is the location of the lesser trochanter. The lesser trochanter is joined by the oblique trochanter crest, forming the posterior margin of the trochanter fossa. In a number of mammalian groups, a third trochanter can be seen as a thin outstanding plate of bone, distal to the greater trochanter, being lateral on the femoral shaft (Sisson and Grossman 1966).

The shaft of the femur broadens towards the distal epiphyseal region. This region has two prominent knuckles of bone, that are articular surfaces known as the medial and lateral femoral condyles. These condyles are separated by the deep inter-condylar fossa. On the anterior aspect, a smooth, wide articular surface, that is concave and elongated transversely, is often referred to as the femoral trochlea. This trochlea serves as a surface in which the knee-cap (patella) articulates (Cornwall 1956; Sisson and Grossman 1966; Bass 1971).

The femur of Rattus is long and slender in form with the third trochanter immediately below the greater trochanter. The carnivores in general possess a femur that is relatively long, slender and straight. The femur of Equus is strongly built, but short in relation to the hind limb as a whole. Even at a neo-natal age, the femur does possess strong muscle markings. The head is large, and the neck thick and indistinct, with the greater trochanter rising above the head. The lesser trochanter can be found along the medial aspect, below the head, as a low ridge. The third trochanter is present, but will develop to a greater extent as the animal ages. The femur of the artiodactyls (Bovidae and Sus) can be distinguished from the horse by the absence of the third trochanter. The femur has a small, rounded head, set on a relatively long neck. The greater trochanter can extend over the femoral head. The condyles once developed are large, allowing a large surface for the patella to articulate.

Views:
a - cranial (anterior)
b - proximal
c - caudal (posterior)
d - distal
e - lateral
f - medial

Landmarks:
1 - Trochanter majus
2 - Trochanter tertius (third trochanter)
3 - Fovea capitis
4 - Fossa trochanteria
5 - Trochanter minor
6 - Fossa plantaris
7 - Trochanter patellaris
8 - Epicondylus lateralis
9 - Patella
10 - Sesmoid bone

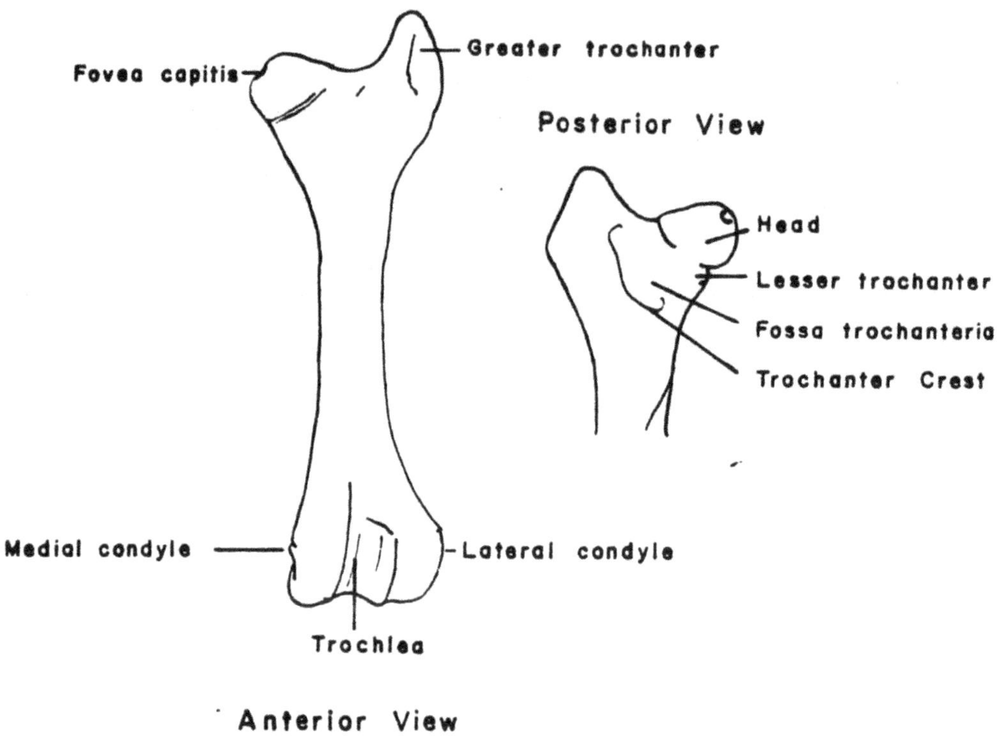

Figure 4-12, The basic architecture of the femur of Bos.

Equus caballus

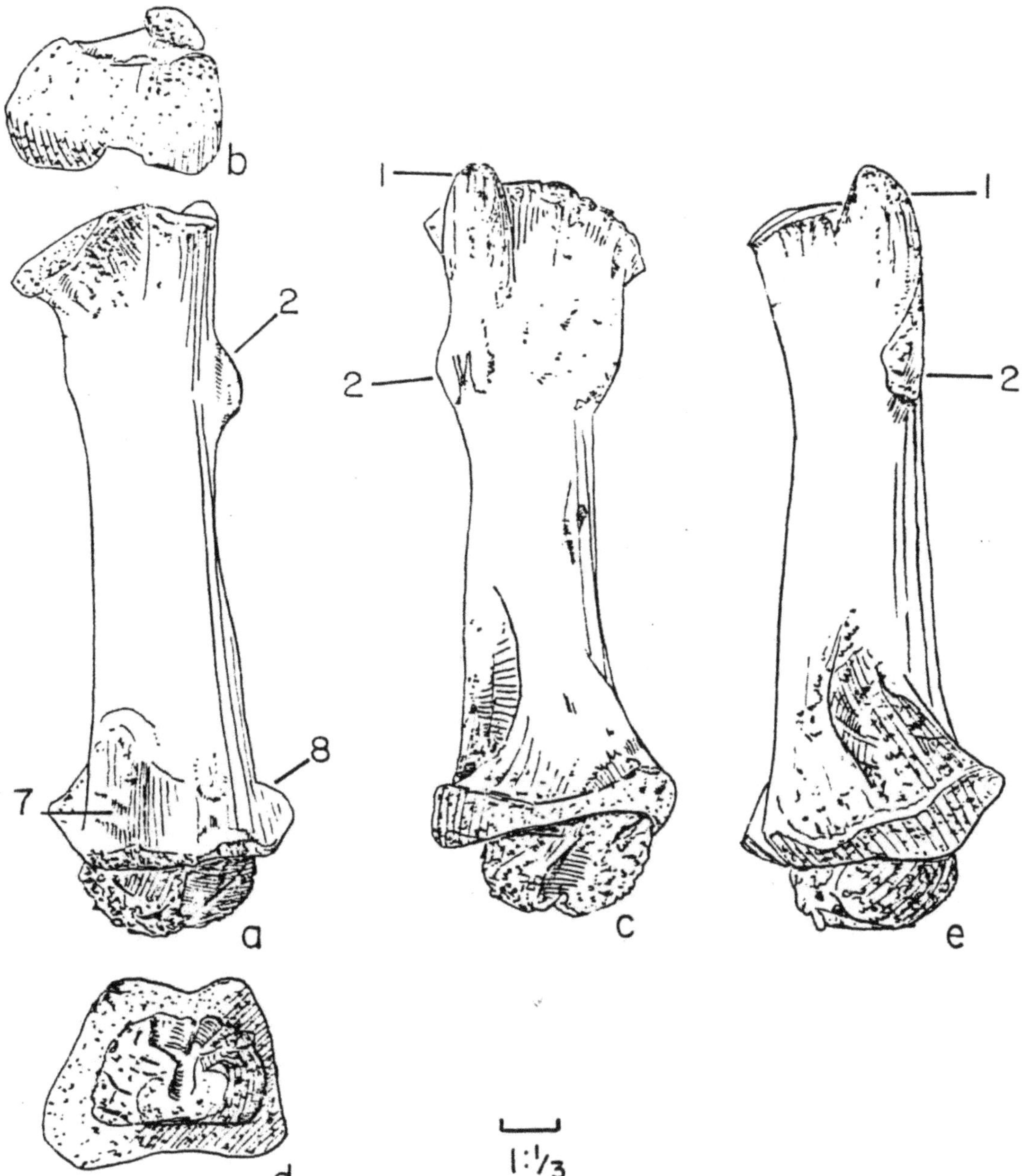

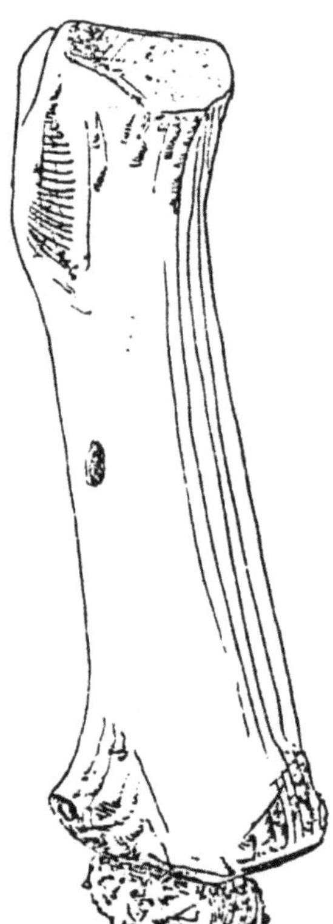

f

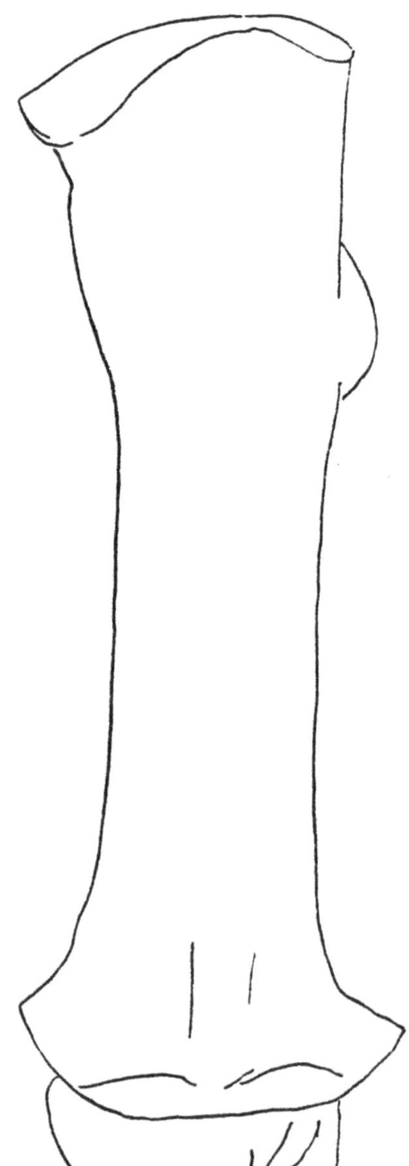

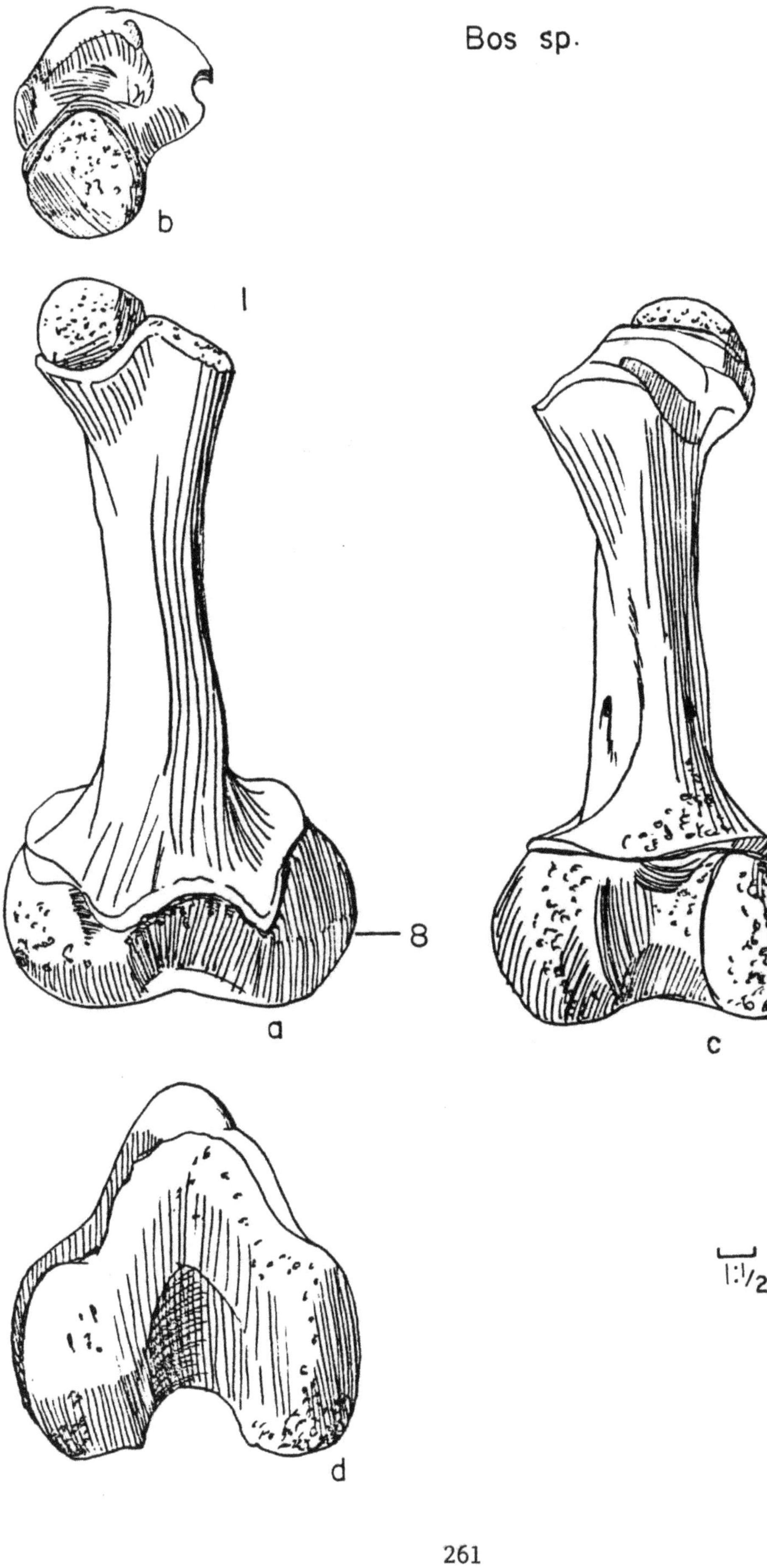
Bos sp.
b
1
8
a
c
d
1:1/2

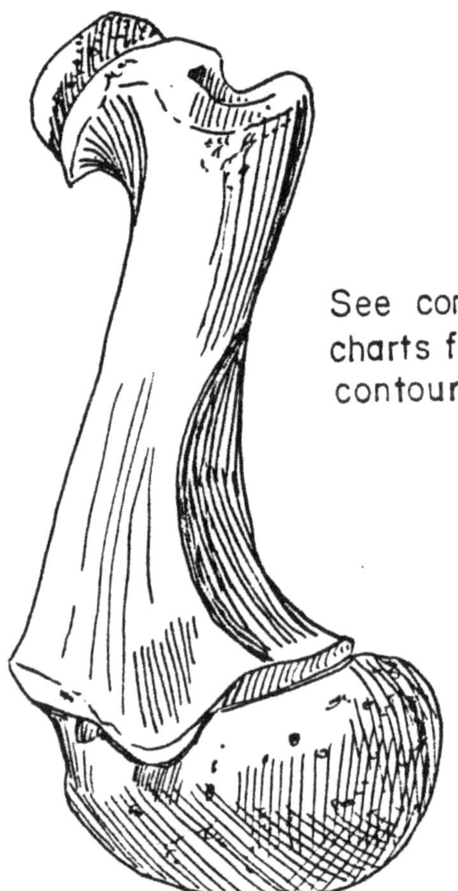

e

See comparison charts for scaled contour drawing.

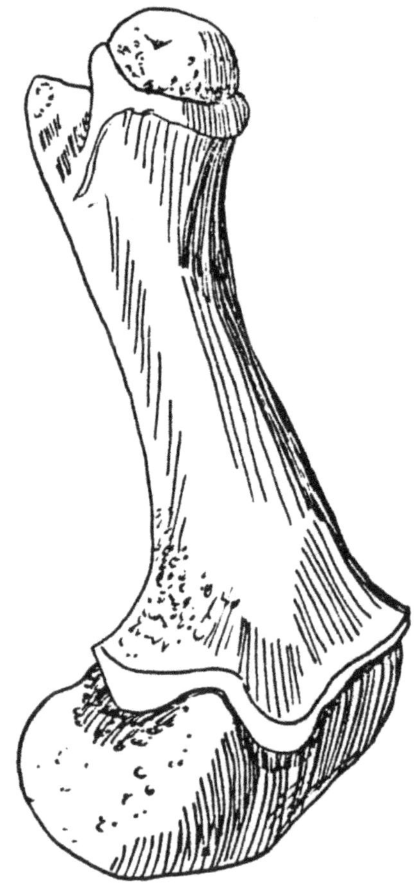

f

Ovis aries

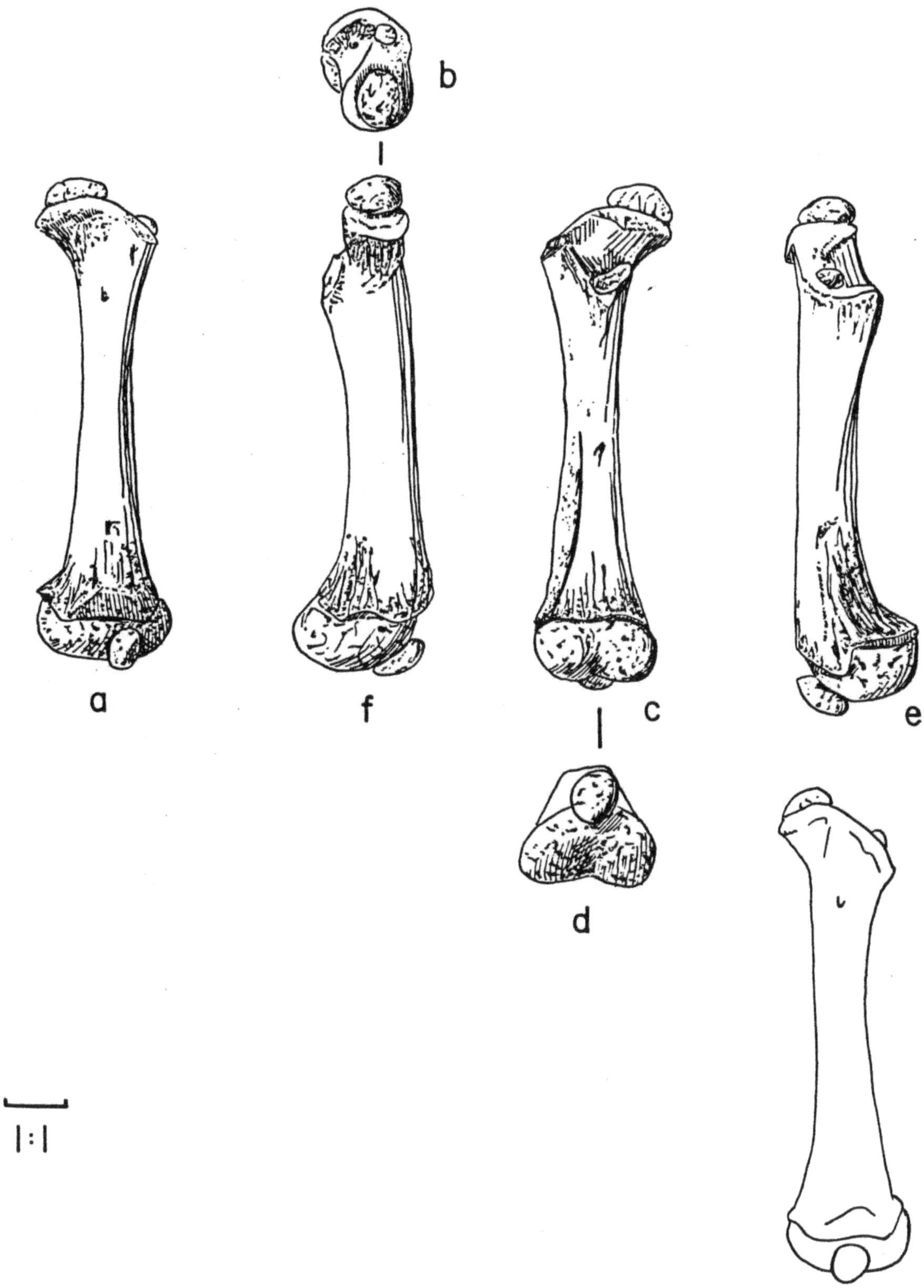

Sus scrofa

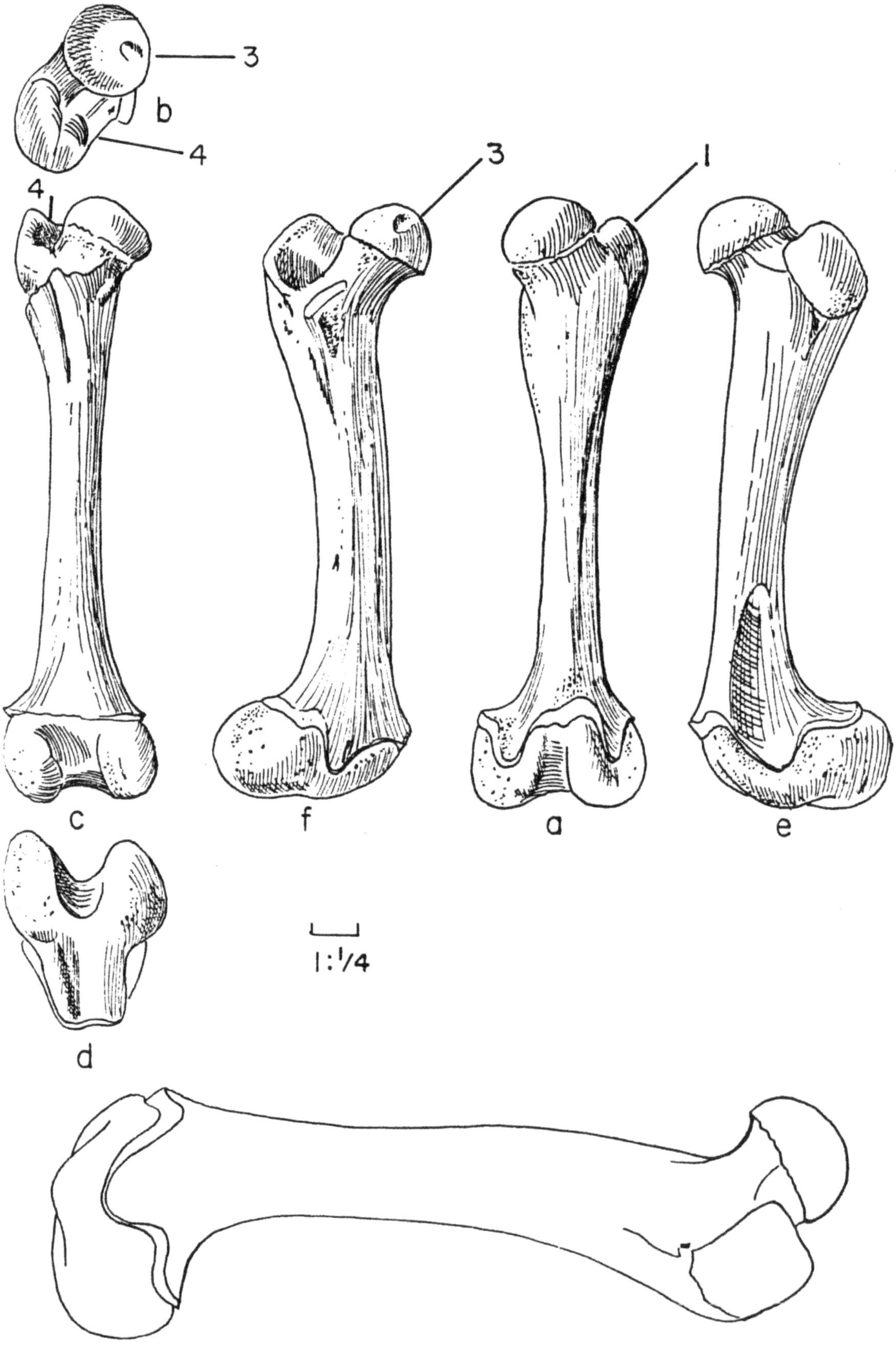

Canis familiaris

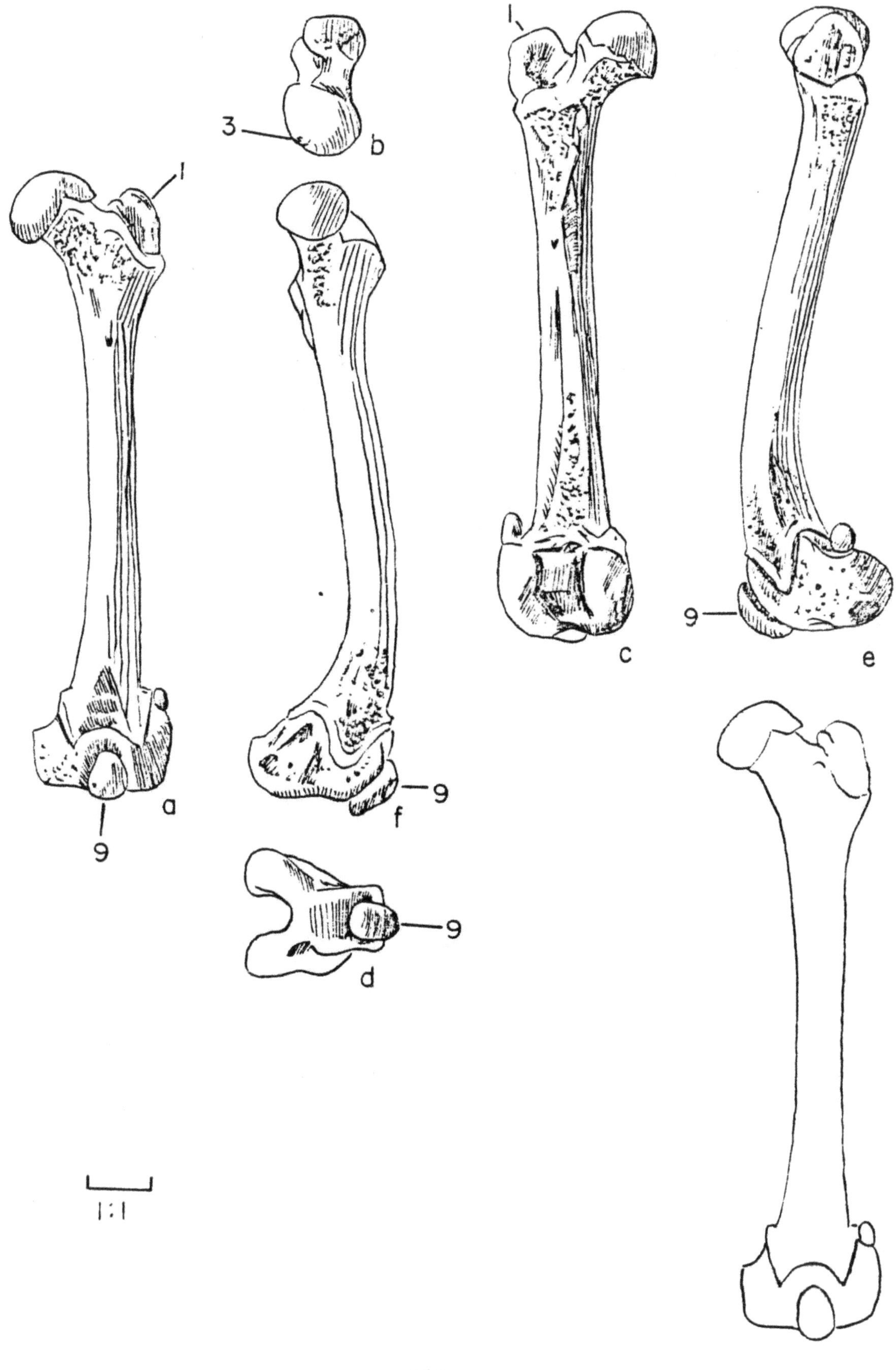

Felis catus L.

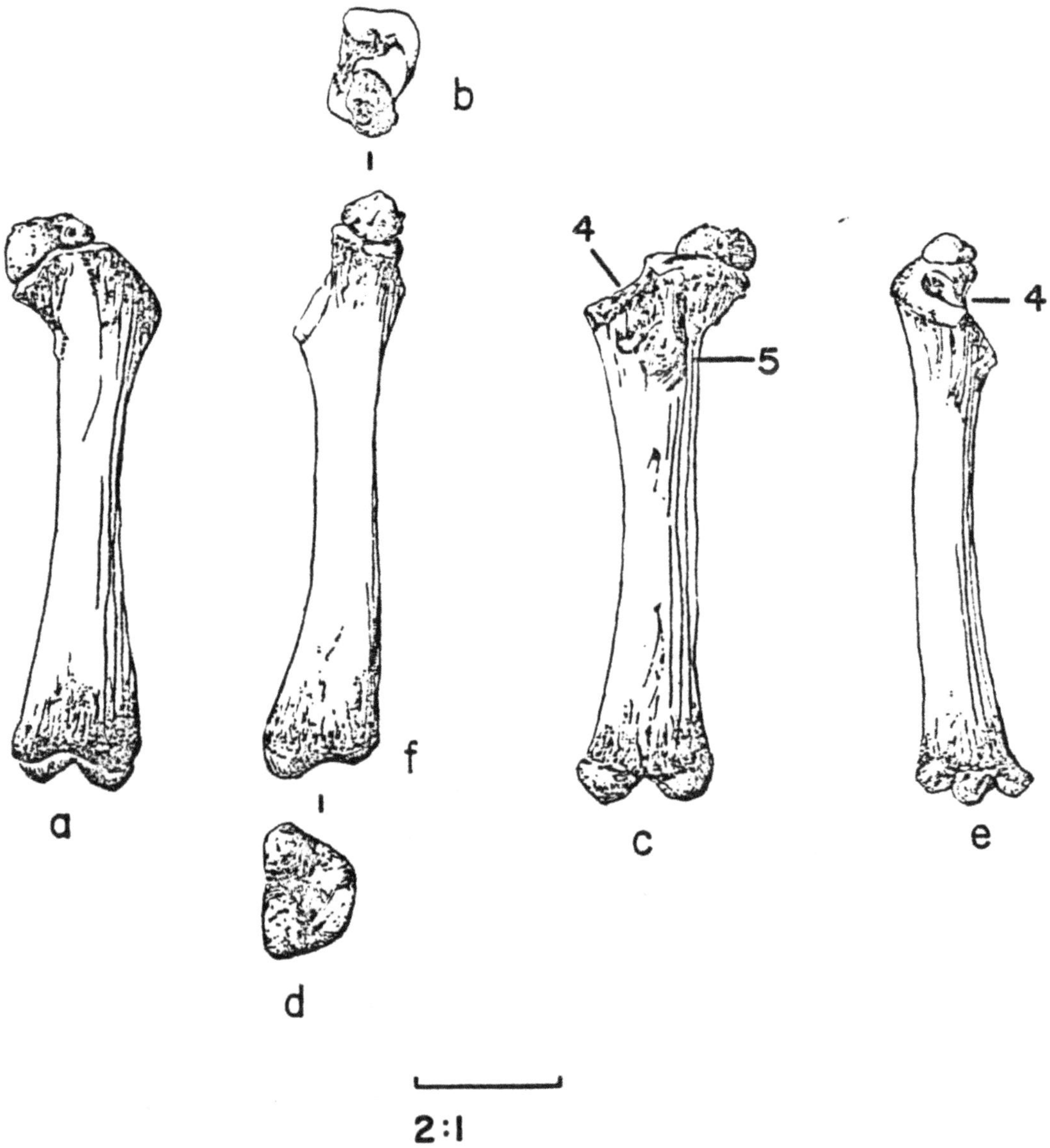

Rattus norvegicus

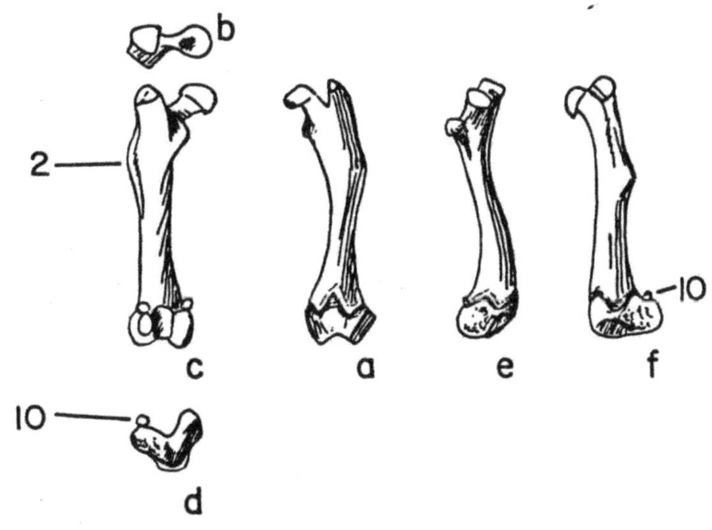

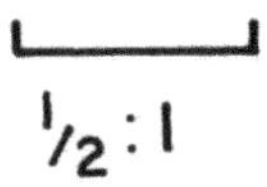

Plates XVI
Tibia

The tibia is the second largest long bone and the heaviest in the body. The proximal articular surface consists of a pair of oval-shaped, concave condylar facets which articulate with the femoral condyles (see figure 4-13). The articular surface is separated by a raised intercondylar area, which plays a large part in maintaining the alignment of the hind limb by permitting free movement in one plane. The sides of the articular facets ar raised into a crest, allowing for articulation of the femoral condyles (Cornwall 1956; Sisson and Grossman 1966; Schmid 1972).

In cross section, the proximal end has a triangular shape. The crista tibiae (tubercle) is the most prominent tuberosity on the proximal, anterior aspect. Characteristic of the tibia, is the deep fovea (groove) seen on the proximal anterior aspect. This fovea affords the attachment of the ligaments for the patella, known as tuberosities tibiae (Schmid 1972).

The shaft of the tibia is the thinnest at its middle, and gradually expands distally. As a consequence, the anterior border is sharp, and there is a flatness of the posterior surface. The anterior border runs somewhat medially from the base of crista tibiae to the distal end. Distally, the tibia expands even more to accommodate the articulation for the proximal surface of the astragalus. Towards the most distal, medial area of the tibia is a process referred to as the medial malleolus, which also articulates with the astragalus.

Common in murid rodents such as Rattus is the distal part of the fibula fusing to the tibia. In carnivores, the tubercle (crista tibiae) is prominent, as is the anterior border running from its base to the distal end. In Equus, the tibia has the characteristic triangular proximal end, with a prominent anterior border extending down the shaft. In Sus, the tibia is slender as compared to ruminants as is the tubercle and anterior border. The distal articulations for the astragalus are shallow, parallel grooves separated sagittally by a ridge. The tibia of ruminants is elongated as compared to Sus. The tubercle is prominent, as is the anterior border. The distal articulation for the astragalus is similar to Sus.

Views:
a - anterior
b - medial

c - posterior
d - proximal
e - distal
f - lateral

Landmarks:
1 - Crista tibiae
2 - No fovea for the ligament of the patella in Tuberositas tibiae
3 - Sulcus muscularis
4 - Ligamentum cruciacun
5 - Caudal margin of Condylus lateralis
6 - Distal articular surface
7 - Upper view of the distal articulation
8 - Malleolus medialis

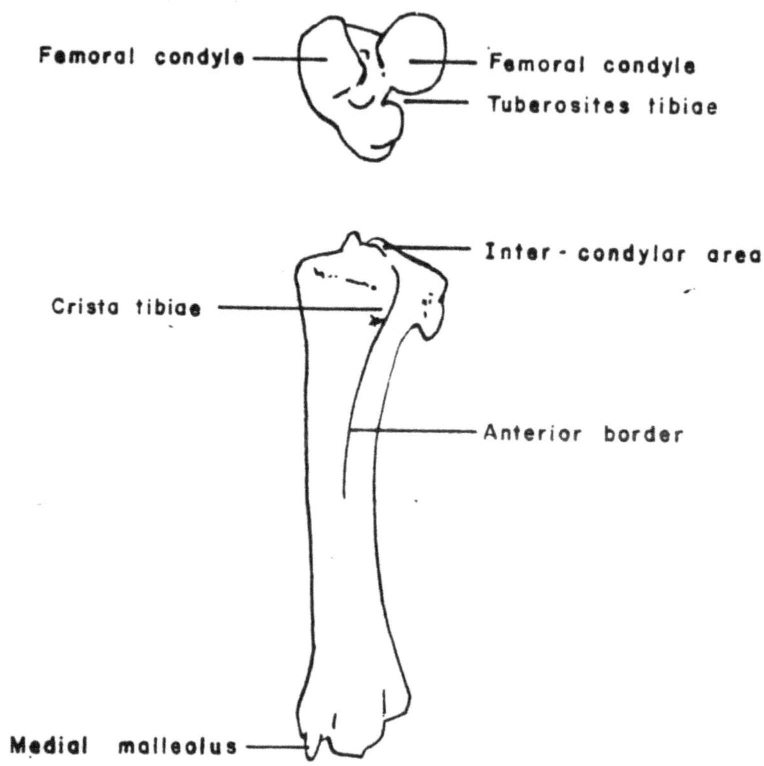

Figure 4-13, The basic architecture of the tibia of Bos.

Equus caballus

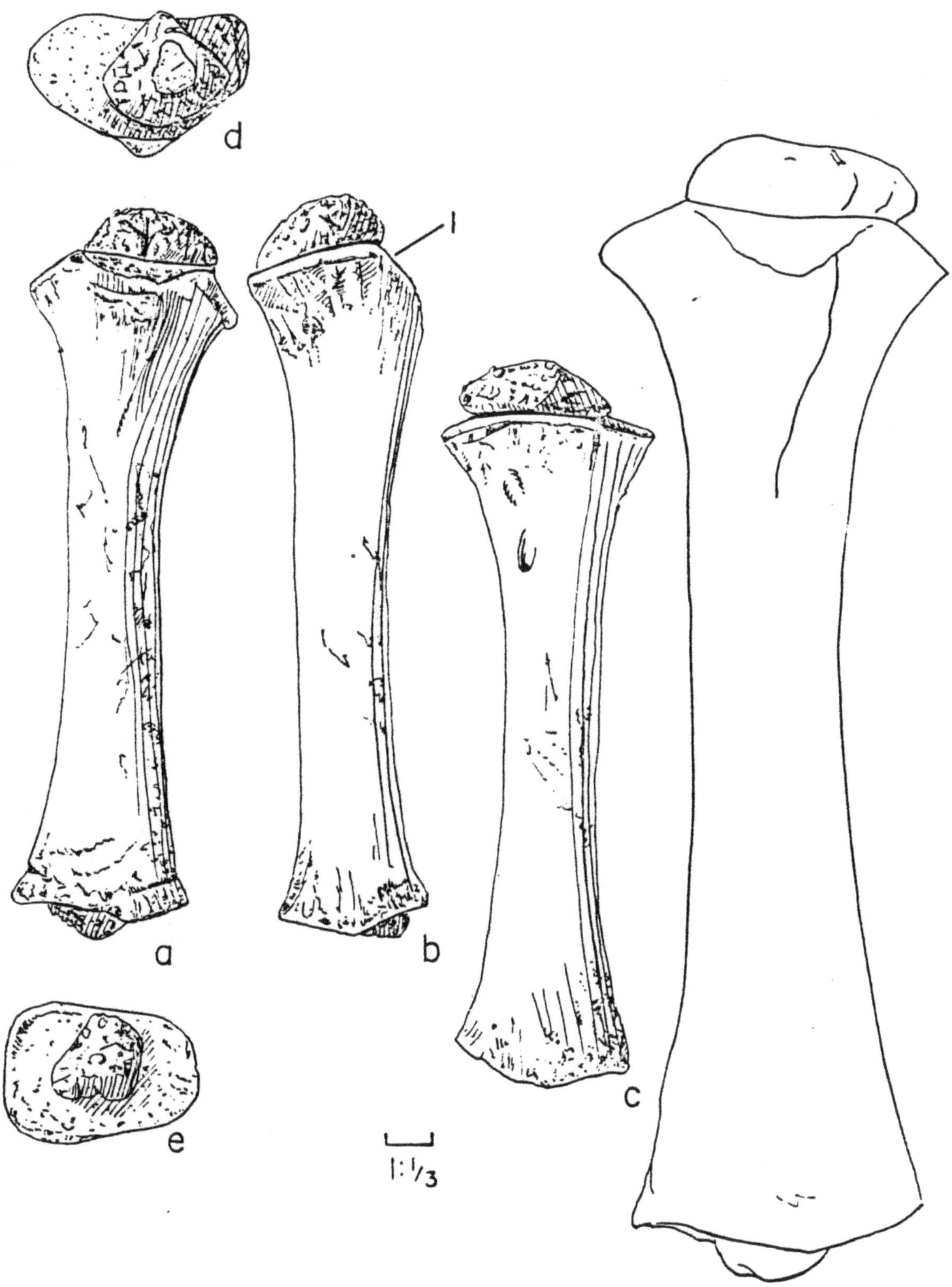

Bos sp.

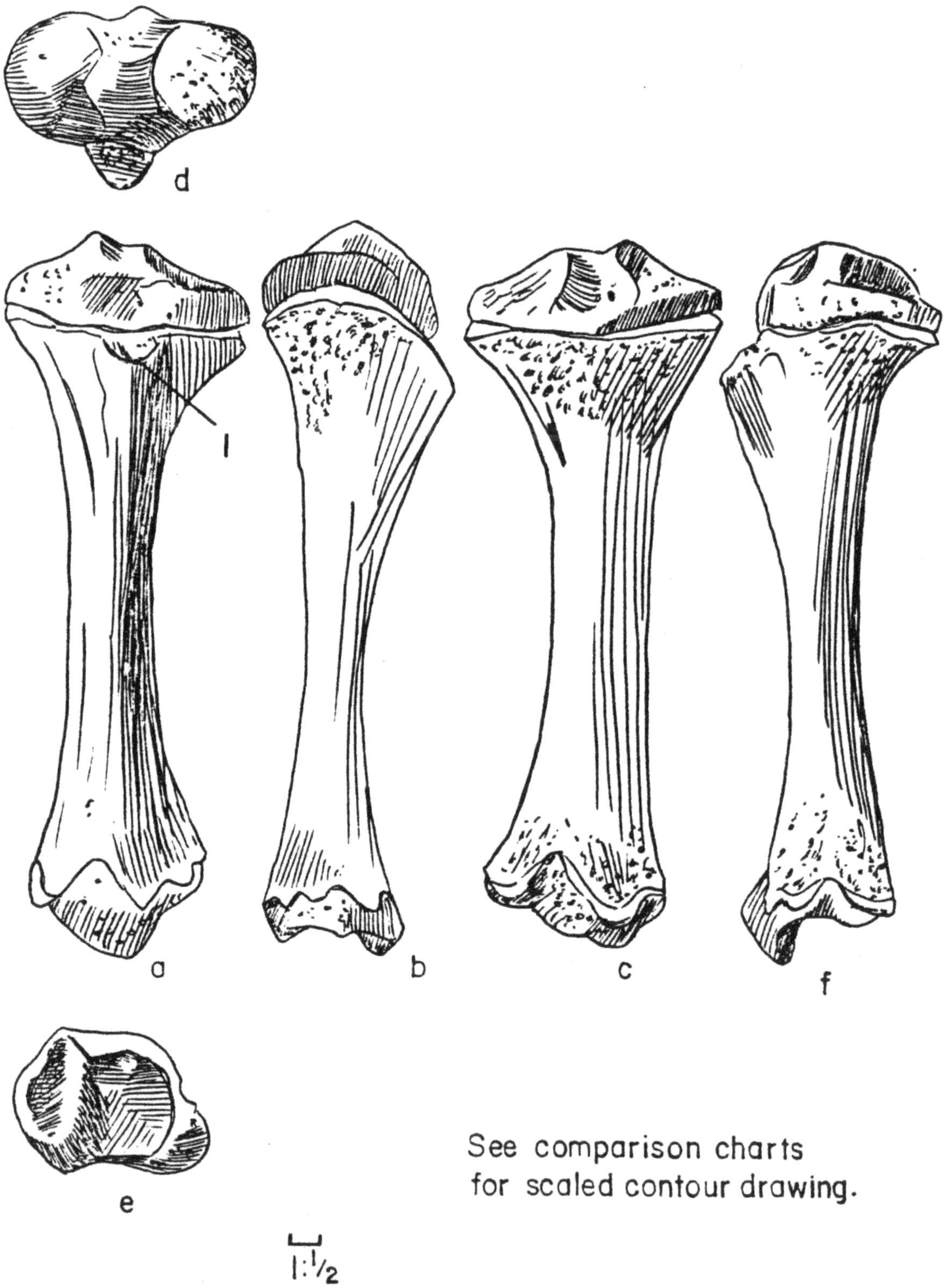

See comparison charts
for scaled contour drawing.

Ovis aries

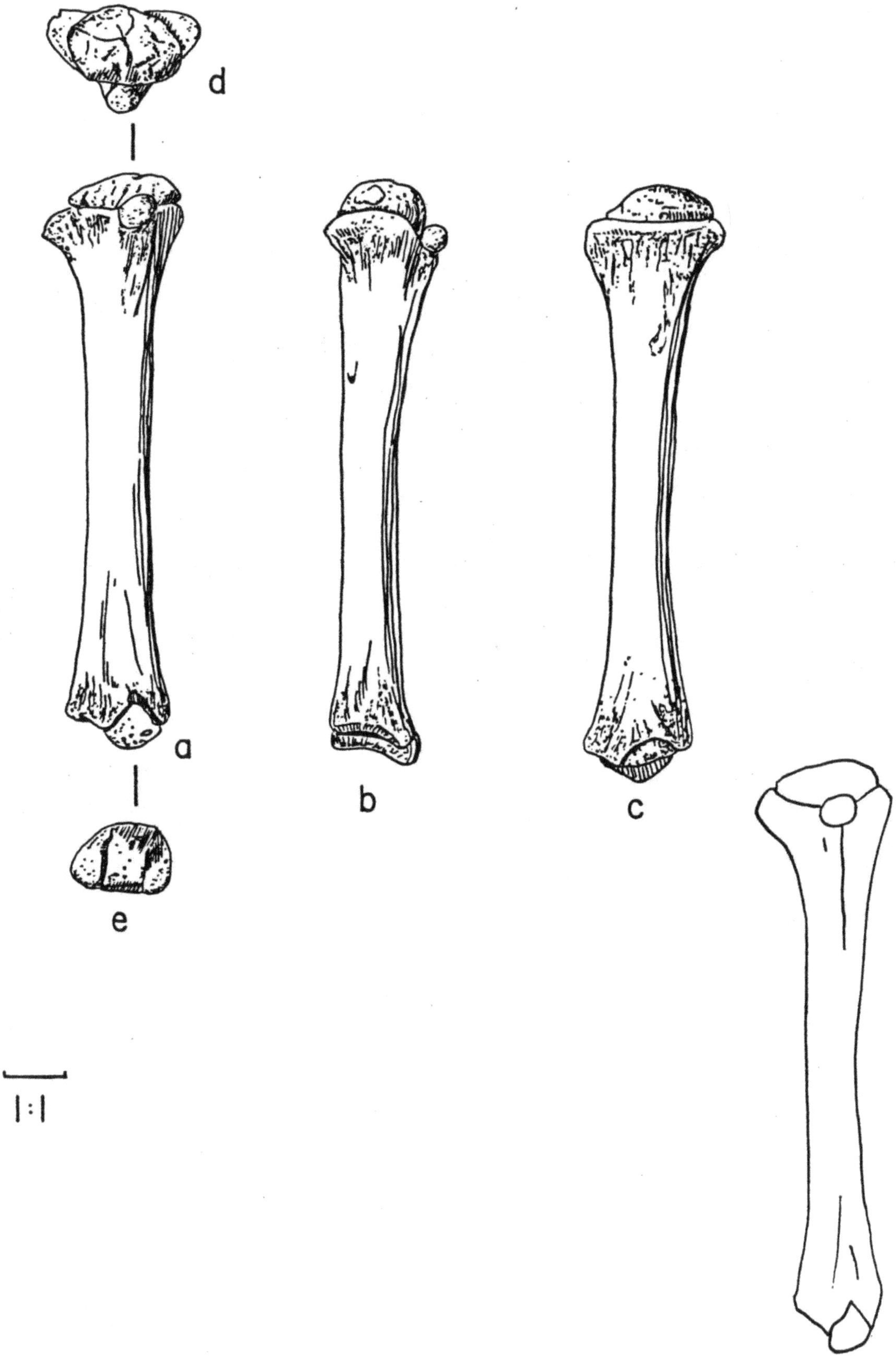

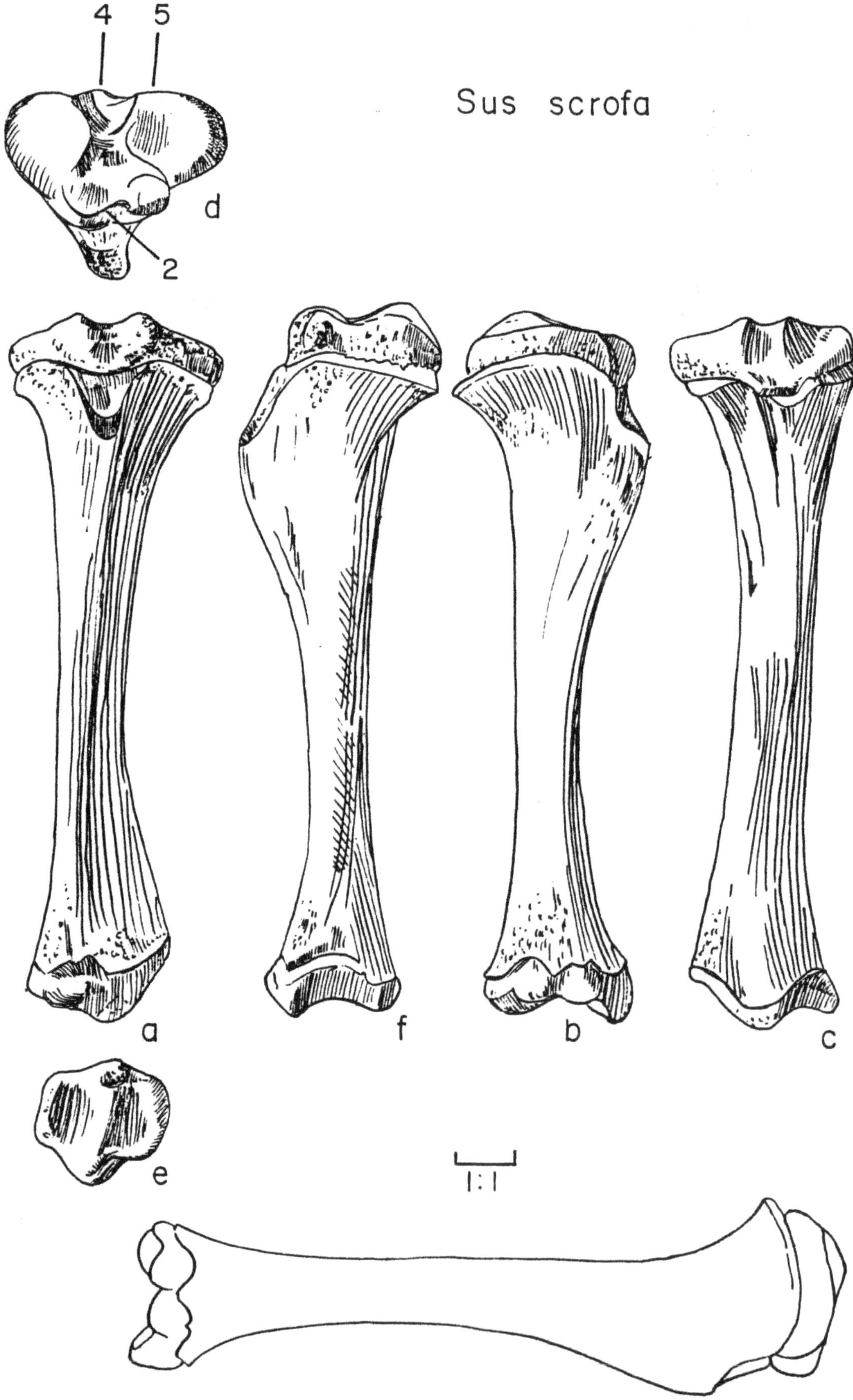
Sus scrofa
4
5
d
2
a
f
b
c
e
1:1

Canis familiaris

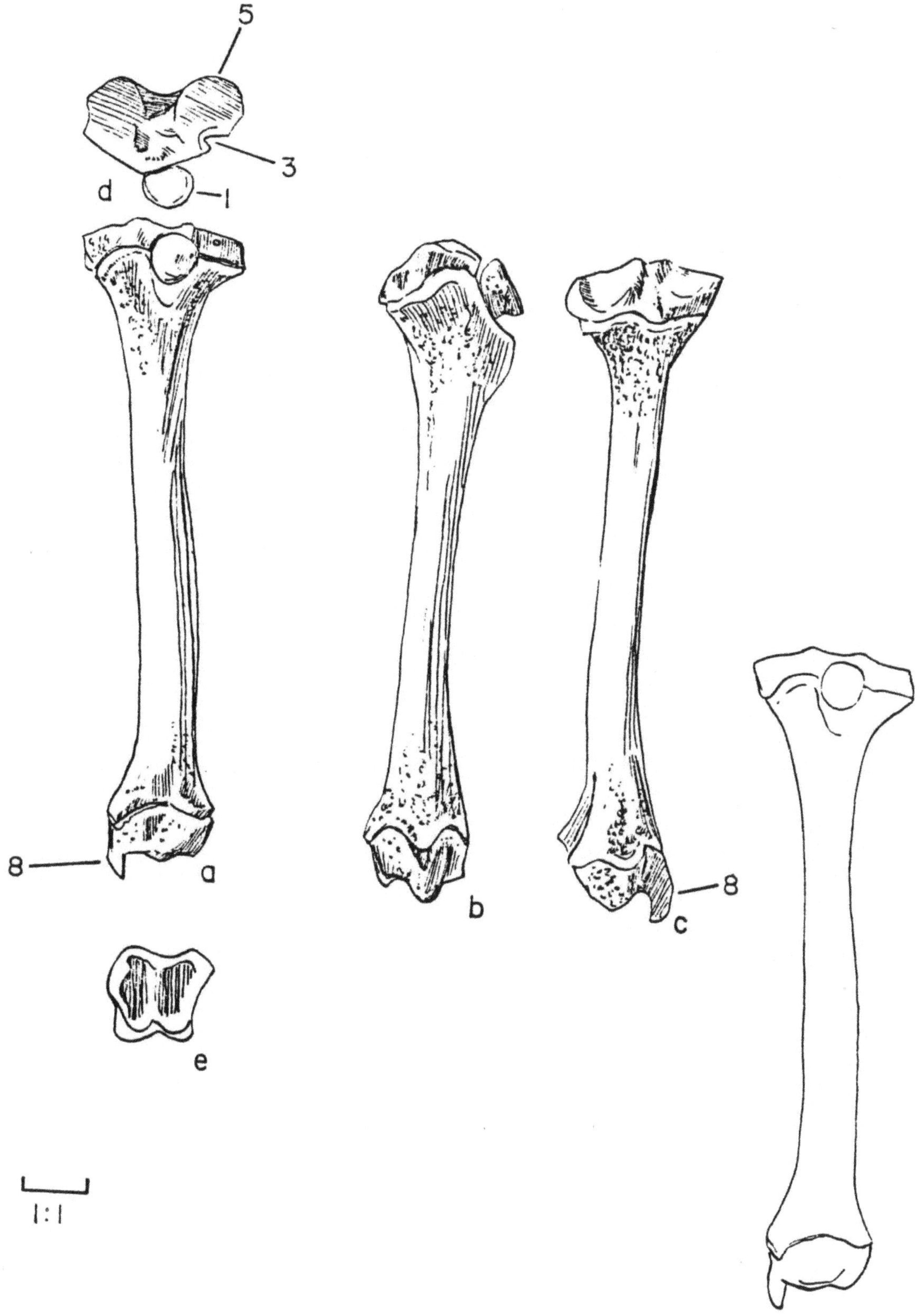

Felis catus L.

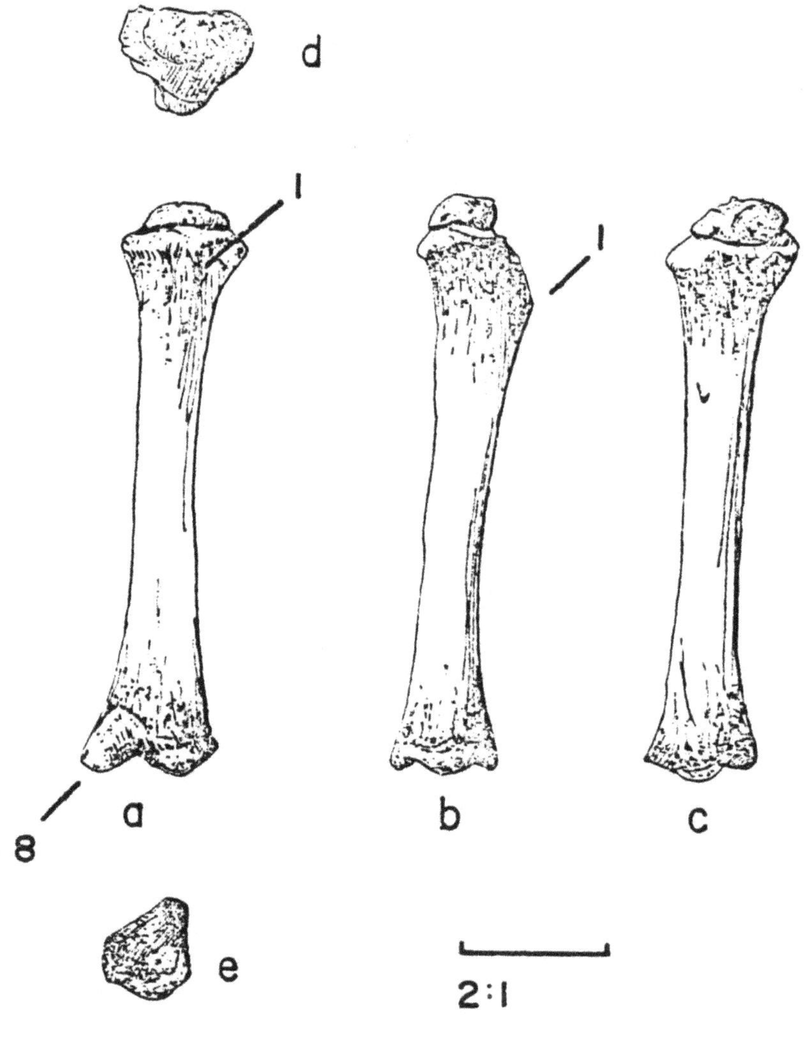

Rattus norvegicus

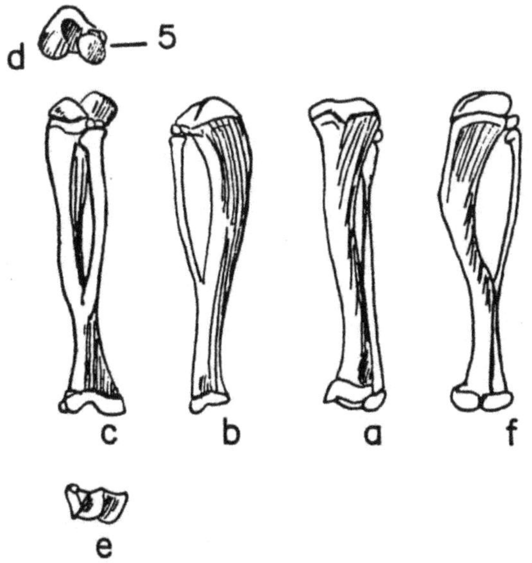

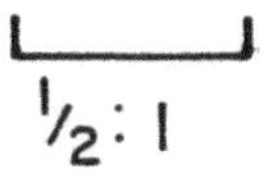

Plates XVII
Fibula

The fibula is a slender long bone that slightly expands proximally or distally to articulate with the lateral side of the tibia, just below the lateral condylar facet of the tibia. The fibula forms no part of the knee-joint. Distally, the fibula also articulates with the lateral side of the astragalus. Together with the tibia's lateral malleolus, the fibula prevents displacement of the astragalus (Cornwall 1956; Sisson and Grossman 1966; Schmid 1972).

The fibula of Rattus (mentioned under the tibia section) fuses distally with the tibia. The fibula in carnivores is distinct, somewhat triangular in shape, and generally well separated from the tibia. Among Canidae the lower half of the fibula lies closely alongside the tibia. The fibula in pigs is well developed, and triangular in cross section. In Equus and ruminants there is no developed fibula. Instead, there is either a small thorn fused on the Condylus lateralis tibiae in mature animals, or a small independent bone (os malleoloare -- not illustrated).

Views:
a - anterior
b - medial
c - posterior
d - proximal
e - distal
f - lateral

Sus scrofa

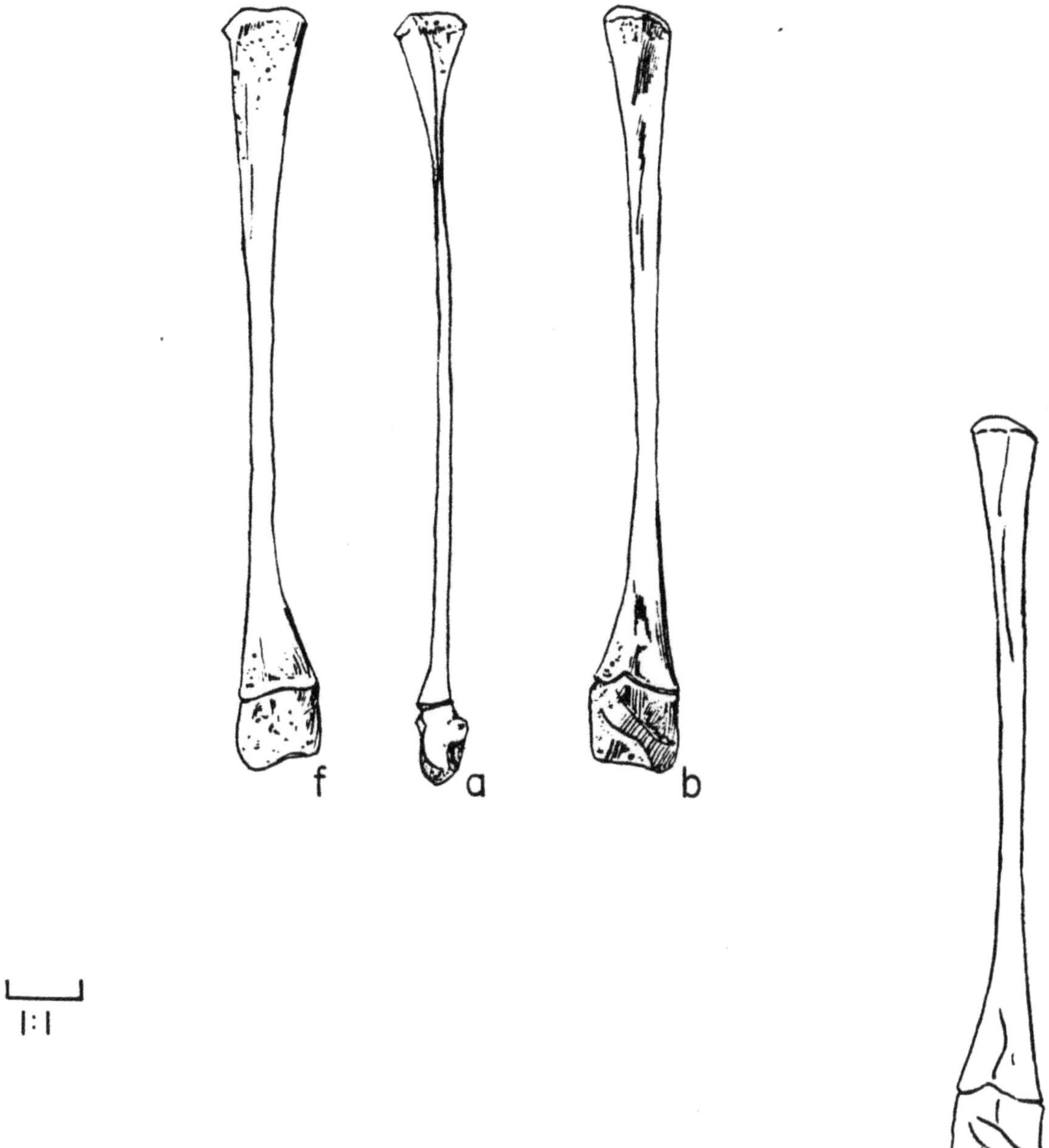

Canis familiaris

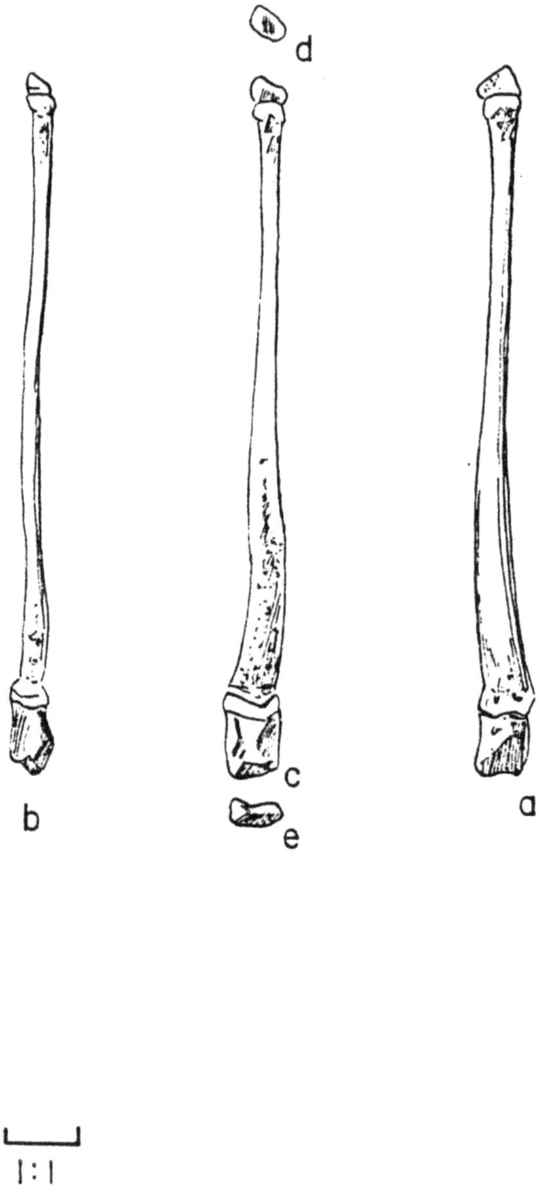

Felis catus L.

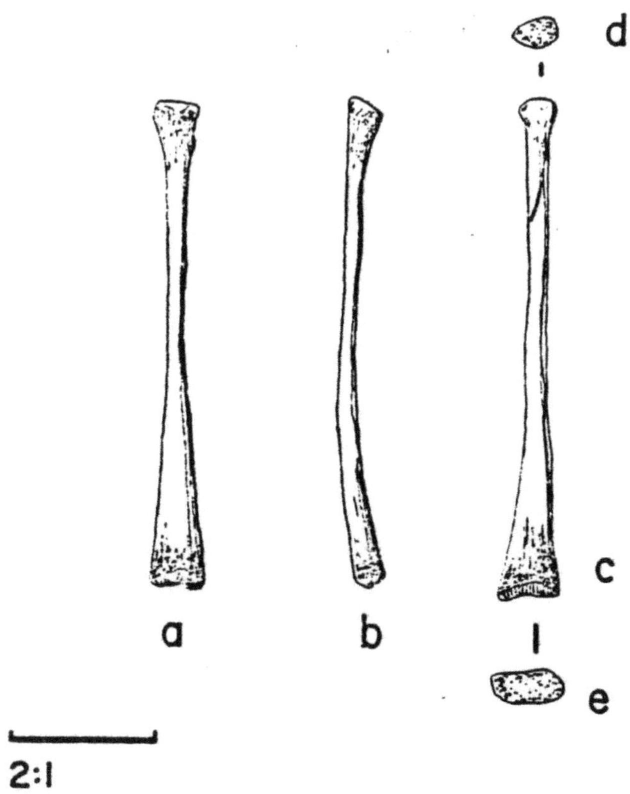

Plates XVIII
Astragalus

The superior articular surface of the astragalus has a grooved pulley-like surface, with lateral and medial facets that articulate with the tibia and fibula. Distally and medially, the bone has a rounded head, which articulates with the navicular bone. The astragalus forms a hinge-joint permitting movement of the pes in only one plane.

Views:
a - anterior
b - lateral
c - medial
d - posterior
e - proximal
f - distal

Landmarks:
1 - Condylus
2 - Collum tali
3 - Facies articularis calcanei
4 - Processes posterior tali

Bos sp.

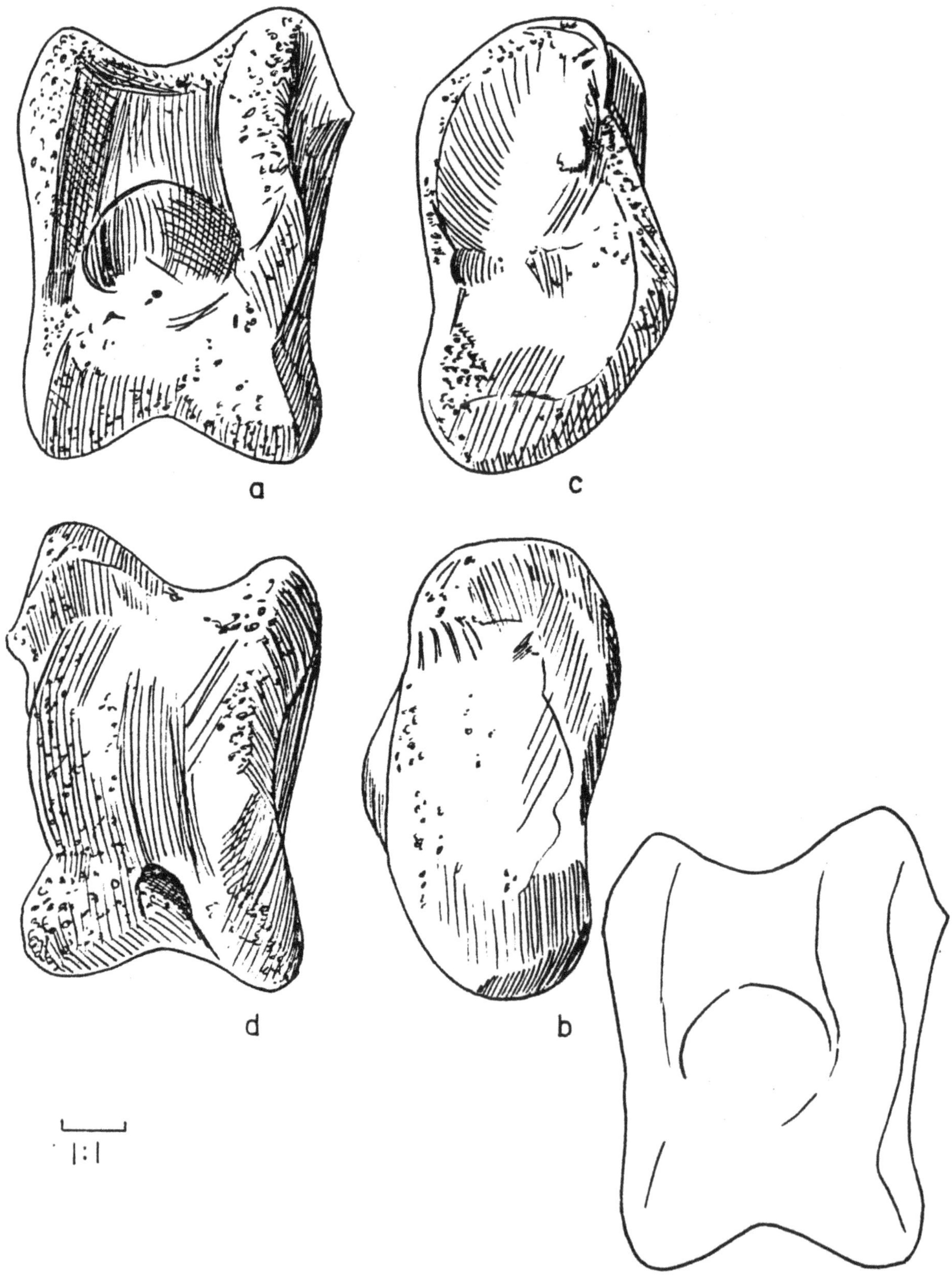

Ovis aries

a c e

1:1

Sus scrofa

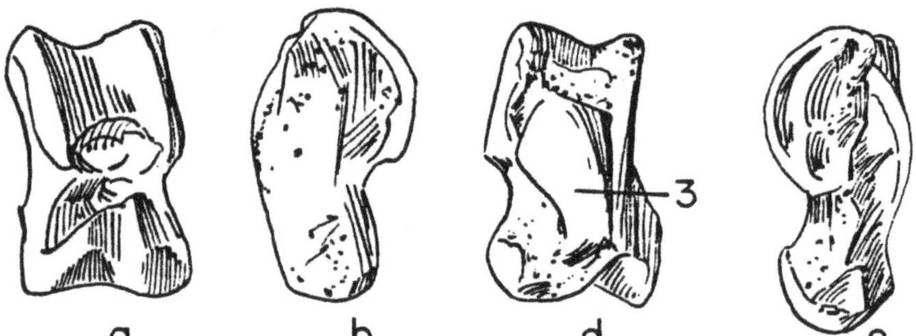

a b d c

1:1

Canis familiaris

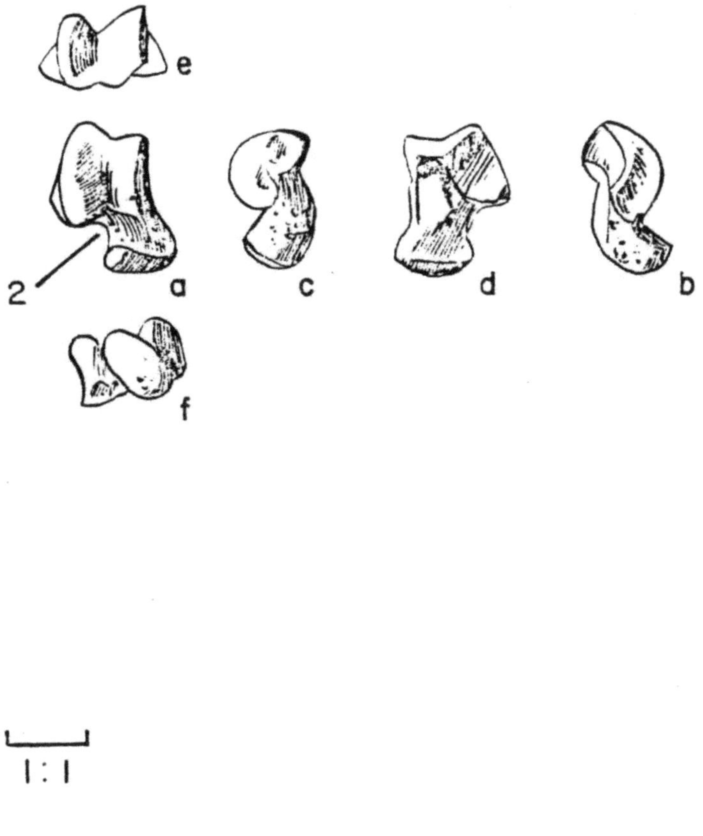

Rattus norvegicus

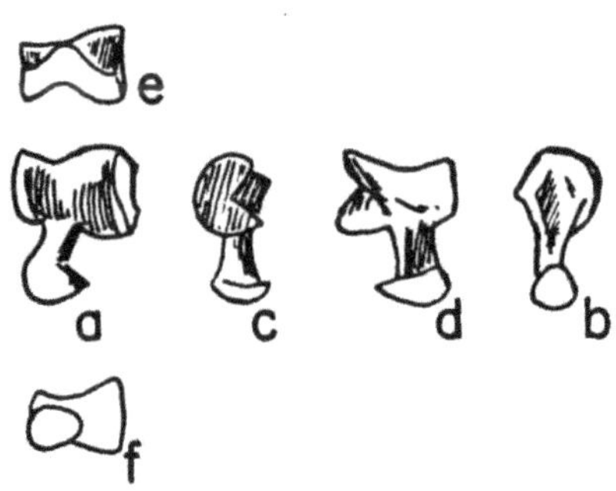

2:1

Plates XIX
Calcaneus

The calcaneus is supported by the astragalus from below and projects posteriorly, as sort of a lever. This tuberosity serves as the attachment to the "Achilles" tendon. These muscles are responsible for extending the pes of the leg, as in a cat's leap. Distally and laterally, the calcaneus extends forward, almost to the level of the head of the astragalus and articulates with the cuboid bone on its rounded surface.

Views:
a - proximal
b - distal
c - anterior
d - lateral
e - medial
f - posterior

Landmarks:
1 - Corpus calcanei
2 - Processus anterior
3 - Sustenaculum tali
4 - Facies articularis talaris
5 - Facies articularis anterior

Equus caballus

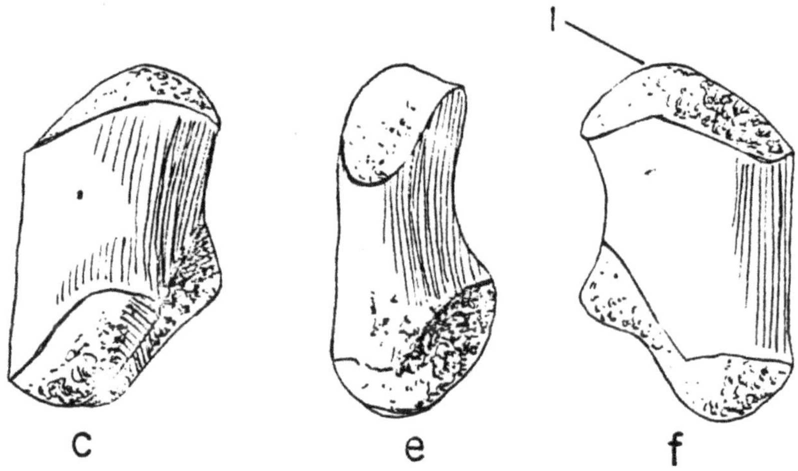

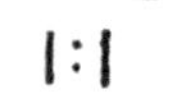

Bos sp.

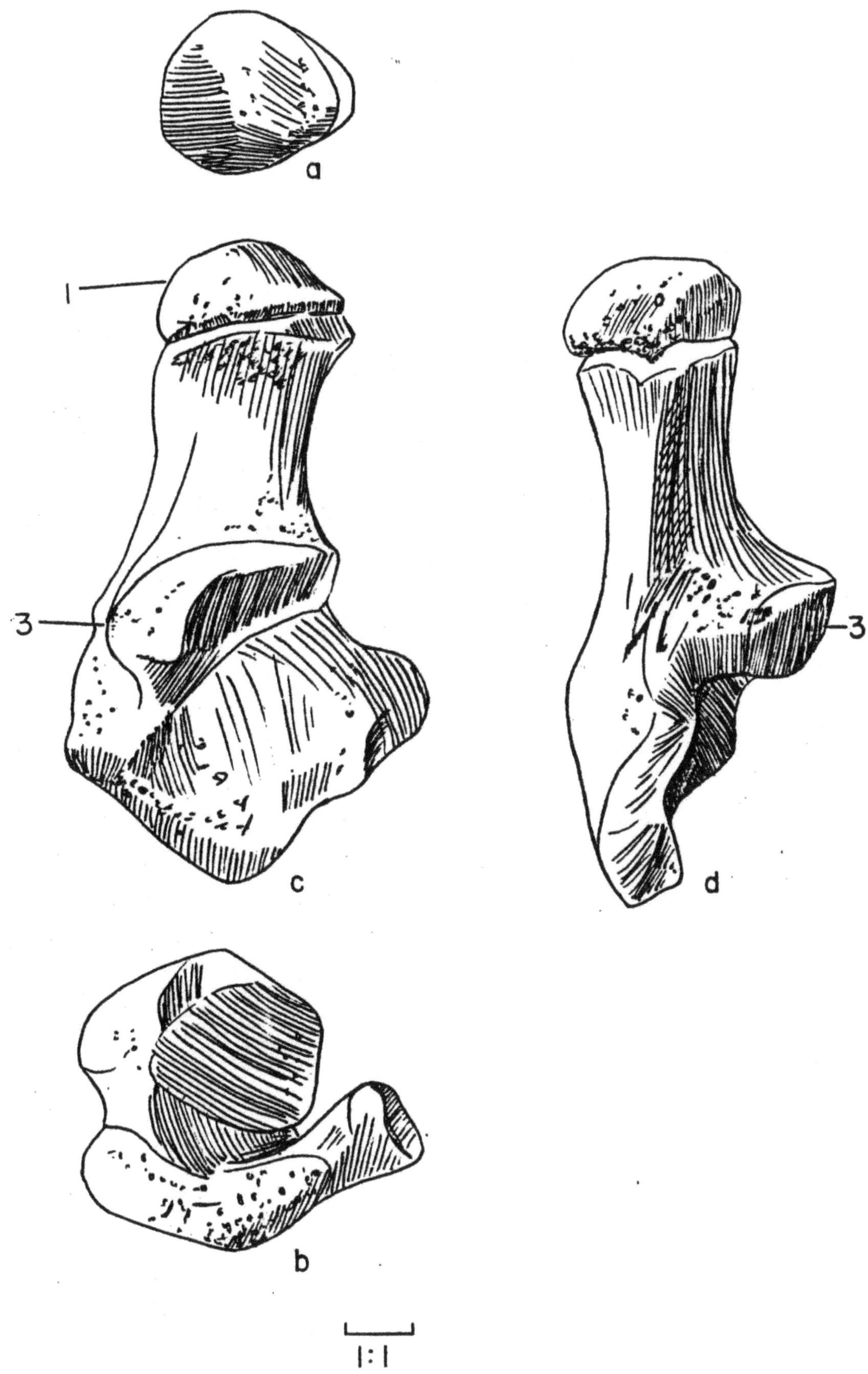

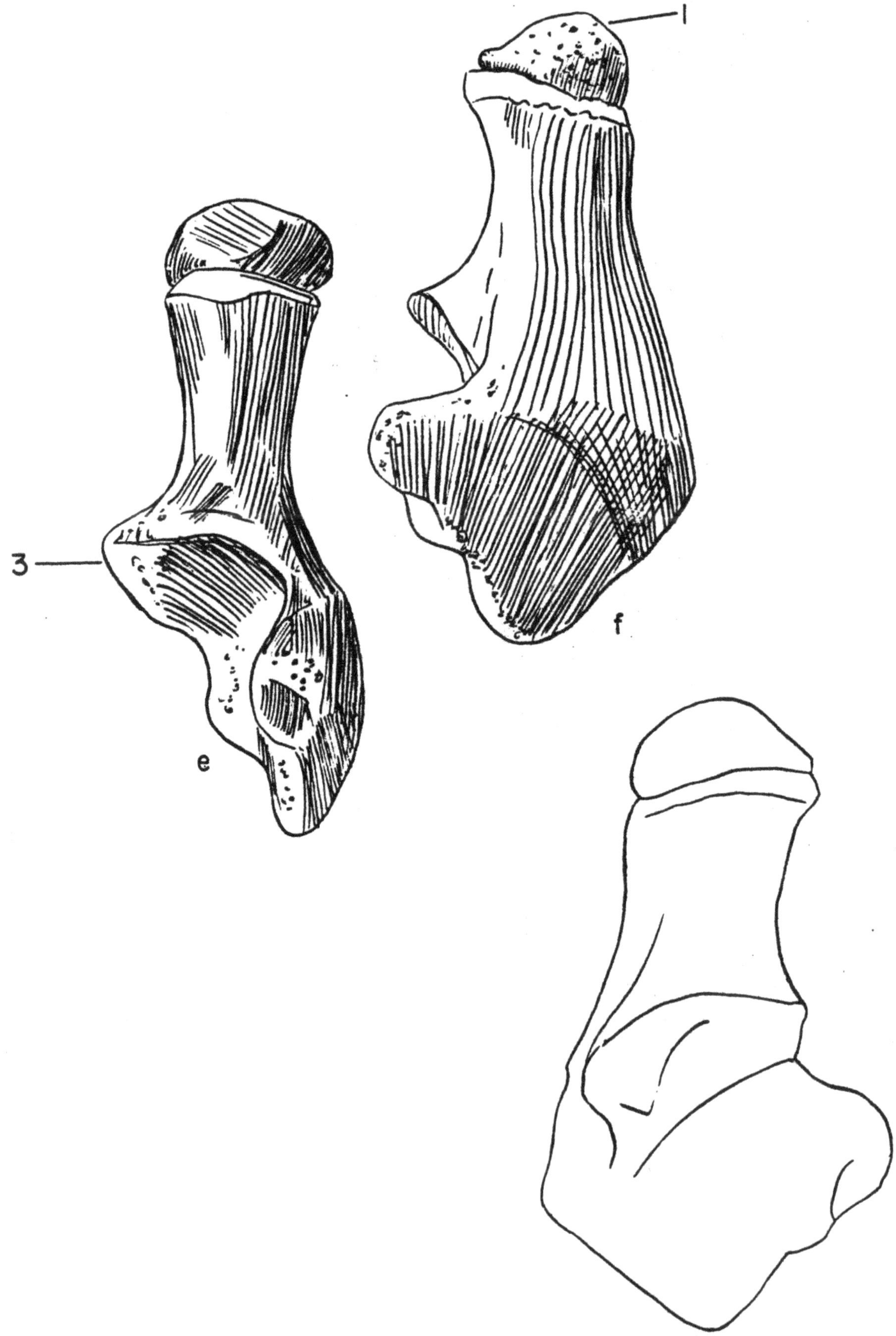
1
3
e
f

Ovis aries

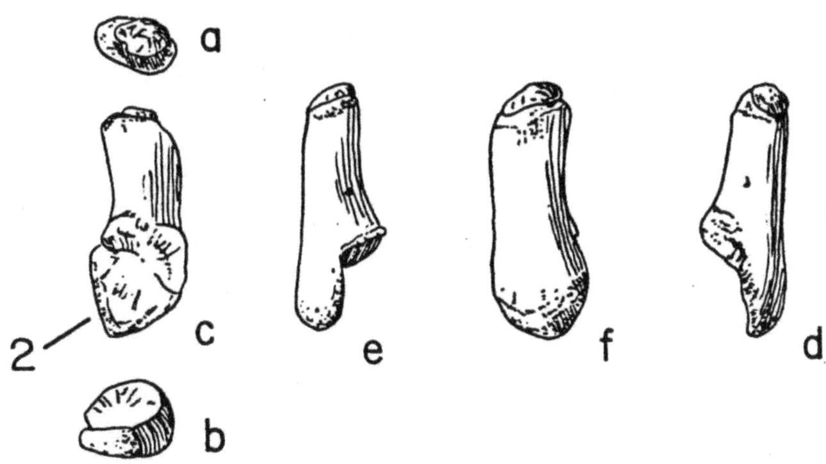

1:1

Sus scrofa

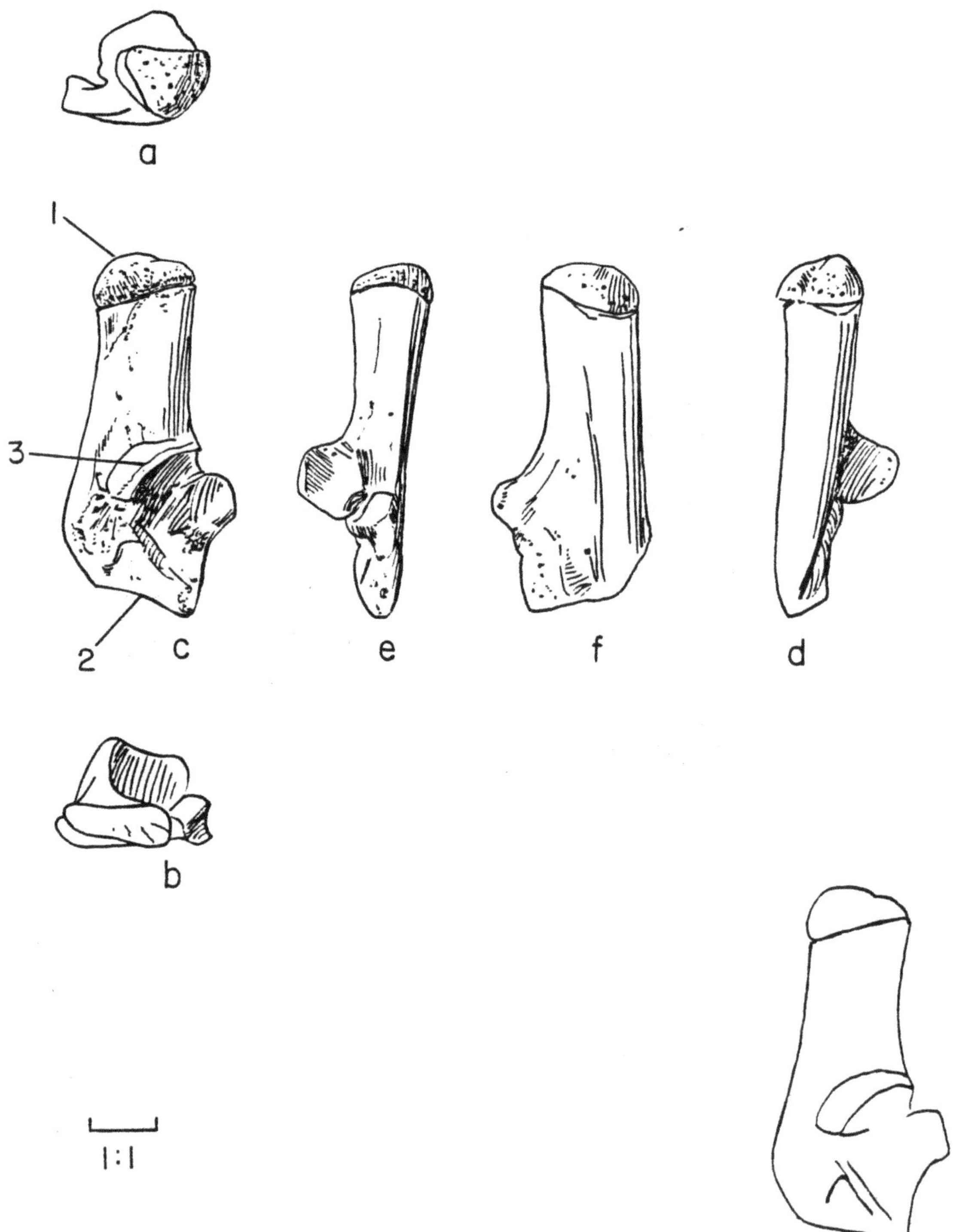

Canis familiaris

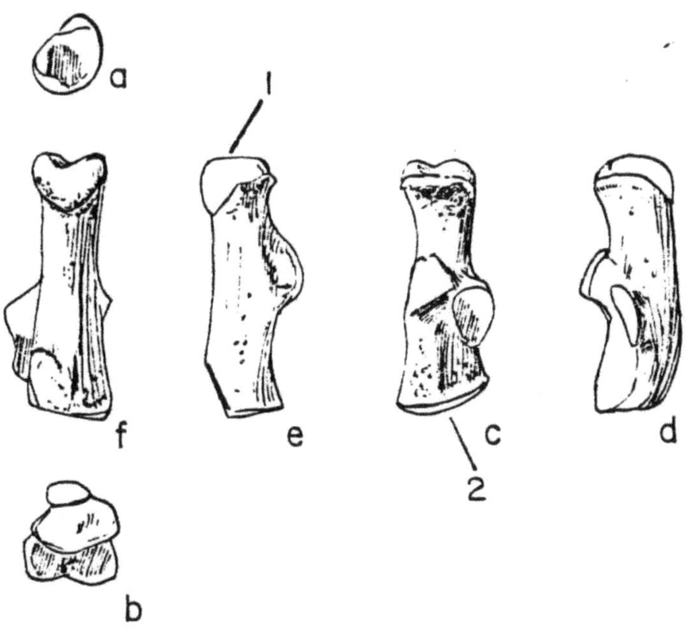

1:1

Felis catus L.

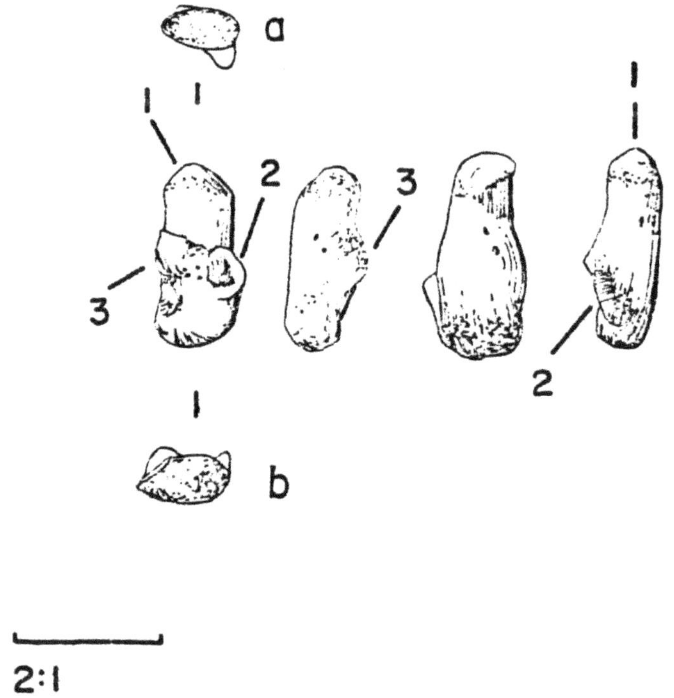

Rattus norvegicus

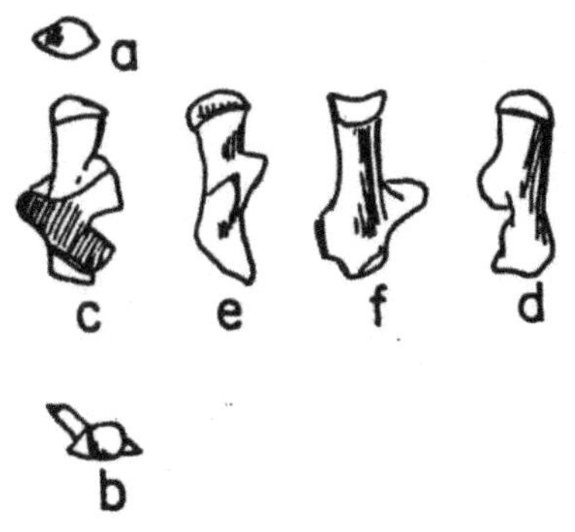

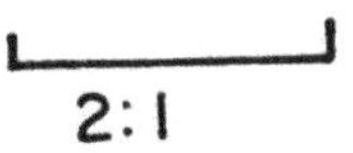

Plates XX
Metatarsus

Similar to the metacarpus, the metatarsus, when fully represented, consists of five bones, numbered I-V from the medial side (see figure 4-14). Their number is frequently reduced among the specialized mammalian quadrupeds. The metatarsal is articulated with the astragalus (tarsal) and with its neighbors, allowing a gliding movement. The distal end of the metatarsal presents a head with a cylindrical articular surface with a raised median sagittal ridge. Amóng the more specialized quadrupeds, this ridge will prevent any lateral displacement with the digits (Cornwall 1956; Sisson and Grossman 1966).

In Canidae and Felidae, the metatarsus is represented by the full complement of bones, being somewhat degenerative in Canidae. In Sus the metatarsal is similar to the metacarpal in that the metatarsals III and IV are much more strongly developed. The metatarsal III of Equus resembles the metacarpal III, but is even more elongated. The "split bones", (metatarsals II and IV) are vestigal, and lie alongside and somewhat posterior to metatarsal III. In Bovidae, the metatarsus like the metacarpus, consists of a "cannon bone" (metatarsus III and IV). Metatrsal I is missing, and II and V are greatly atrophied and rarely present.

Views:
a - anterior (dorsal)
b - proximal
c - lateral
d - posterior (volar)
e - distal
f - medial

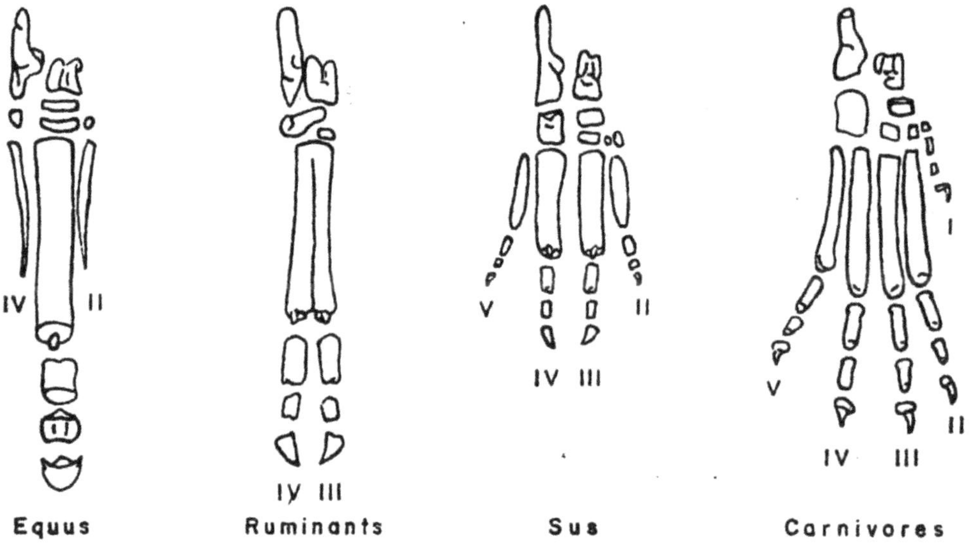

Figure 4-14, Basic pattern of the pes in different mammalian groups.

Equus caballus

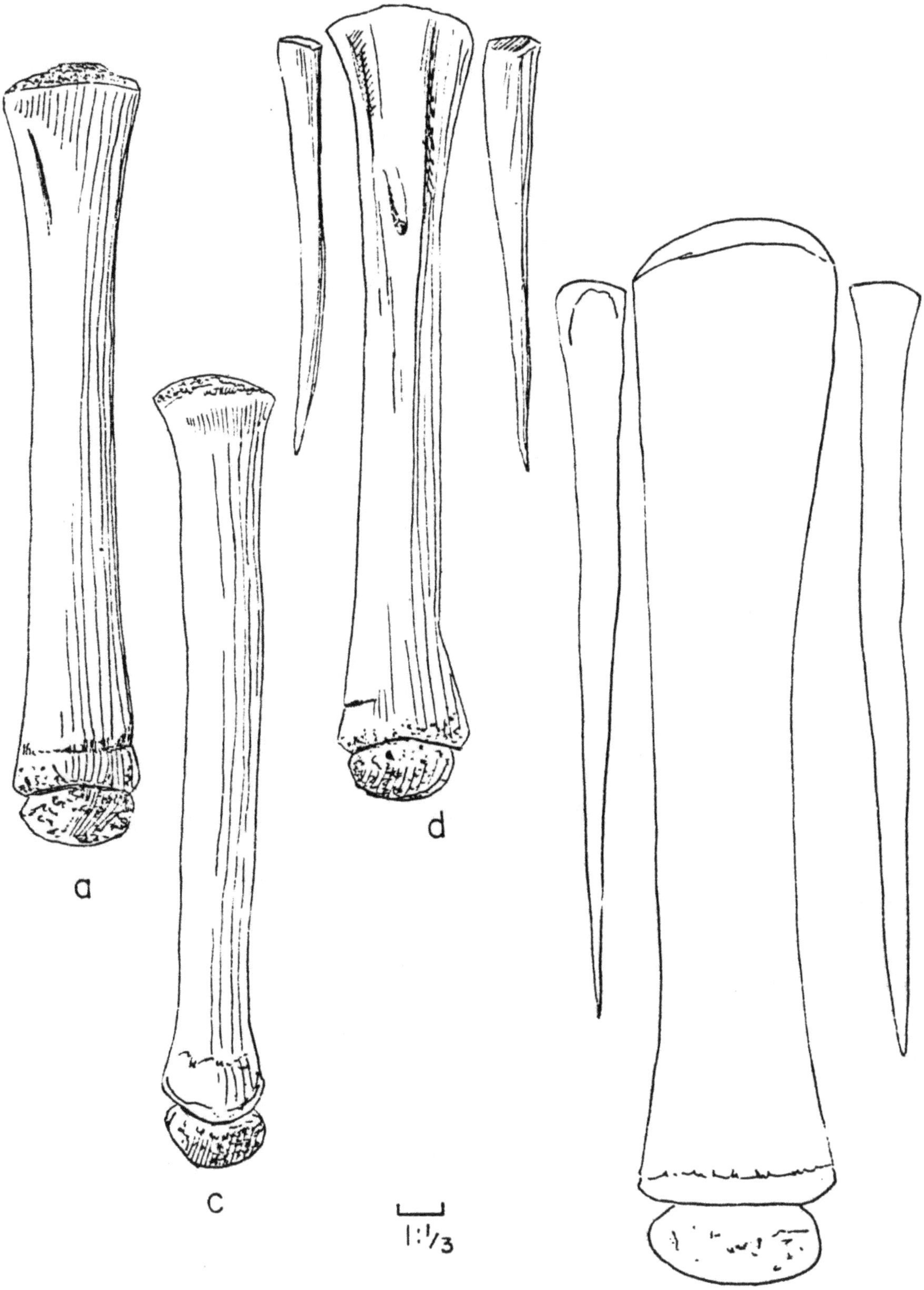

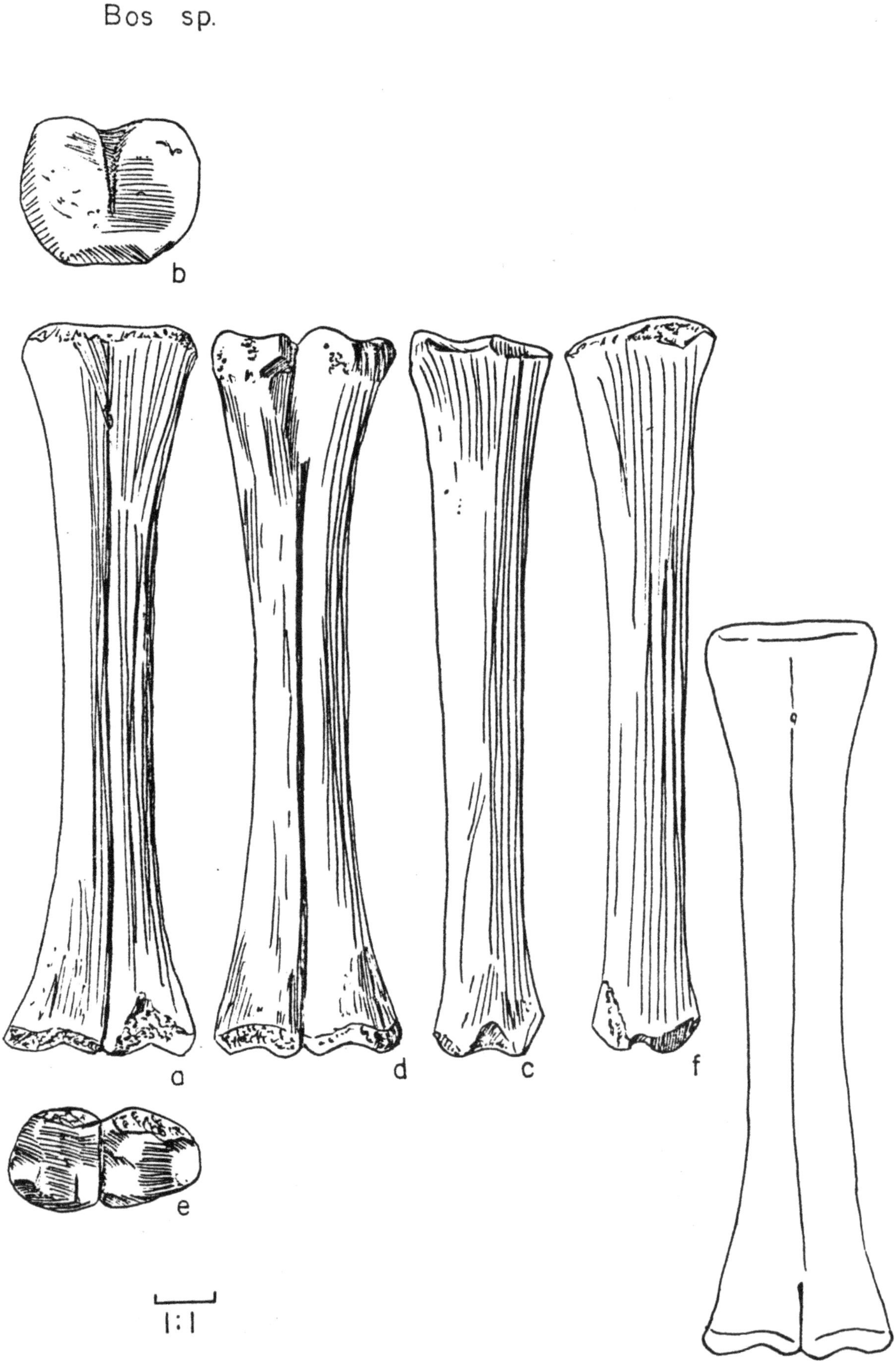
Bos sp.
b
a
d
c
f
e
1:1

Bos sp.

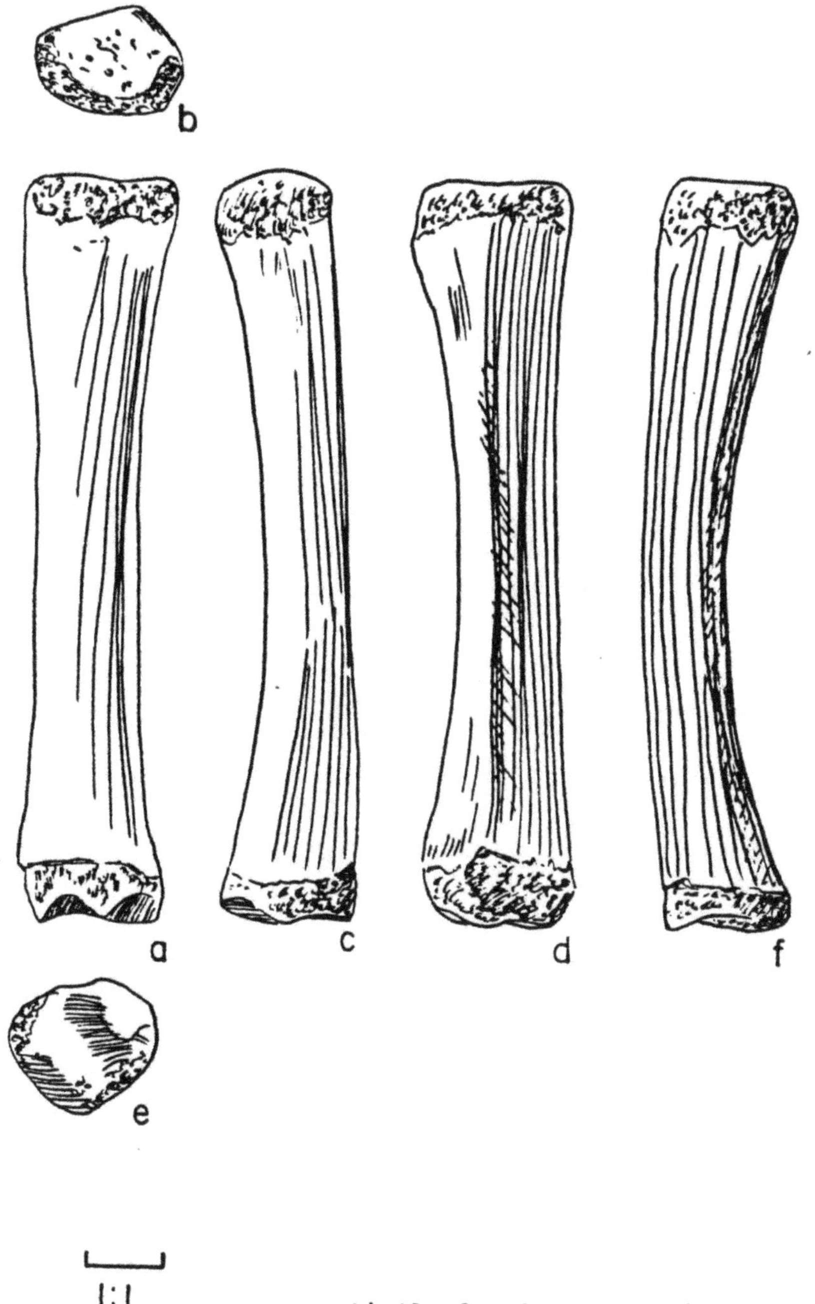

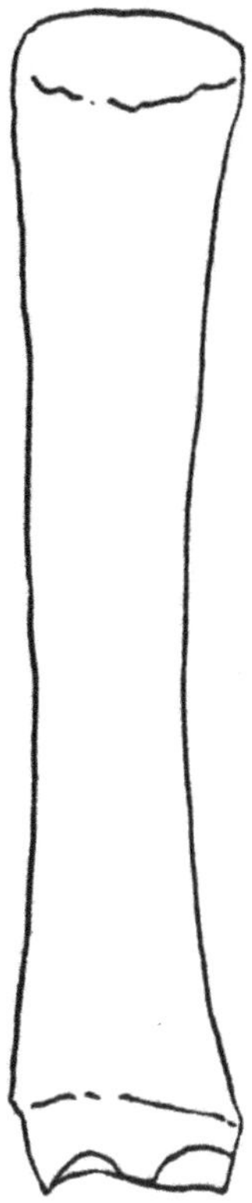

Half of a Neo-natal metapodial prior to fusion with the opposite half.

Bos sp.

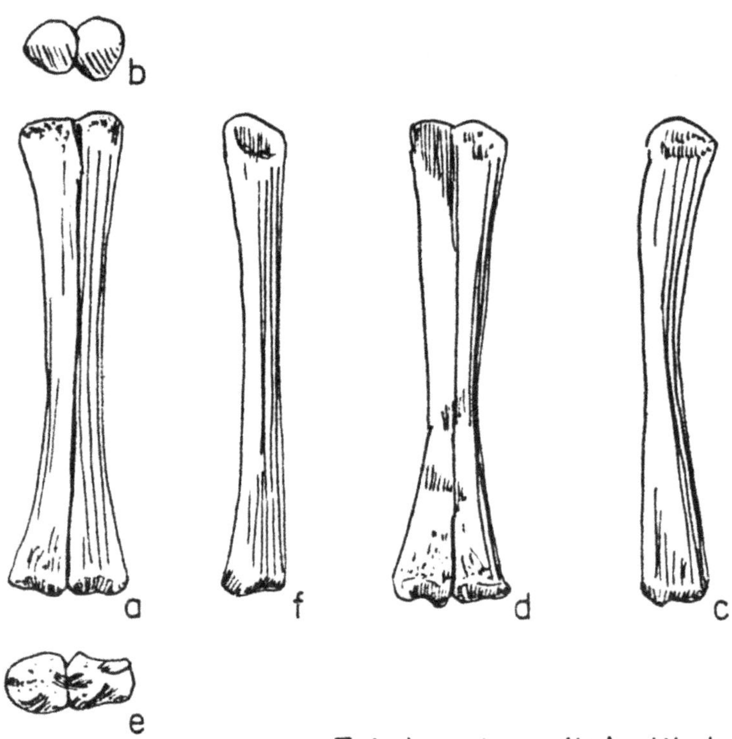

1:1

Fetal metapodial with both halves present. Each half is separate, but will become fused during a later juvenile stage.

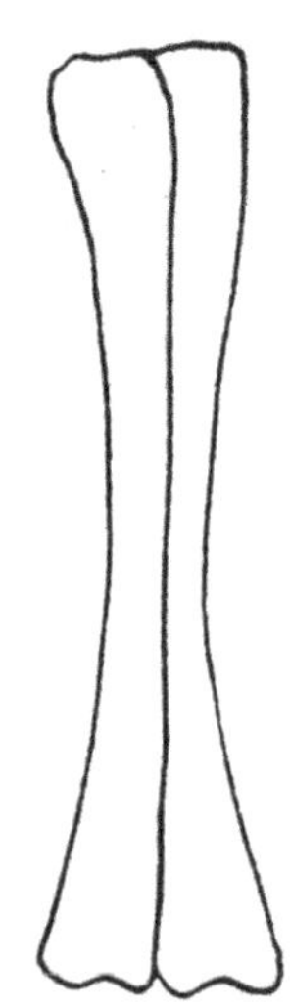

Ovis aries

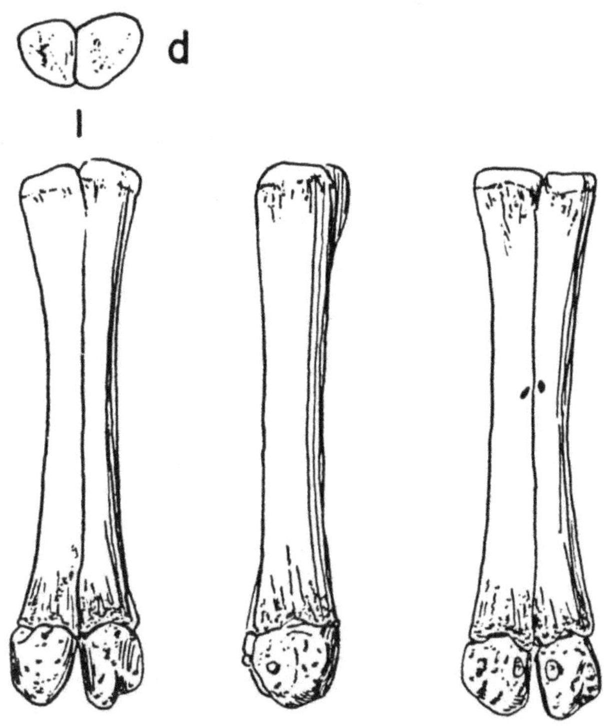

a c b

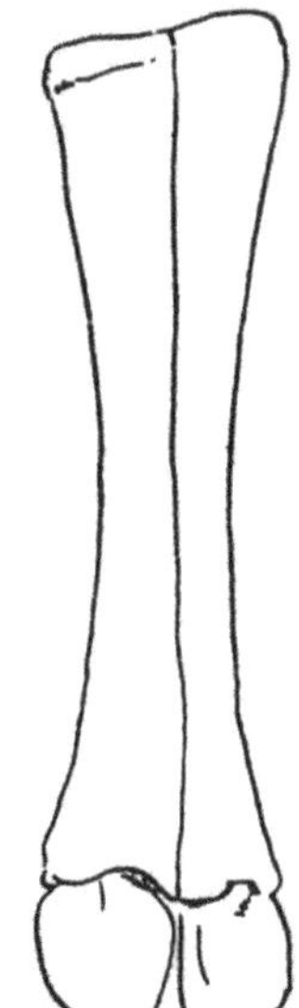

1:1

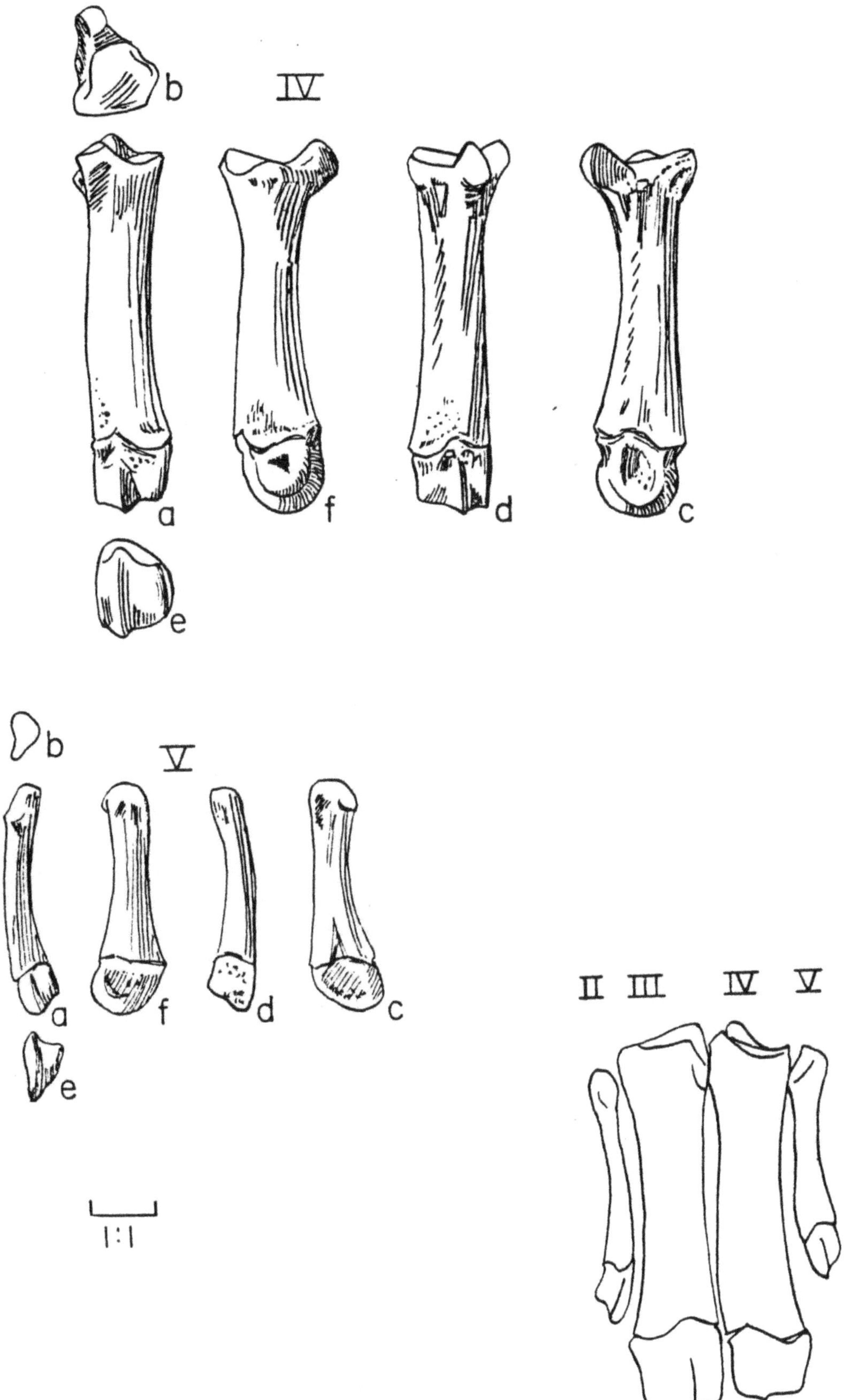
b
IV
a
f
d
c
e
b
V
a
f
d
c
e
1:1
II III IV V

Canis familiaris

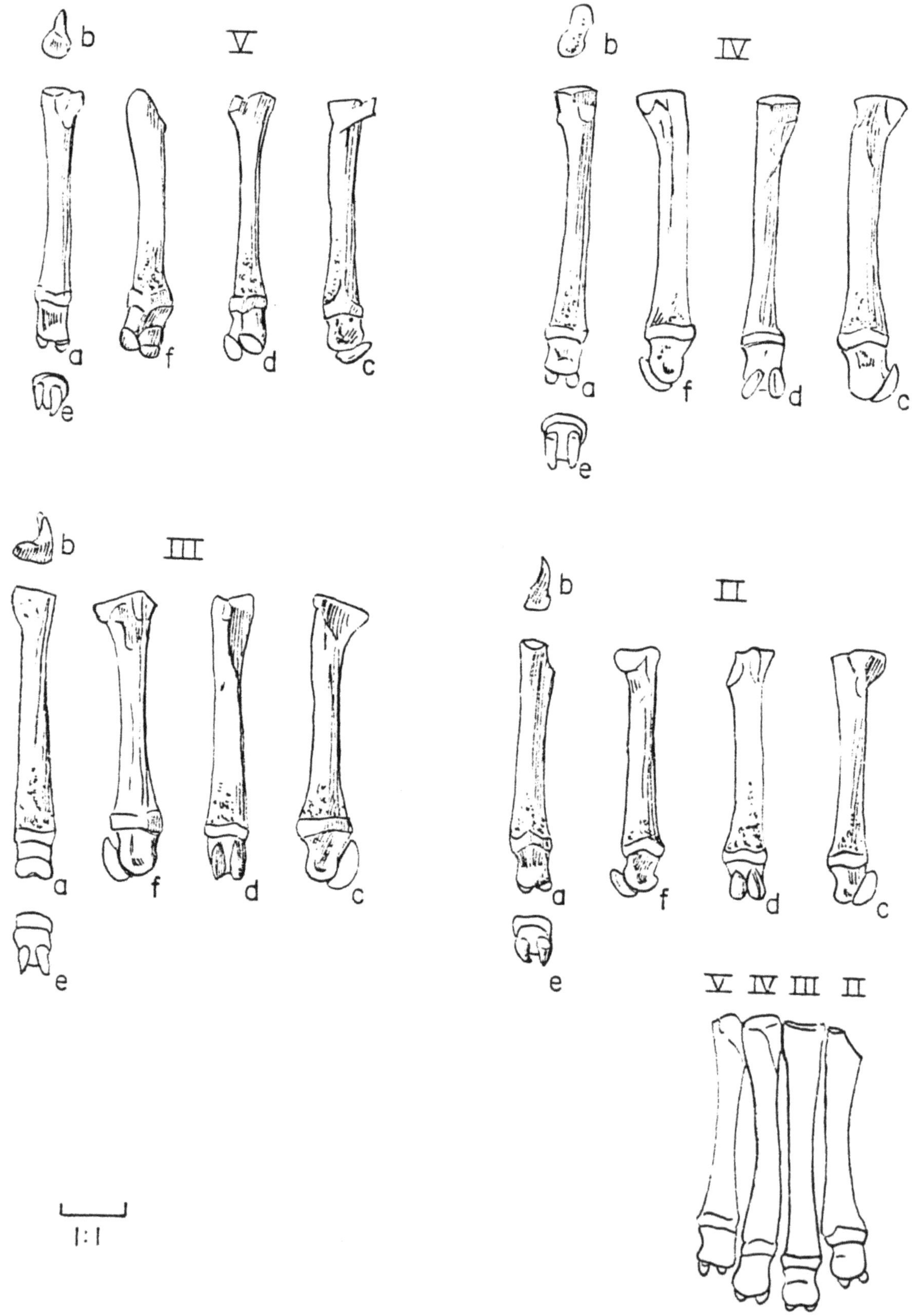

Plates XXI
Phalanges (Digits)

The phalanges consist of five digits, phalanges II-V, having three digits each, and phalanx I having only two. The phalanges are usually referred to as the 1st (proximal), 2nd (intermediate), and the 3rd (distal, terminal, or ungual referring to the nail). The first digit articulates with the distal end of the metapodial. The third digit is often modified as a bony core, forming either a hoof or claw (Cornwall 1956; Sisson and Grossman 1966).

Often the pattern, or reduction of the phalanges are of importance in classification. The carnivores (Canidae and Felidae) possess what Cornwall (1956:150) termed as "waisted" proximal articular faceted phalanges. The proximal facet of the phalanx is deeply hollowed, and the distal end has a marked median ridge. The third phalanx, carries a strong claw, with a thin lamina of bone sheathed over the core of bone. The phalanges of the main digits in Sus are short, stout, and less "waisted" than that of the carnivores. Phalanx sets II and V in the pig are clearly degenerative. The third phalanx in Sus, is a simple, conical-like ossicle. The equine first phalanx is fairly long, wide and flattened on its palmar surface, while the second phalanx is compressed. The third phalanx represents a specialized digit, forming the core of the hoof. The phalanges in Bovidae are represented by the III and IV sets. They are similar in nature to Sus, in that the first is longer, the second is compressed, and the third phalanx is a simple, conical-like ossicle.

Views:
a - dorsal
b - proximal
c - lateral
d - distal (plantar)
e - posterior (plantar)

Equus caballus

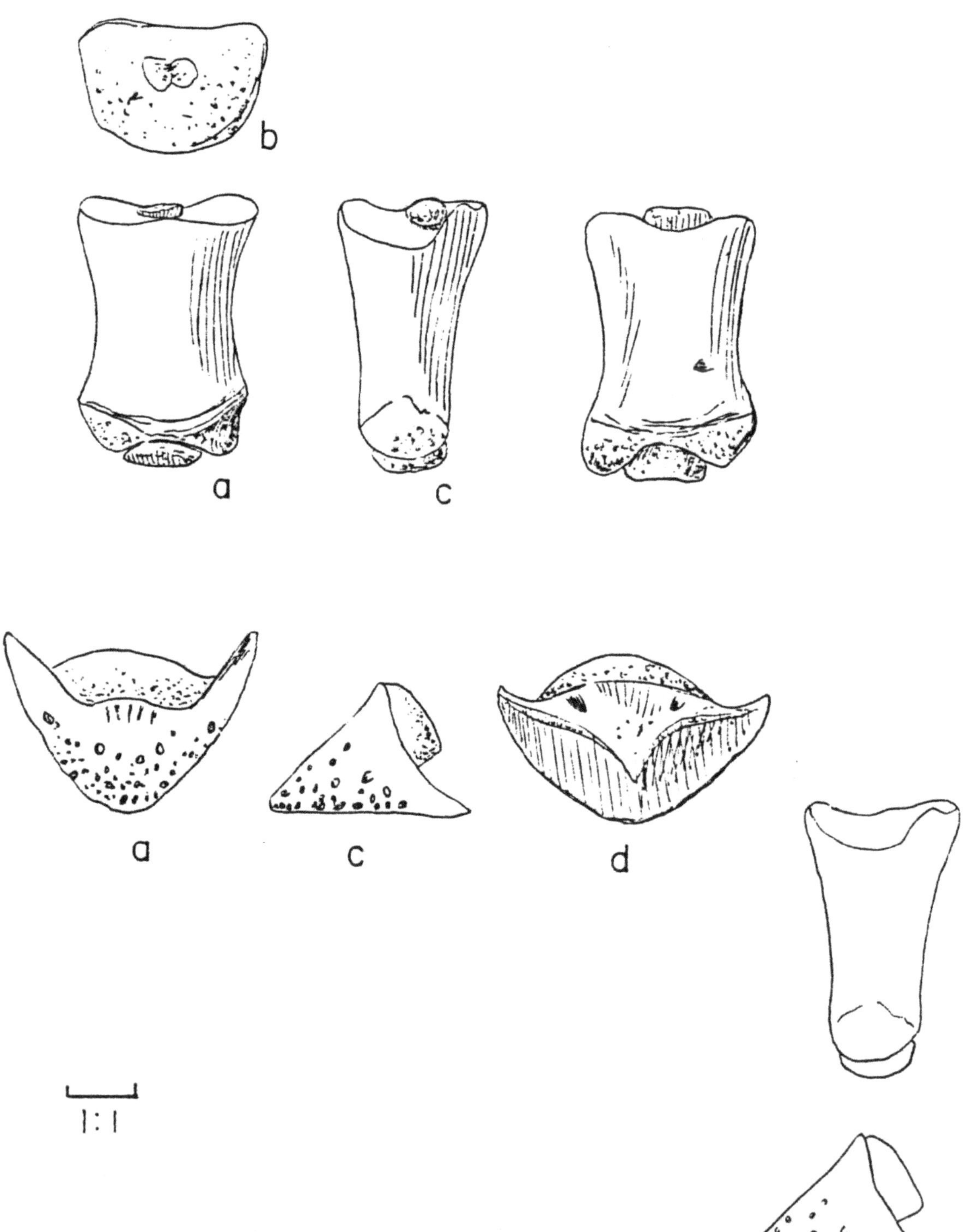

Bos sp.

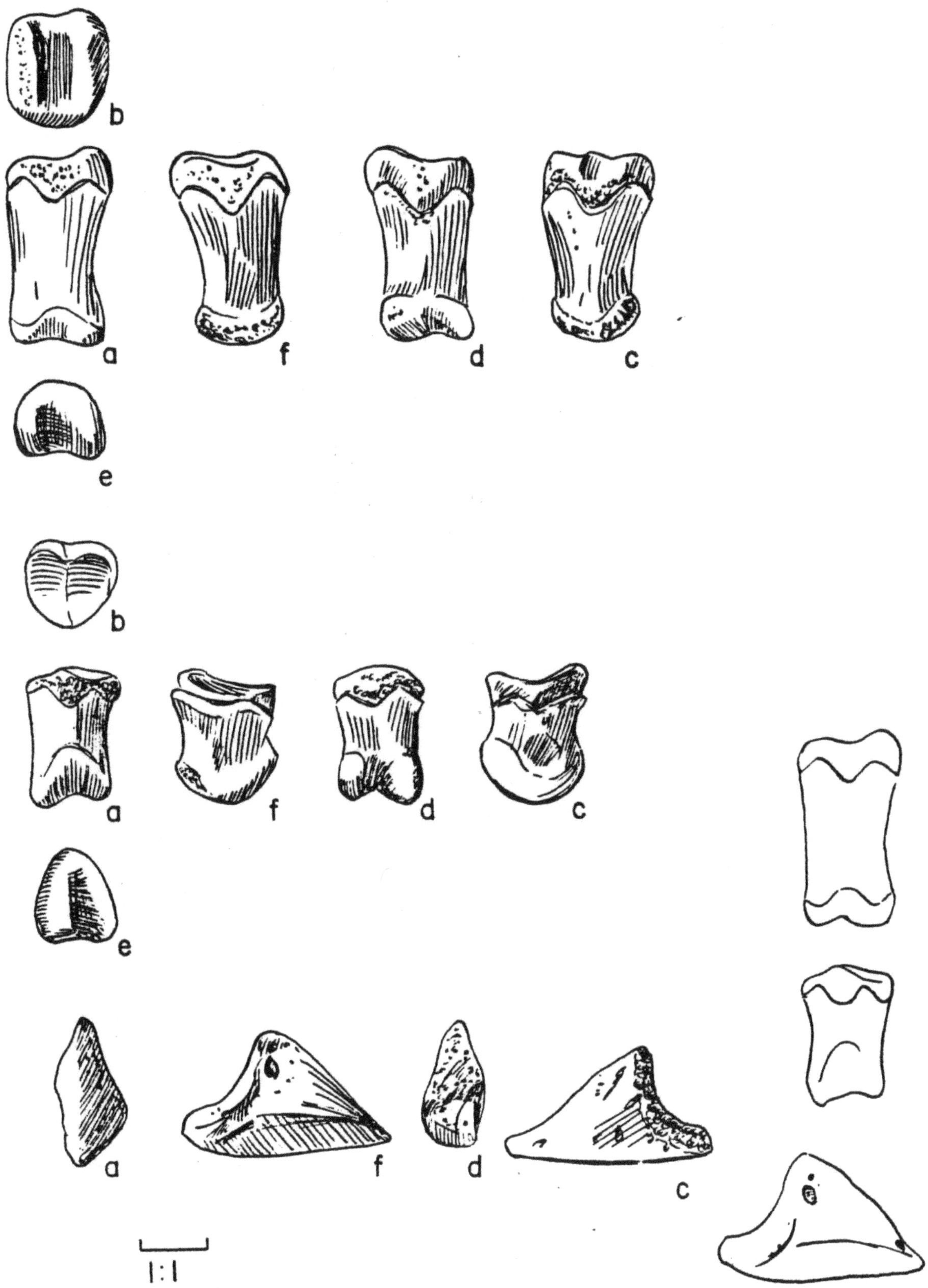

Ovis aries

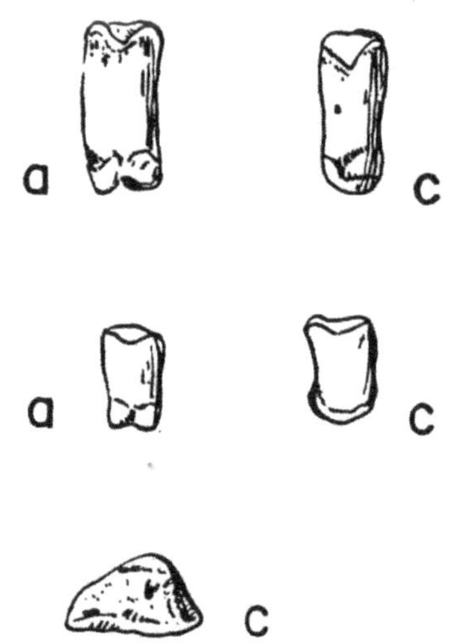

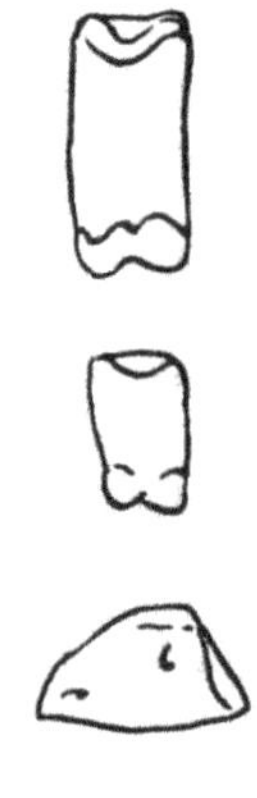

1:1

Sus scrofa

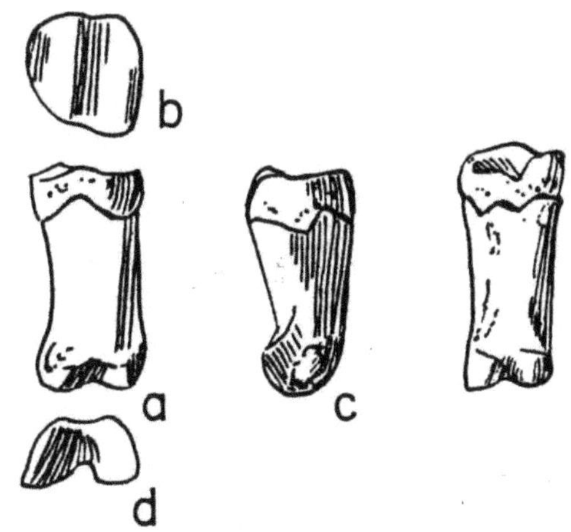

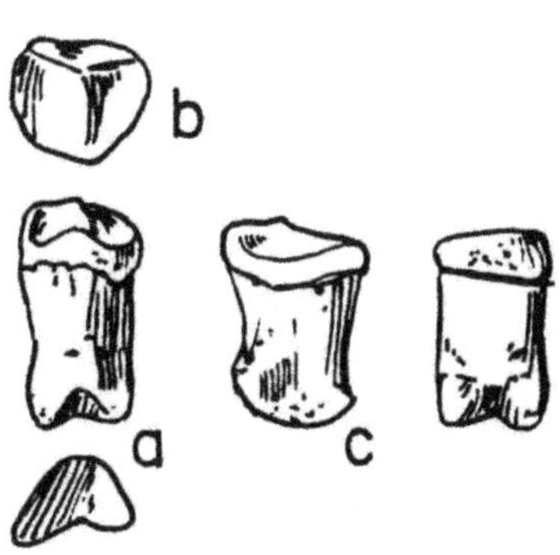

1:1

Canis familiaris

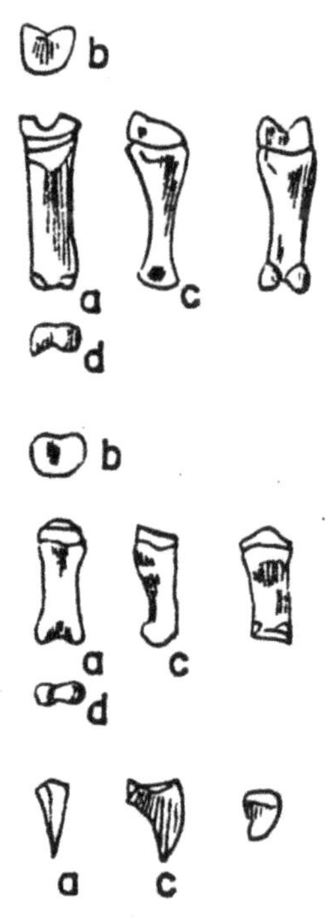

1:1

Contour Drawings of Bone Elements to Scale:

The intent of these charts is to expedite the rough sorting of juvenile faunal material. Each chart is a comparison of one bone element's size in relation to the same bone of another species represented in this volume. The bone elements are portrayed by a contour drawing at each elements natural size. The use of natural size is necessary in order to demonstrate this size relationship.

Initial sorting can be done by comparing bone fragments to the contour illustrations. A fragment must first be identified by bone element. The fragment is then compared to the contour drawing of the same element for a general match. Once a match is made, the bone fragment can be further compared to the more detailed illustrations of the same bone element. Finally, when the pre-sorting is done, the fragment is compared to an actual bone for a definitive identification.

a - *Equus caballus*
b - *Bos sp.*
c - *Ovis aries*
d - *Sus scrofa*
e - *Canis familiaris*
f - *Felis catus*
g - *Rattus norvegicus*

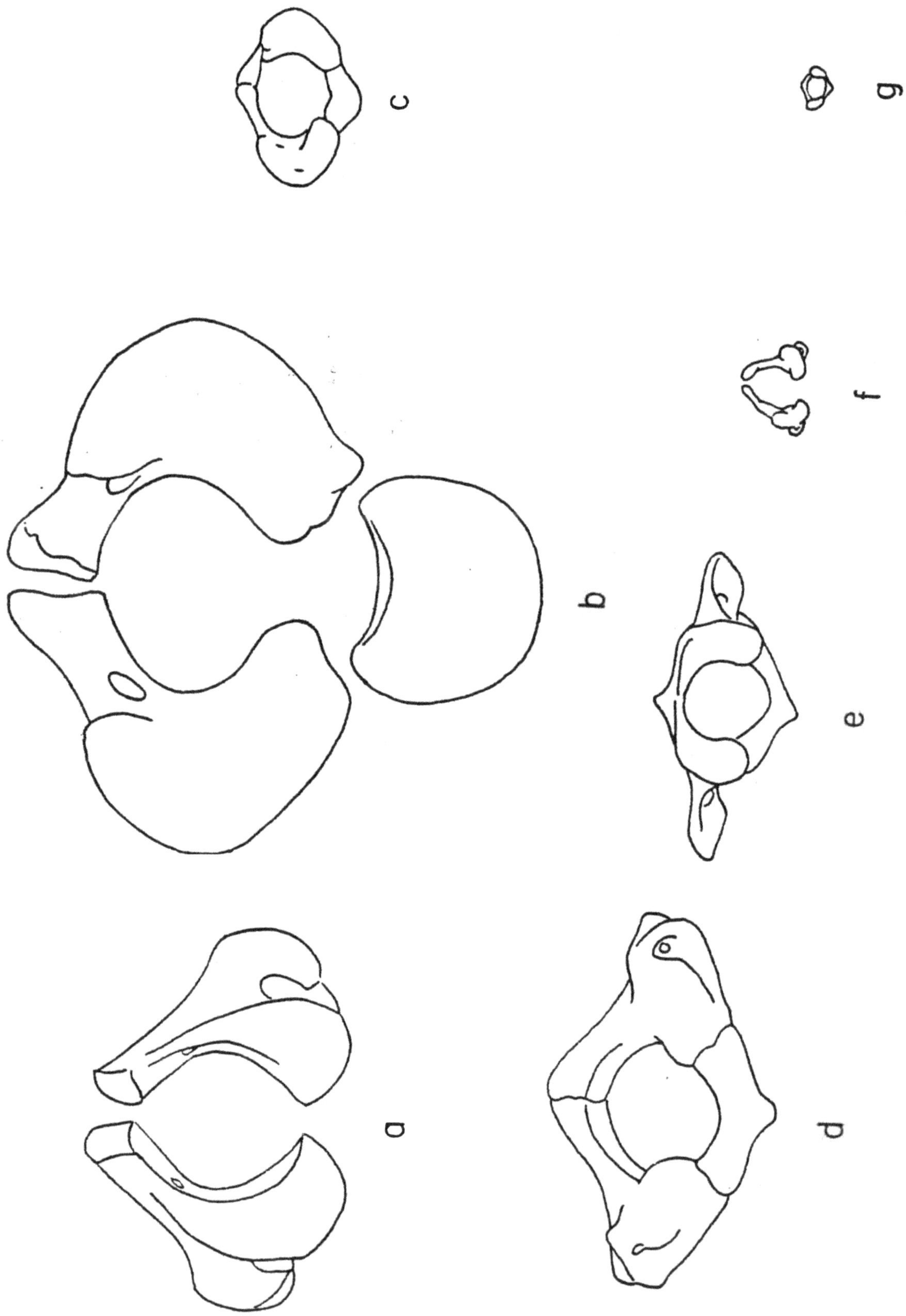
a
b
c
d
e
f
g

a
b
c
d
e
f
g

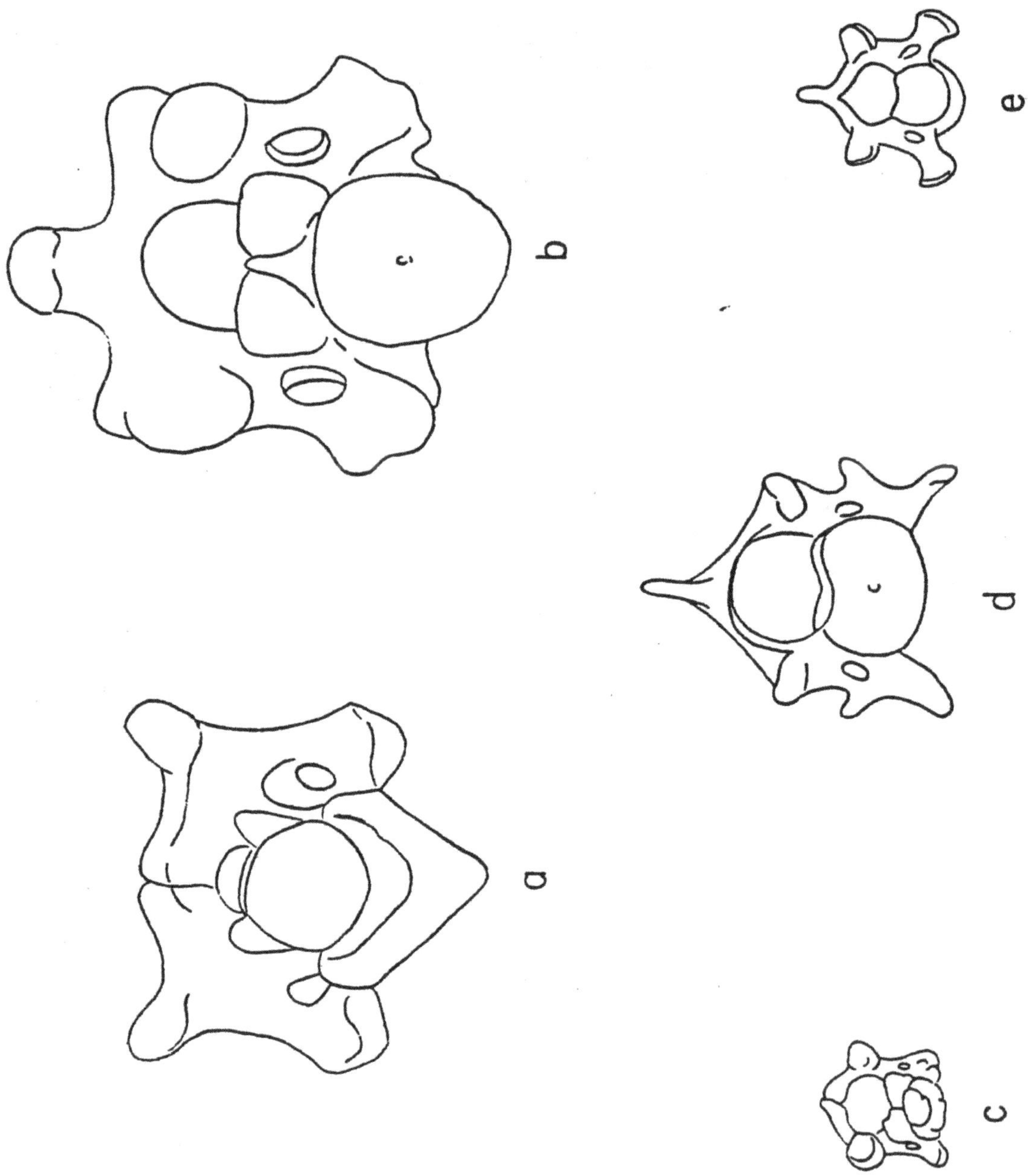
a
b
c
d
e

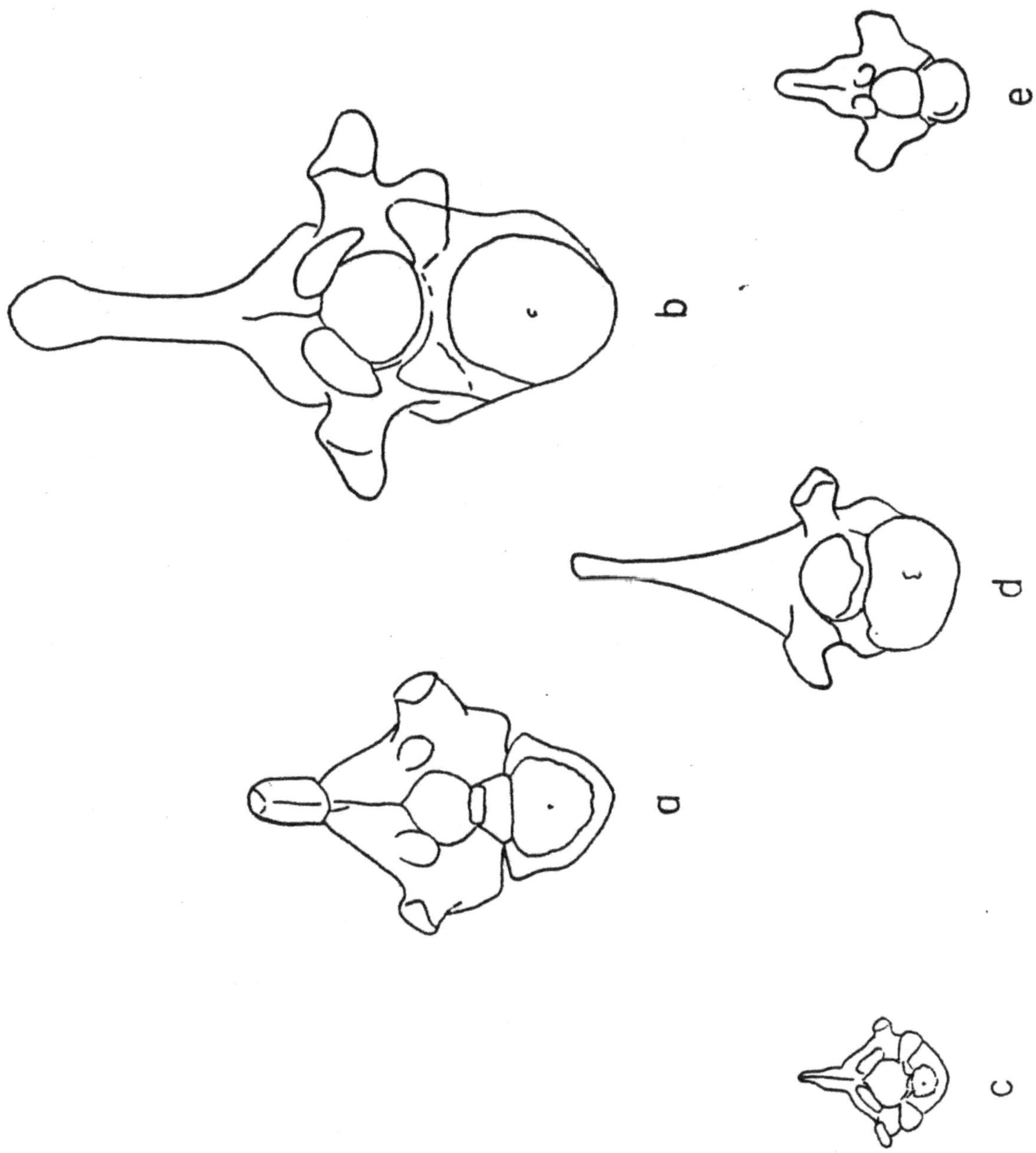
b
e
d
a
c

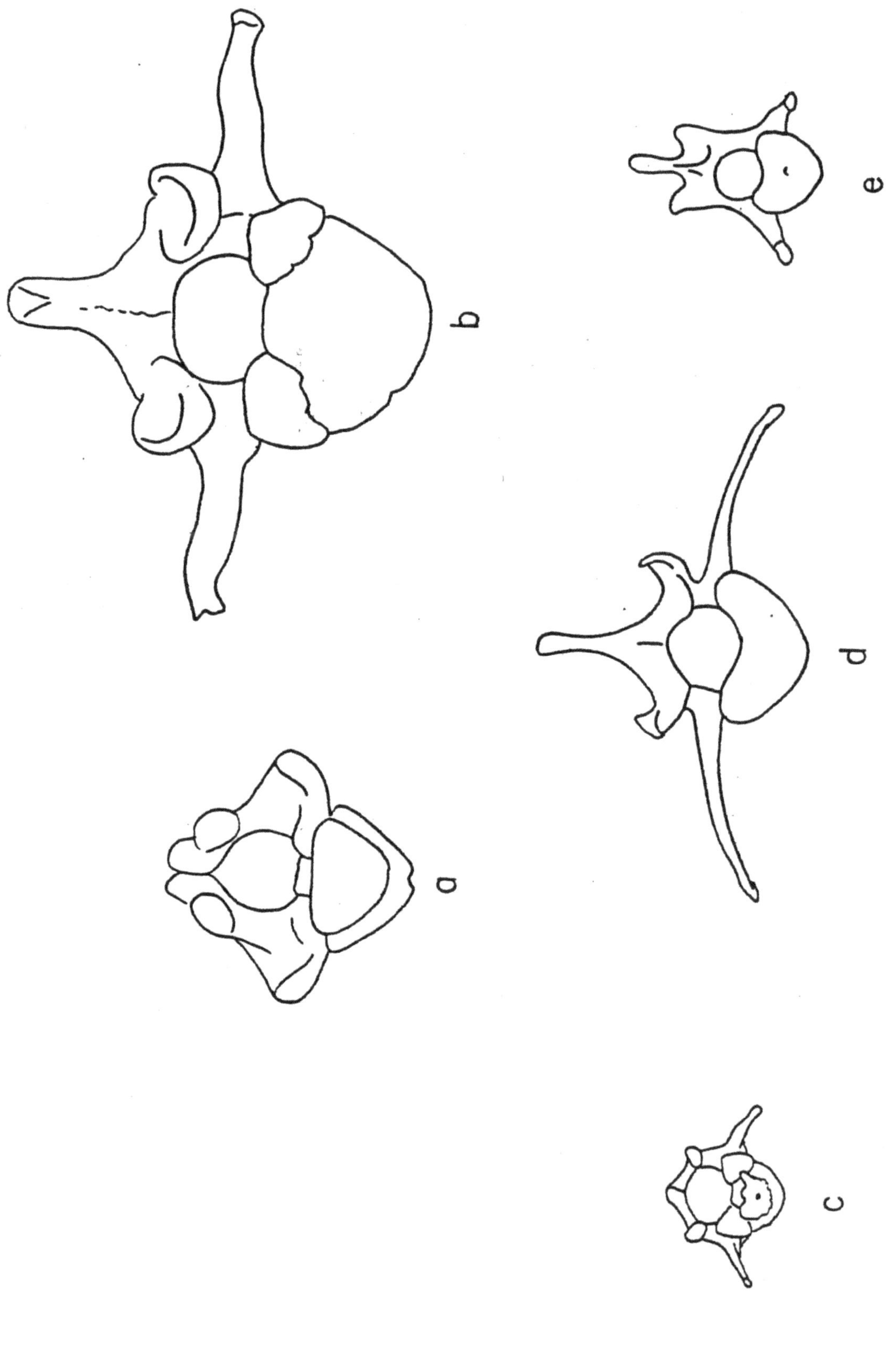
a
b
c
d
e

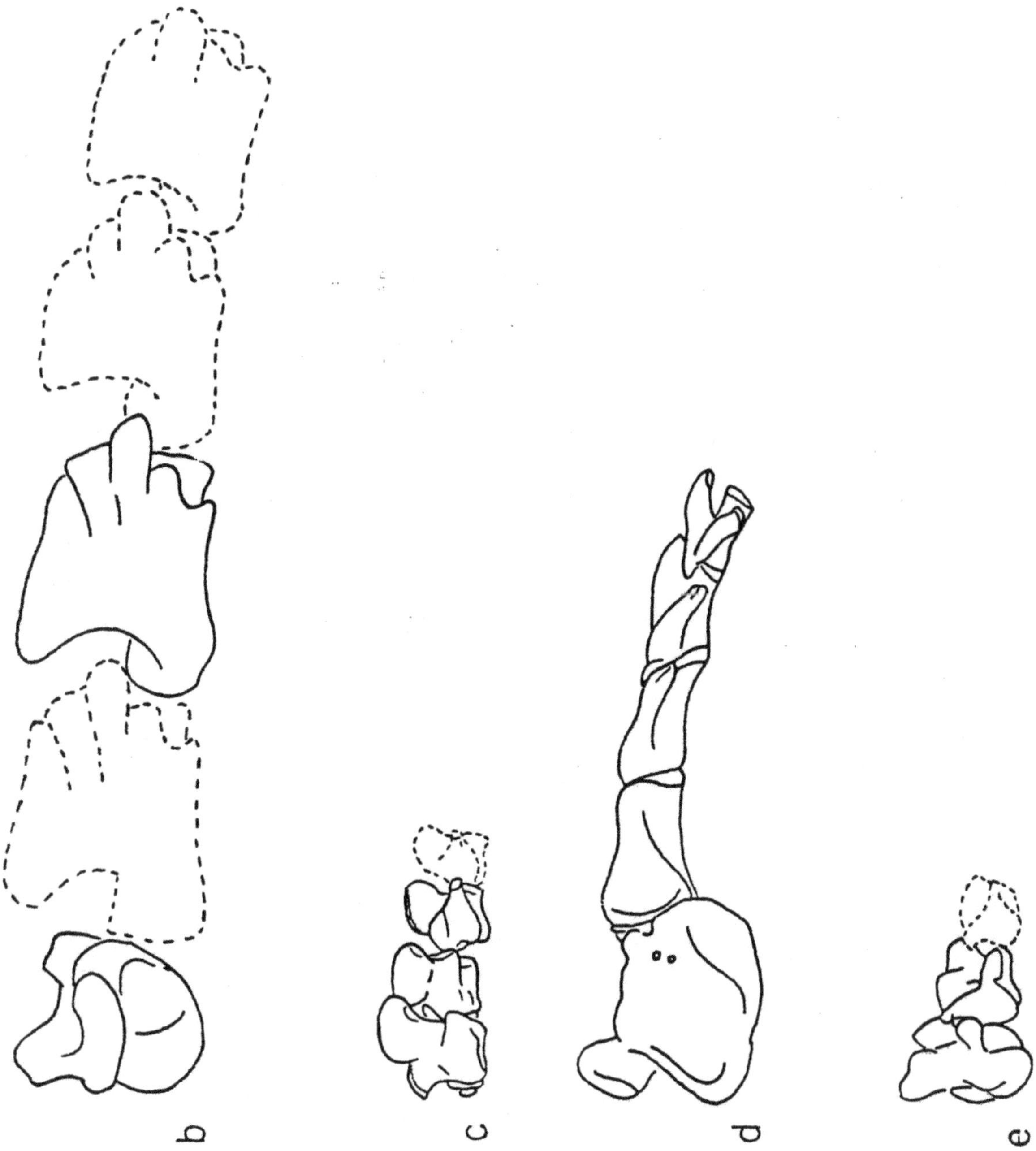
b
c
d
e

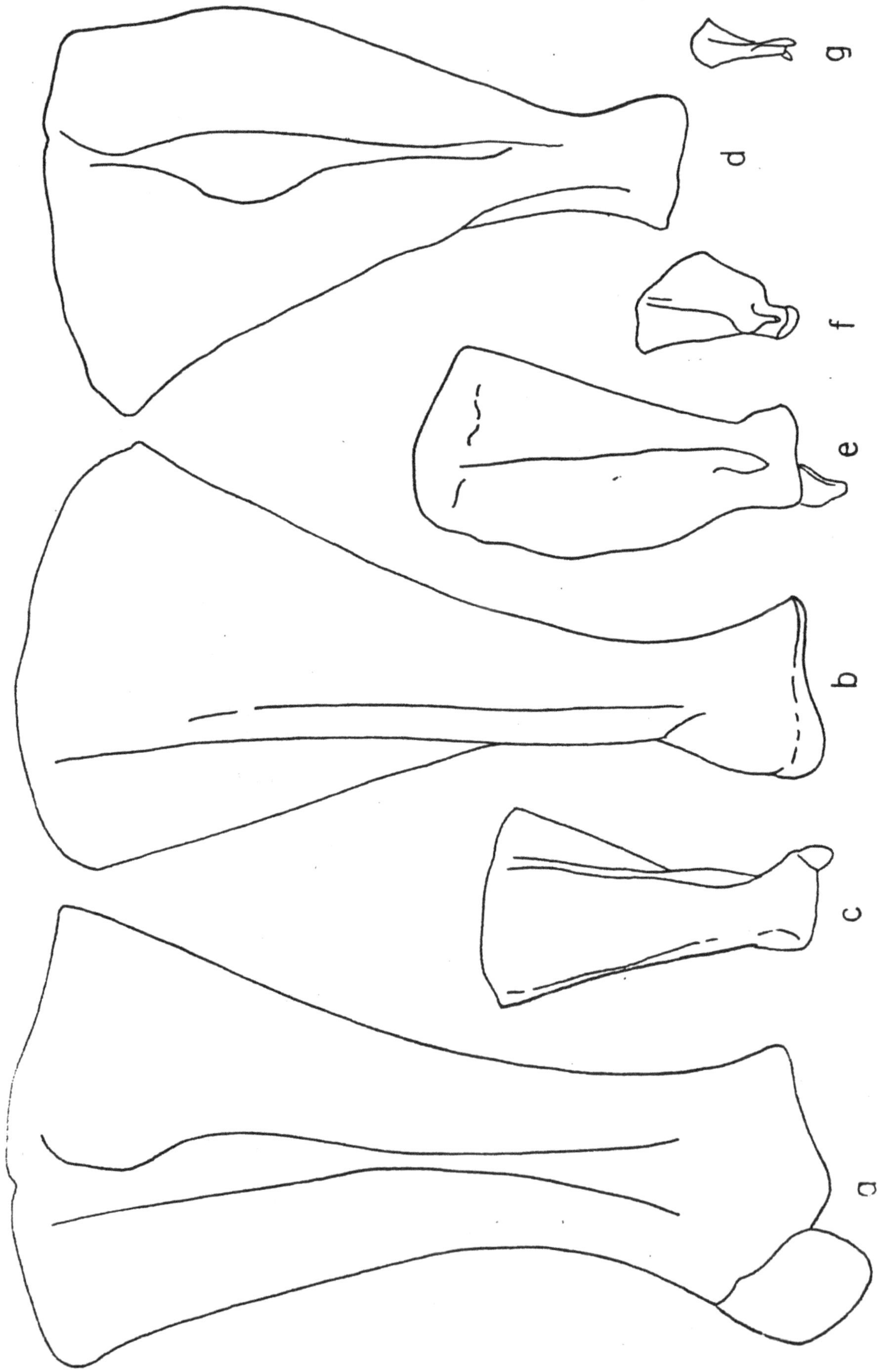
a
b
c
d
e
f
g

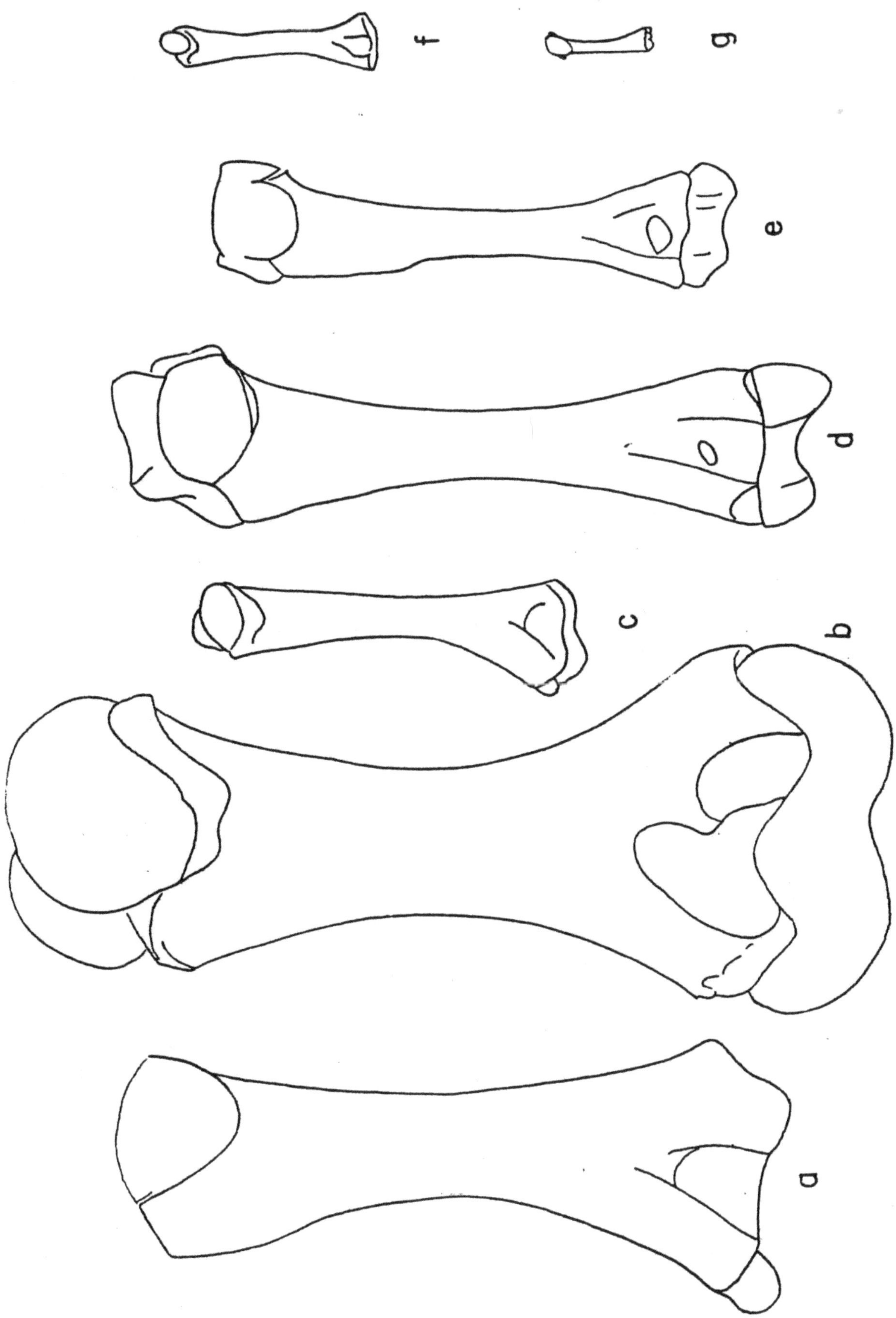
a
b
c
d
e
f
g

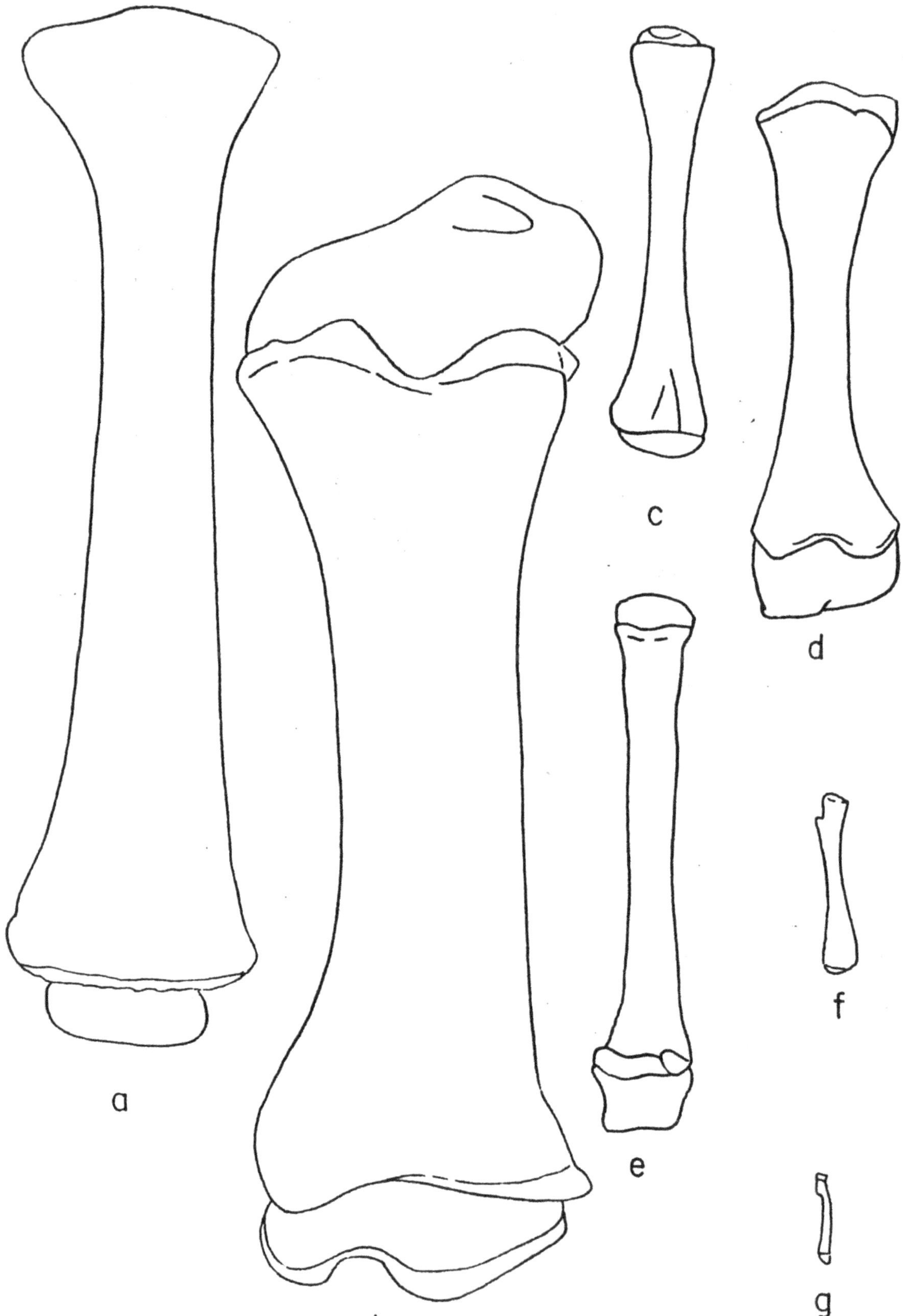
a
b
c
d
e
f
g

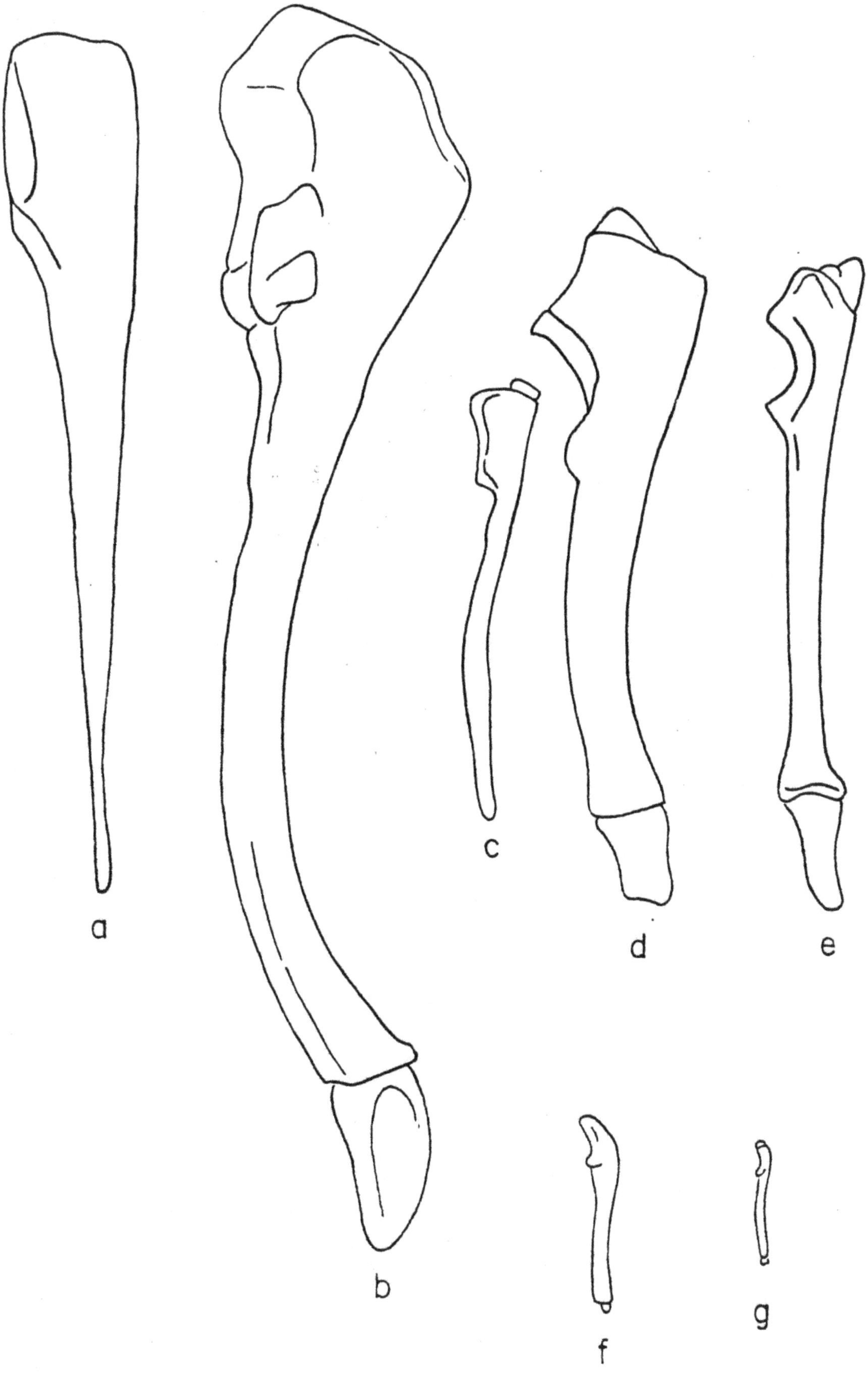
a
b
c
d
e
f
g

Metacarpus

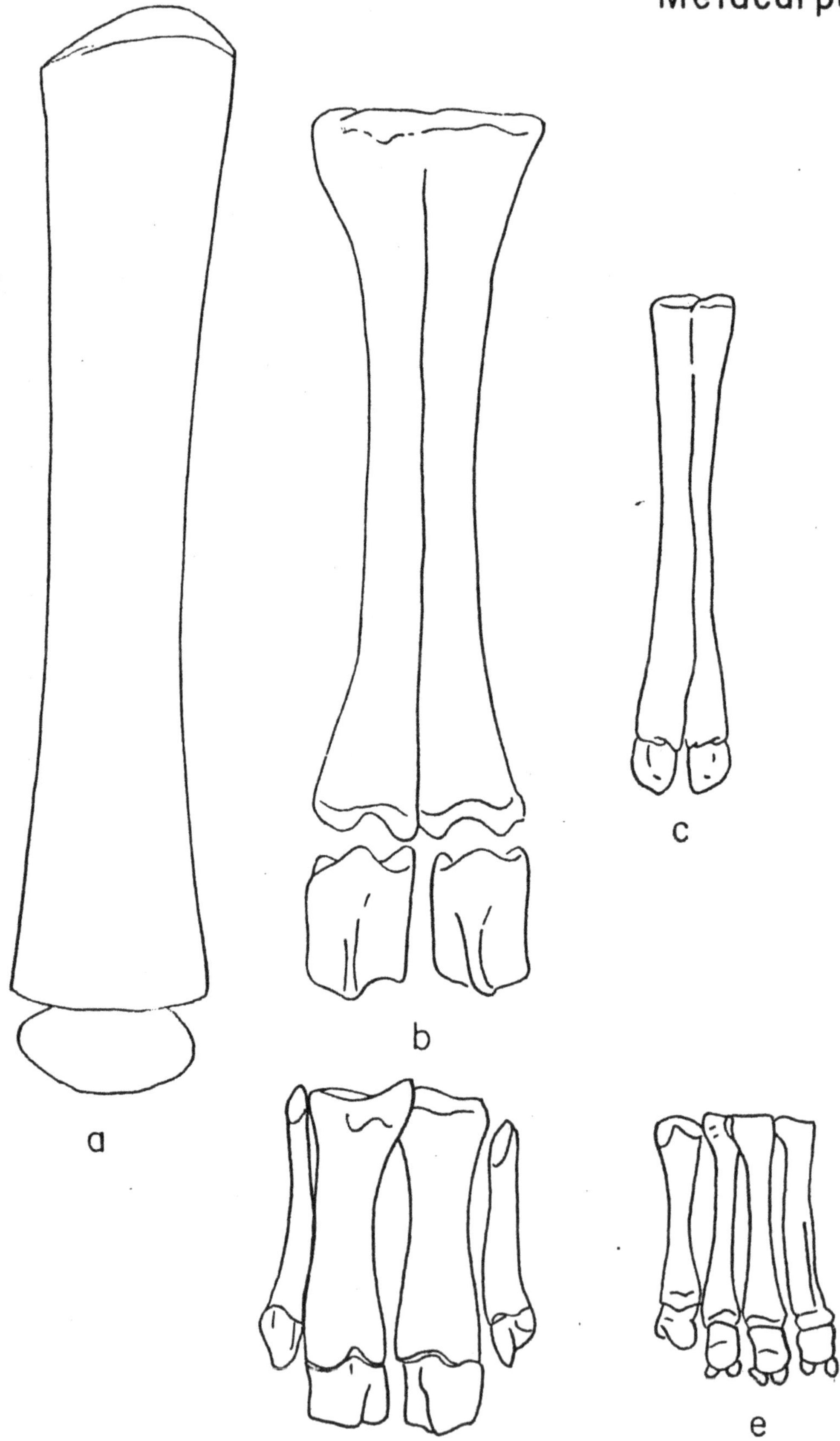

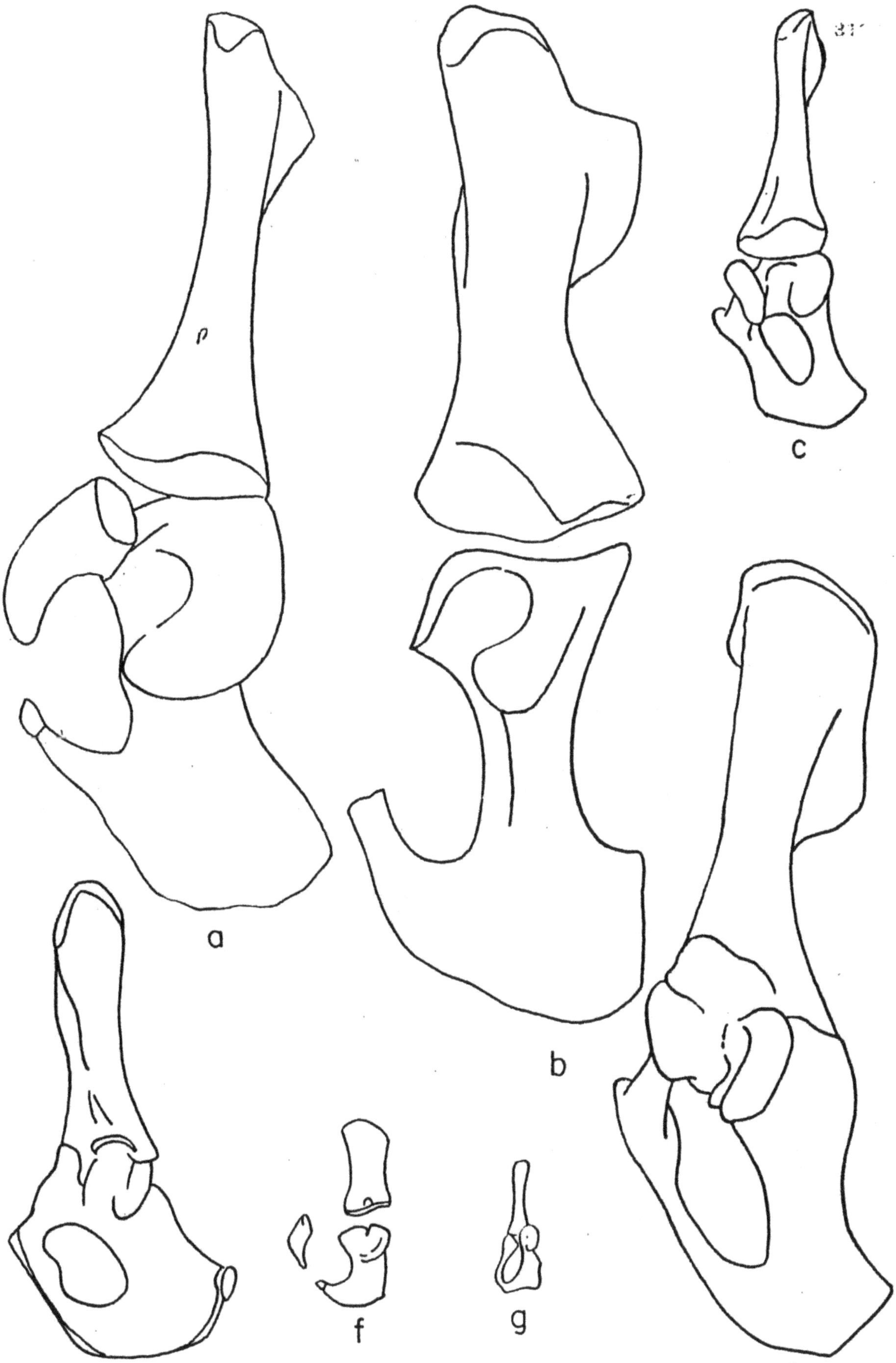
a
b
c
d
e
f
g

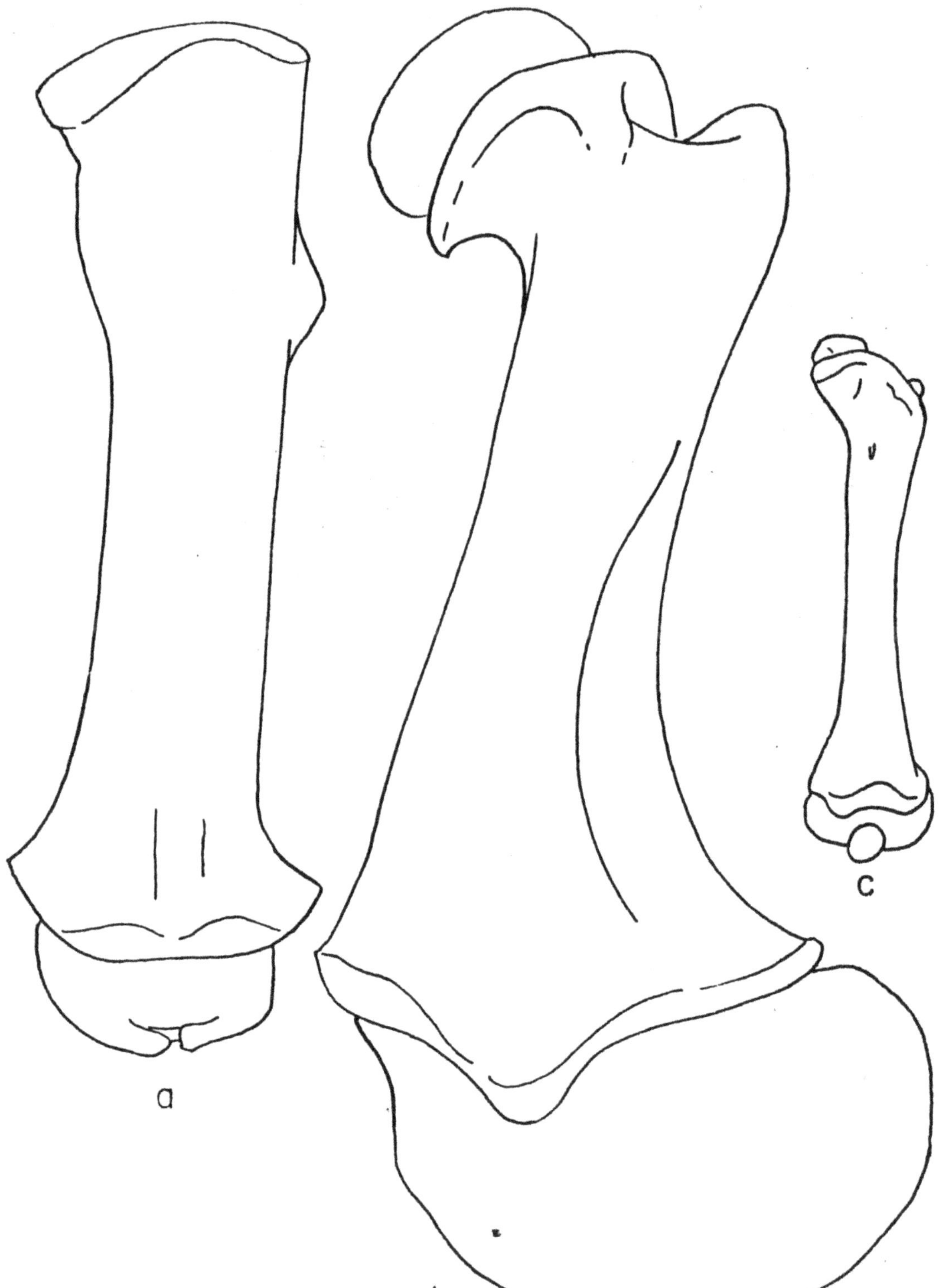
a
b
c

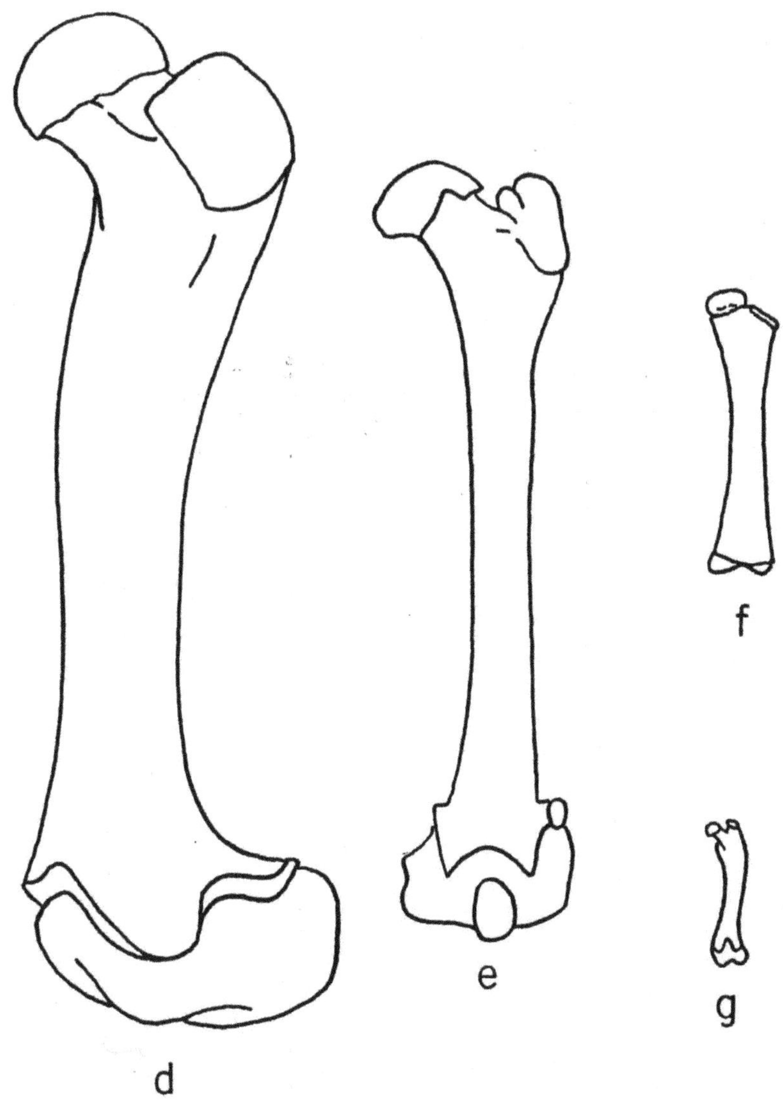
d
e
f
g

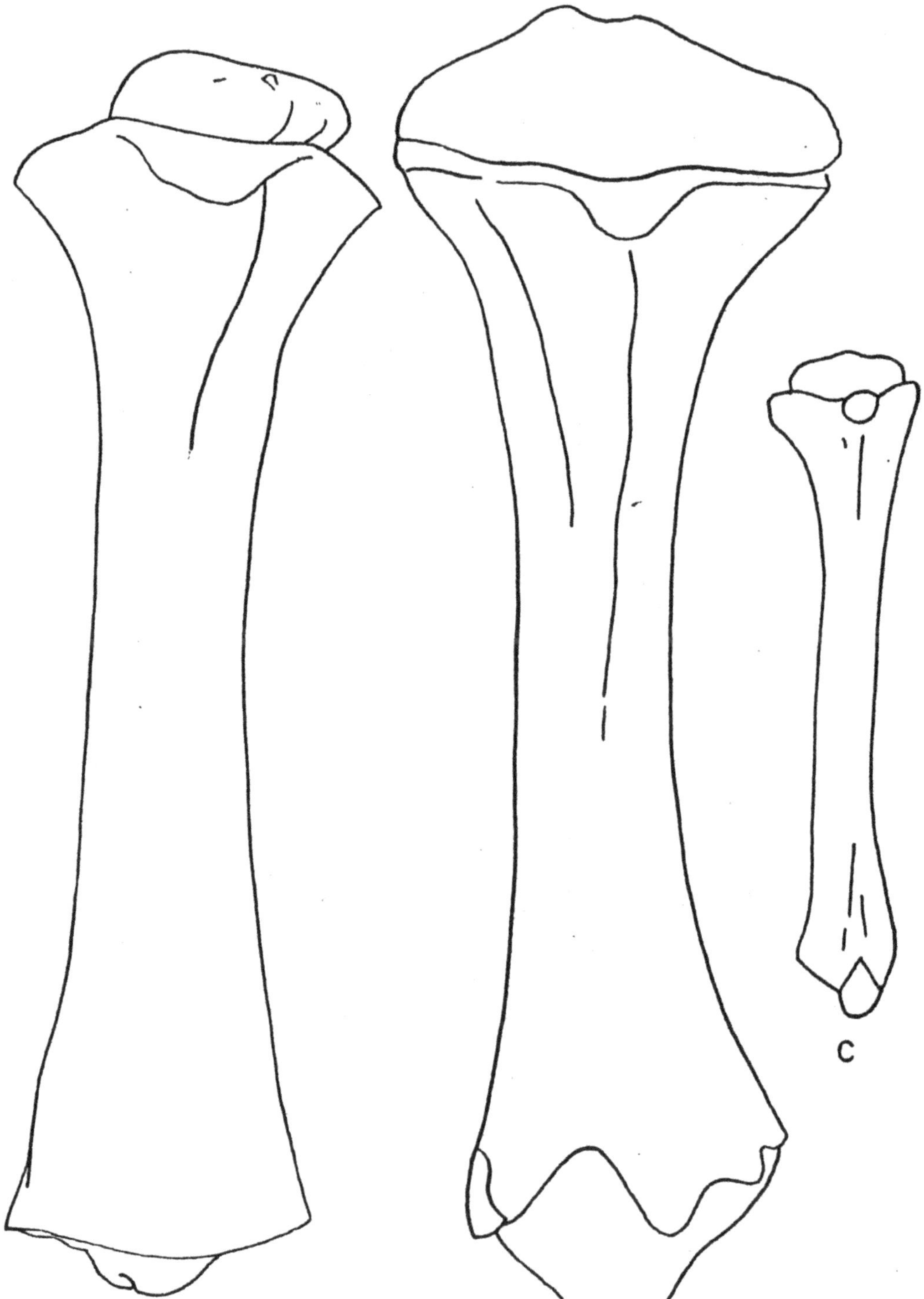

a b c

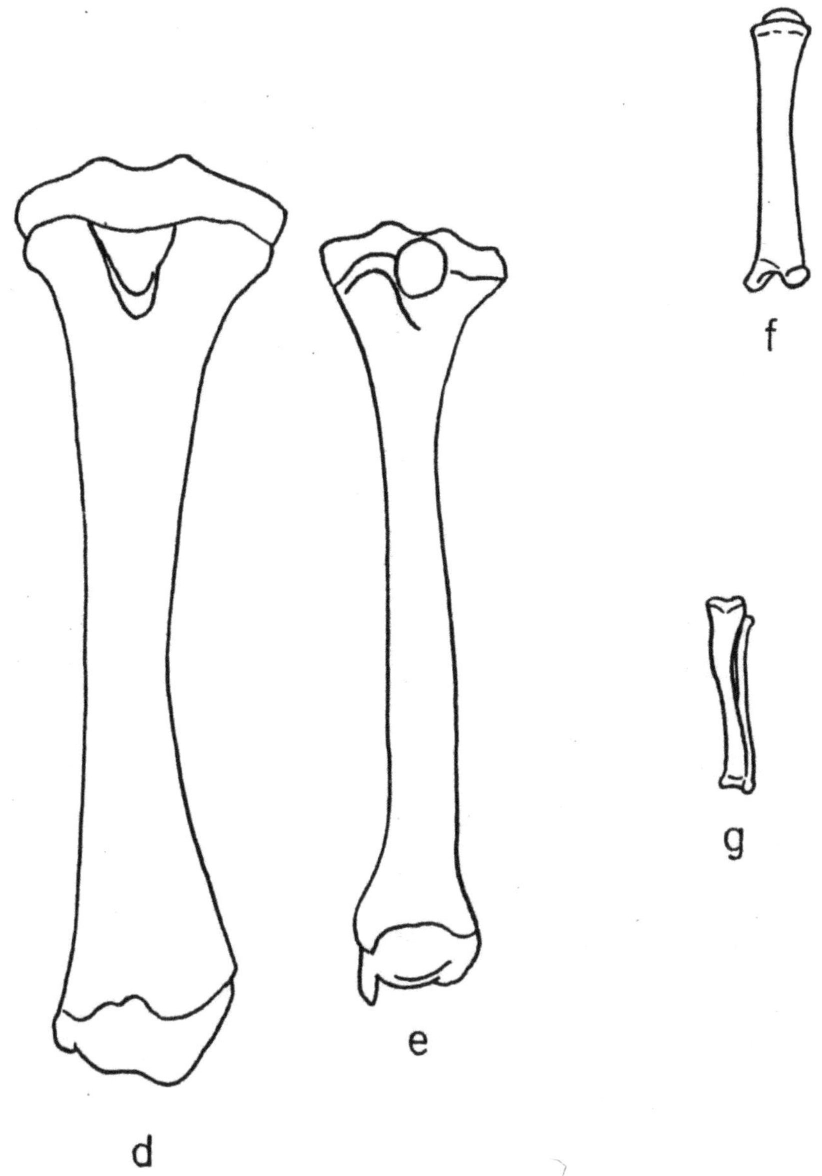
d
e
f
g

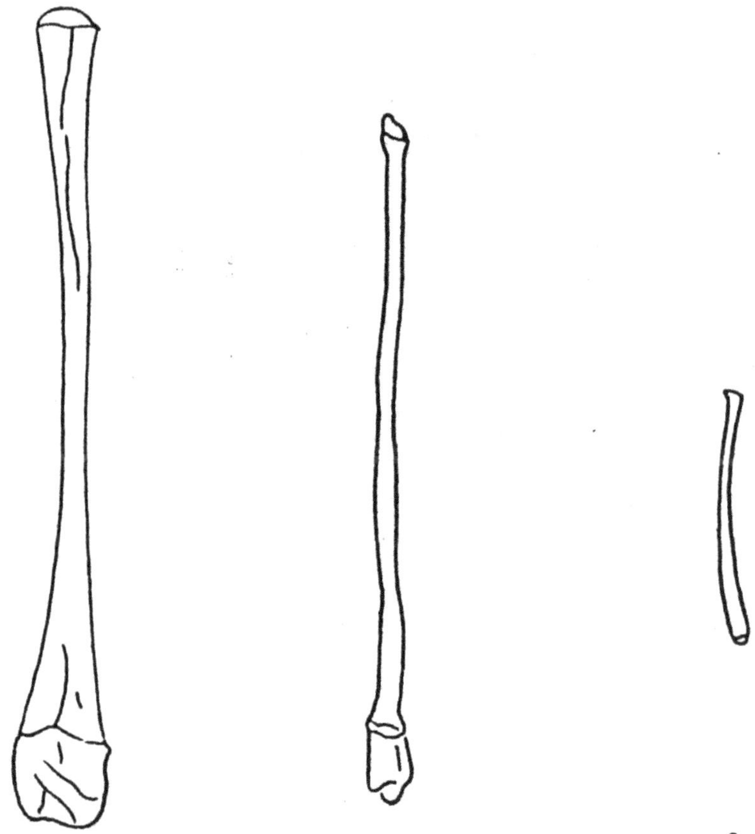

d e f

Metatarsus

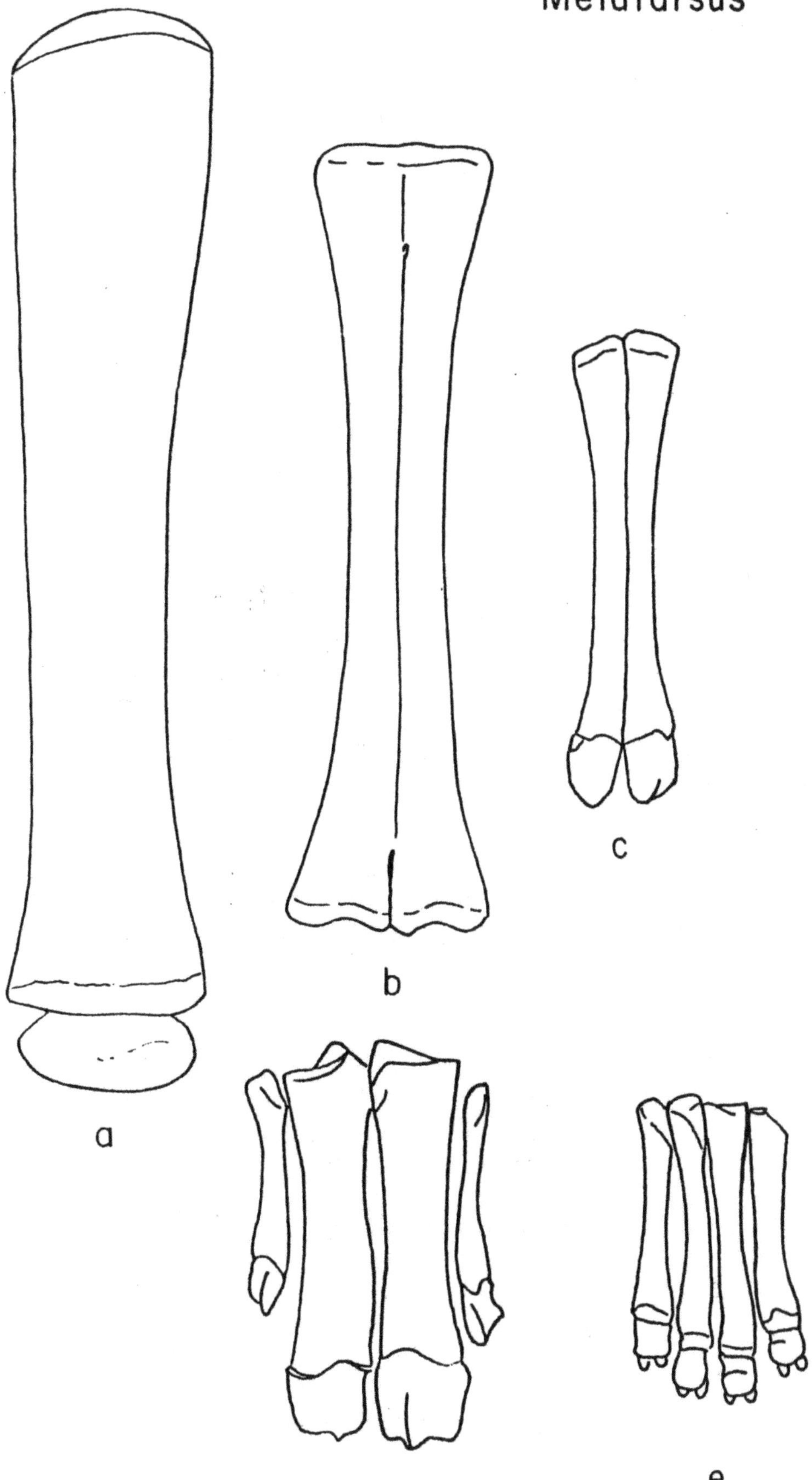

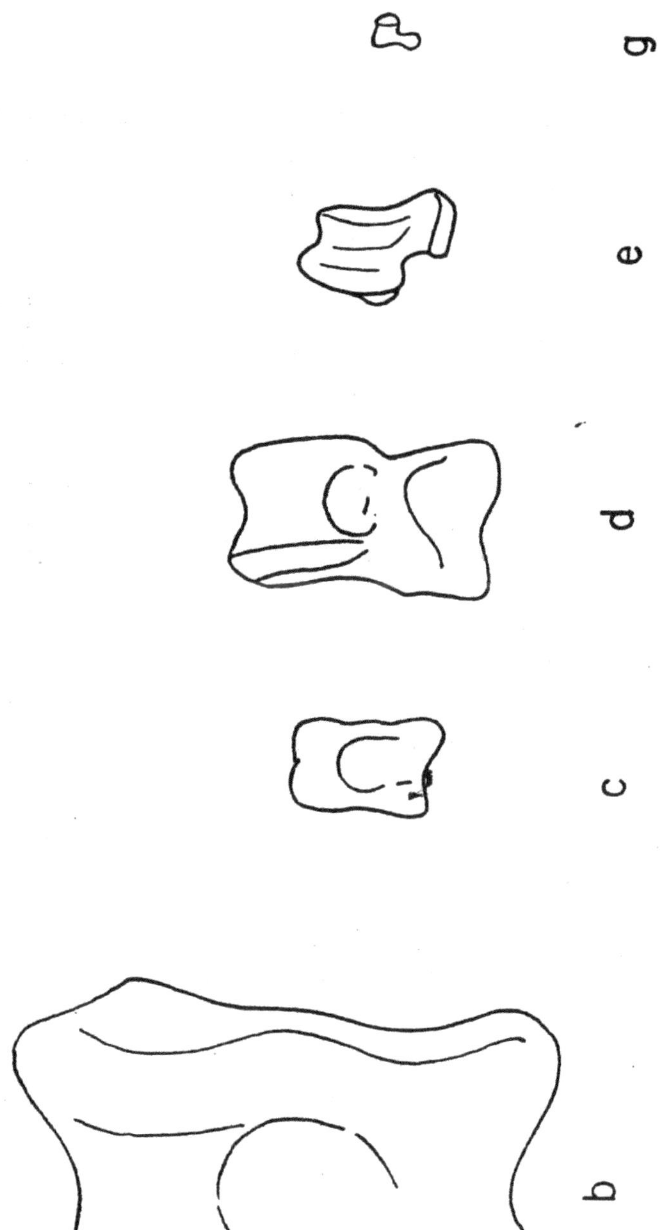
b
c
d
e
g

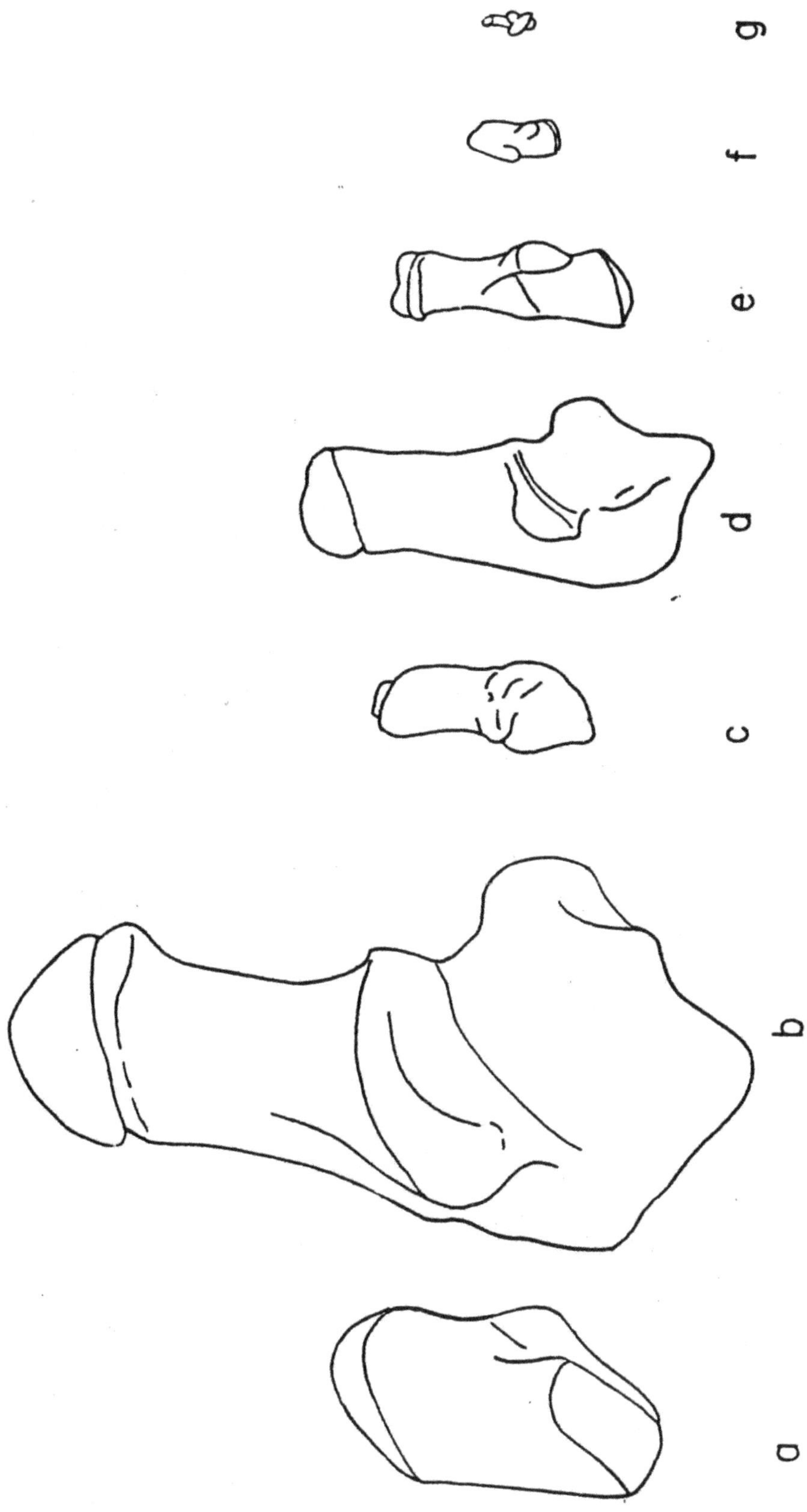
a
b
c
d
e
f
g

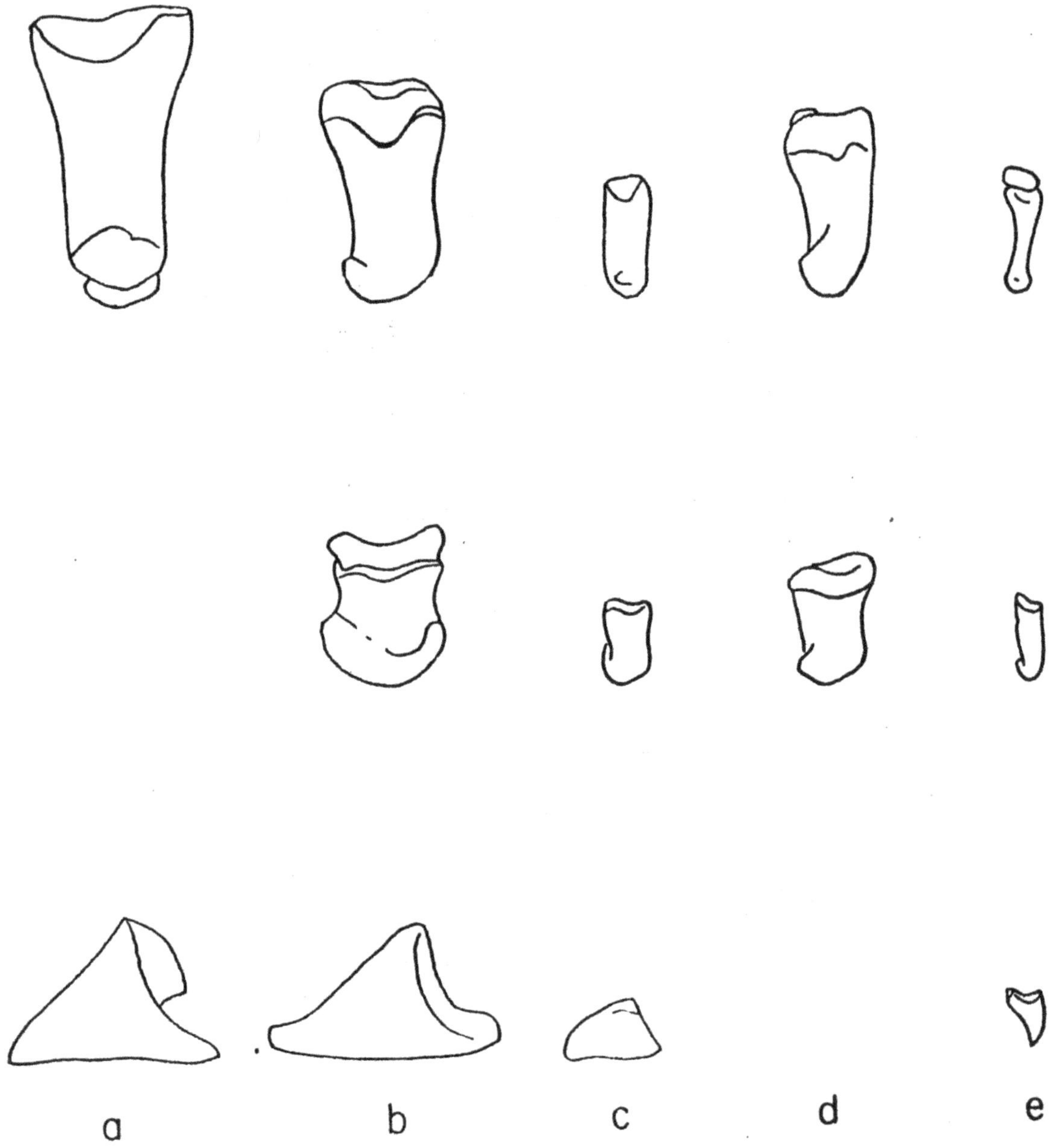

Appendix A

MATERIAL CHOSEN FOR ILLUSTRATION

The sample of osteological material used in this work incorporated as many neo-natal/juvenile skeletons or bone elements as were available. The faunal material chosen for the illustrations was obtained from three comparative collections: the Department of Anthropology, Bioarchaeology Laboratory at Hunter College (City University of New York -- HC), the Department of Mammalogy at the American Museum of Natural History (AMNH), and the author's comparative collection.

The material used from the collections at the American Museum represent a small sample of complete disarticulate skeletons. The sample comprises of Equus (AMNH 204001), a specimen of Ovis (AMNH 146549), one of Sus (AMNH 54851), one of Canis (AMNH 146522), and seven specimens of Rattus (Rattus norvegicus) -- AMNH 238556, 238558, 238623; Rattus rattus -- AMNH 42661, 210313, 240709, 2420709). However, only those specimens that were once part of the S.H. Chubb collection of Equus (AMNH 204001), Ovis (AMNH 146549), and Canis (AMNH 146522) were curated with reliable documentary data on age and sex. These specimens in particular were used as a means to evaluate some of the relative aging tables presented in chapter 3. The rest of the skeletal material was aged on the basis of the aging tables in chapter 3. The ages given to these specimens represent the earliest and latest possible ages as indicated by the physiological events of dental eruption, wear, and epiphyseal fusion. The ages assigned to these specimens were checked by an independent observer to test for accuracy.

The material drawn from the American Museum consists of:

AMNH No.	Species	Background Information
204001	Equus caballus	S.H. Chubb Collection, 1902, complete skeleton, aged between the 10th fetal month to birth, male.
146549	Ovis aries	S.H. Chubb collection, complete skeleton, aged between birth to 6 weeks, male.
54851	Sus scrofa	India, complete skeleton, aged between 6 but less than 12 mos.,

146549	Ovis aries	S.H. Chubb collection, complete skeleton, aged between birth to 6 weeks, male.
54851	Sus scrofa	India, complete skeleton, aged between 6 but less than 12 mos., female.
146522	Canis familiaris	S.H. Chubb collection, complete skeleton, aged between 5-7 mos., female.
238623	Rattus norvegicus	New York, Rockland Co. Haverstraw, 1974 complete skeleton, aged approximately 3 mos..

The material obtained from the author's collection was prepared so that the age could be documented. In the case of the cat skeleton (Felis catus L.), each was aged on the basis of breeders records, recording the age of the individual at birth and death. The cattle sample Bos sp. was aged on the basis of information supplied by Victory Veal, Inc. (823 Washington Street, New York, N.Y.C.) and verified by the use of the aging tables in chapter 3.

The material obtained from the author's collection consists of:

No.	Species	Background Information
3-7	Felis catus	East Rockaway, N.Y., 1970, complete skeleton, aged between 4-6 weeks, female.
3-135	Felis catus	East Rockaway, N.Y., 1981, complete skeleton, aged between 6-8 weeks, female.
3-210	Bos sp.	Obtained at Victory Veal Inc., N.Y., 1983, complete cervical vertebrae series, thoracic vertebrae series, an incomplete lumbar vertebrae series (missing L6), and ribs, aged 6-12

mos..

3-211 Bos sp. Obtained at Victory Veal Inc., N.Y., 1983, left thoracic limbs consisting of two humeri, radii, ulnae, and carpals, aged between 6-12 mos..

3-212 Bos sp. Obtained at Victory Veal Inc., N.Y., 1983, pelvic limbs consisting of femora, tibiae, astragali, calcanae, and tarsals, aged between 6-12 mos..

The faunal material used from the collection at Hunter College are not modern comparative material, but archaeological specimens. The two collection used are those from the Storaborg site in southwestern Iceland (cf. Russell, Amorosi and McGovern 1986), and the Clibberswick site located in Baltasound, Unst, Shetland, United Kingdom (Bruns, Amorosi and McGovern n.d.).

The neo-natal/juvenile sample from Storaborg comprises 356 whole or fragmentary bone elements. This collection is composed of 22 cataloged and 82 uncataloged specimens of Bos, 250 uncataloged specimens of Ovis/Capra, and 2 uncataloged specimens of Sus. The Clibberswick neo-natal/juvenile sample contains 25 uncataloged specimens, of which 22 are Bos, 2 are Ovis/Capra, and 1 is Sus. From the Storaborg sample, only 17 specimens were chosen for illustration. This material was aged on the basis of the aging criteria found in chapter 3 and then checked by an independent observer for accuracy.

Those specimens obtained for illustration at Hunter College are:

HC No.	Species	Background Information
M-1	Bos sp.	Storaborg, Iceland, Field No. STB 1980, H14-9, between floor, east of fireplace. Left scapula approximately aged between 7-10 mos..
M-3	Bos sp.	Storaborg, Iceland, Field No. STB 1979, H14. First sacral vertebra, aged to a

		neo-natal phase.
M-4	Bos sp.	Storaborg, Iceland, Field No. STB 1980, H15 floor. Third sacral vertebra, aged to a neo-natal phase.
M-5	Bos sp.	Storaborg, Iceland, Field No. STB 1981, H18 19x. First phalanx, approximate age 18-24 mos..
M-6	Bos sp.	Storaborg, Iceland, Field No. STB 1980, H19 floor. First phalanx, approximate age 18-24 mos..
M-7	Bos sp.	Storaborg, Iceland, Field No. STB 1981, "G" Trench leveling to 25 cm.. Second phalanx, approximate age 15-18 mos..
M-9	Bos sp.	Storaborg, Iceland, Field No. STB 1981, "F" Trench A2 4b. Third phalanx, approximate age 18-24 mos..
M-10	Bos sp.	Storaborg, Iceland, Field No. STB 1979, H14 floor. Left ilium, approximate age is less than 6 mos..
M-12	Bos sp.	Storaborg, Iceland, Field No. STB 1981, H18 25x. Right ilium, approximate age is less than 6 mos..
M-14	Bos sp.	Storaborg, Iceland, Field No. STB 1981, "G" Trench SE 80-90 cm.. Left ischium, approximate age is less then 6 mos..
M-15	Bos sp.	Storaborg, Iceland, Field

		No. STB 1981, H18 17x. Metapodial half, approximate age is less than 24 mos..
M-16	Bos sp.	Storaborg, Iceland, Field No. STB 1981, H18 19x. Distal epiphysis of a metapodial (metacarpus right side), approximate age is less than 24 mos..
M-17	Bos sp.	Storaborg, Iceland, Field No. STB 1980, H18 floor. Metacarpus, approximate age is less than 24 mos..
M-18	Bos sp.	Storaborg, Iceland, Field No. STB 1981, H18 19x. Distal epiphysis (metacarpus right side), approximate age is less than 24 mos..
M-20	Bos sp.	Storaborg, Iceland, Field No. STB 1979, H14 5y. Metatarsus, approximate age is less than 24 mos..
M-22	Bos sp.	Storaborg, Iceland, Field No. STB 1979, H9 floor. Metapodial, aged to be a terminal fetus to birth.

In order to test the accuracy of the illustrations, each drawing was then compared to the samples from Storaborg, Clibberswick, and an unproveniened archaeological collection obtained by the author from Con-Edison workers in the Lower Manhattan (New York City) area. The Manhattan collection consists of a small sample of neo-natal/juvenile Bos, totaling 9 specimens. The Storaborg and Clibberswick collections were used as control samples of known identified specimens. The Manhattan assemblage was used as an uncontrolled sample, since the author has prior knowledge of the Storaborg and Clibberswick collections, which might serve as a source for bias.

Each illustration was compared to a bone element to test for the genera identification and to see if the landmarks indicated in the illustration could be ascertained. In the case of the Storaborg and

Clibberswick assemblages, the degree of accuracy was high, with only a 1% difference in cases of fragmentary material. The identification of the Manhattan assemblage yielded similar results by both the author and an independent observer.

Appendix B

A LISTING OF OSTEOLOGICAL ATLASES, MANUALS AND KEYS

Appendix B is a general cross-listing of osteological atlases, manuals, and keys useful for the zooarchaeologist. Readers should note that this list is not comprehessive, and emphasis is placed on Circumpolar (Arctic) Studies. The reference citations can be found in the references cited section.

CLASS - AVES

Anserifomes:
Woolfender 1961

Galliformes:
Butenbieck and Wissdorf 1982
Hargrave 1970
Hargrave and Emslie 1979
Holman 1961
Koch 1973
McKusick 1986
von den Driesch 1976

General:
Cohen and Serjeantson 1986
Cowles 1987
Gilbert, Smith and Savage 1981
Holman 1961
Koch 1973
Olsen 1979a
Schmid 1972
Woolfender 1961

North America:

Emslie 1982
Gilbert, Smith and Savage 1981
Hargrave and Emlsie 1979
Olsen 1979a

North Atlantic:
Cohen and Serjeantson 1986
Cowles 1987

GENERAL OSTEOLOGY

Olsen 1961a and b
Parker 1988
Reynolds 1913
Romer 1949

CLASS - MAMMALIA

Africa:
Walker 1985

Canidae:
Hildrebrand 1949, 1954
Howard 1949

Domesticates:
Boessneck 1970
Brown and Gustafson 1979
Ellenberger and Dittrich 1959
Getty 1975
Grigson 1971, 1975, 1976, 1978
Hildrebrand 1949
Howard 1949
Lawrence 1951
McGovern and Amorosi 1986
Olsen 1951
Prummel 1987a and b
Schmid 1972
Sisson and Grossman 1966
von den Driesch 1976
Walker 1982

General:
Cranial:

Burt and Grossenheider 1976
Glass 1951
Grigson 1974, 1975, 1976, 1978
Hall 1981
Hildebrand 1949
Hillson 1986
Howard 1949
Mead 1975
Peyer 1968

Cranial and Post-Cranial:
Boessneck 1970
Brained 1939
Getty 1975
Gilbert 1980
Olsen 1979a, 1980, 1982
Prummel 1987a and b
Sisson and Gossman 1966
Smith 1979
Spiess 1976
Spiess and Hedden 1983 (see appendix, Table 8)
Torres, Enciso and Porras 1986
von den Driesch 1976

General Mammalian Osteology:
Chaplin 1971
Cornwall 1956
Davis 1987
Ellenberger and Dittrich 1959
Flower 1885
Hesse and Wapnish 1985
Klein and Cruz-Uribe 1984
Lavocat 1966
Lawlor 1979
Ryder 1969
Wolf-Heidegger 1961

Post-Cranial:
Bangs 1895
Brown and Gustafson 1979
Brown and Twigg 1969
Carter 1976
Hildebrand 1954, 1955
Kasper 1980
King 1983
Lawrence 1951

McGovern and Amorosi 1986
Olsen 1951
Robinette and Stains 1970
Schimd 1972
Walker 1985
Walker 1982
Wolsan 1982

Hominidae:
Bass 1971
Brothwell 1981
Carlsen 1987
Steele and Bramblett 1988
Shipman, Walker and Bichell 1985
Ubelaker 1978
Warwick and Williams 1973

Meso-America/ South America:
Olsen 1982
Torres, Enciso and Porras 1982

North Atlantic/ Europe:
Chaplin 1971
Davis 1987
Lavocat 1966
Wolsan 1982

North America:
Bangs 1895
Burt and Grossenheider 1976
Brown and Gustafson 1979
Glass 1951
Hall 1981
Hildrebrand 1949
Howard 1949
Gilbert 1980
Olsen 1979a, 1980
Smith 1979

Pinnipedia:
Kasper 1980
King 1983
Robinette and Stains 1970
Spiess 1976
Spiess and Hedden 1983

Rodentia:
Amorosi 1988
Brown and Twigg 1969

Ruminats:
Boessneck 1970
Brown and Gustafson 1979
Getty 1975
Grigson 1974, 1975, 1976, 1978
Hildrebrand 1955
Lawrence 1951
McGovern and Amorosi 1986
Olsen 1951
Prummel 1987a and b
Sisson and Grossman 1966
von den Driesch 1976

MEASUREMENMT

Morales and Rosenlund 1979
Spiess 1976
von den Driesch 1976

SUPERCLASS - PISCES

Cyrinid:
Harrington 1955

Gadidae:
Amorosi 1988
Canon 1987
Frost 1981
Morales and Rosenlund 1979
Mujib 1967
Schultz and Welander 1935

General:
Brinkhuizen and Clason 1986
Canon 1987
Casteel 1972, 1973, 1974b, 1976a and b
Clothier 1950
Gregory 1933
Hakonen 1986
Morales and Rosenlund 1979

Olsen 1968

Salmonidae:
Beachman 1985
Canon 1987
Casteel 1974a
Jollie 1984
Kazakov, Dorofeeva, Kozlov and Ll'eukova 1982
Norden 1961
Parker 1873
Pichugin 1983
Tchernavin 1938
Traquair 1895
Wilson and Veilleux 1982
Vladykov 1962

CLASS - REPTILIA

Gaffney 1979
Olsen 1968
Romer 1956

ZOOARCHAEOLOGICAL TEXTBOOKS

Chaplin 1971
Cornwall 1956
Davis 1987
Hesse and Wapnish 1985
Klein and Cruz-Uribe 1984

REFERENCES CITED

Adams, L., and S.G. Watkins
1967 Annuli in Tooth Cementum Indicate Age in California Ground Squirrels. Journal of Wildlife Management 31(4): 836-839.

Aitken, R.J.
1975 Cementum Layers and Tooth Wear as Criteria for Aging Roe Deer. Journal of Zoology, London 175: 75-128.

Amorosi, T.
1988 Non-Metric Trait Distinction for the Family Gadidae and the Genus Rattus. Presented at the Norse of the North Atlantic Conference, Bioarchaeological Seminar, Bowdoin College, Maine April 17-21. Unpublished Ms. on file at the Bioarchaeological Laboratory, Department of Anthropology, Hunter College, City University of New York, 695 Park Ave., New York, N.Y. 10021.

n.d. A Report on the Small Mammal Assemblage from Selected Features at 175 Water Street, New York City. Report under preparation.

Andrews, A.H.
1973 A Survey of the Relationship Between Age and the Development of the Anterior Teeth in Cattle. Veternary Record 92: 275-282.

1982 The Use of Dentition to Age Young Cattle. In Ageing and Sexing Animal Bones from Archaeological Sites. B. Wilson, C. Grigson, and S. Payne, eds. pp. 141-153. British Archaeological Reports, British Series 109.

Armitage, P.
1982 A System for Ageing and Sexing the Horn Cores of Cattle from British Post Medieval Sites (with Special Reference to Unimproved British Longhorn Cattle). In Ageing and Sexing Animal Bones from Archaeological Sites. B. Wilson, C. Grigson, and S. Payne, eds. pp. 37-54. B.A.R., Brit. Series 109.

Arey, L.B.
1974 Developmental Anatomy. 7th edition. Philadelphia: W.B. Saunders Company.

Banfield, A.W.F.
1960 The Use of Caribou Antler Pedicels for Age Determination. Journal of Wildlife Management 24: 99-102.

Bangs, O.
1895 The Geographical Distribution of the Eastern Races of the Cotton-Tail (Lepus sylvaticus Bach.) with a Description of the Northern Hare (Lepus americanus Erxl.) in the East. Proceedings of the Boston Society of Natural History XXVI: 404-414.

Bass, W.M.
1971 Human Osteology: a Laboratory and Field Manual of the Human Skeleton. Columbia, Missouri: Archaeological Society.

Beachman, T.D.
1985 Variation in the Number of Vertebrae and Gill Rakes of Sockeye Salmon, Oncorhynchus nerka, in North America. Environmental Biology of Fishes 14(2/3): 97-105.

Benzie, D. and J.C. Gill
1974 Radiography of the Skeletal and Dental Condition of the Soay Sheep. In Island Survivors: the Ecology of the Soay Sheep of St. Kilda. P.A. Jewell, C. Milner, and S. Morton Boyd, eds. London: Athlone Press.

Binford, L.R.
1981 *Bones: Ancient Man and Modern Myths*. New York: Academic Press.

Boessneck, J.
1970 Osteological Differences Between Sheep (*Ovis aries* L.) and Goat (*Capra hircus* L.). In *Science in Archaeology*. D. Brothwell, and E. Higgs, eds. pp. 331-358. New York: Praeger Publisher.

Boessneck, J. and A. von den Driesch
1975 Tierknochhenfunde von Korucutepc bei Elazig in Ostanatolien. In *Studies in Ancient Civilization*. Korucutepe, 1. Amsterdam: M.N. van Loon.

Borque, B.J., K. Morris and A. Spiess
1978 Determing the Season of Death of Mammal Teeth from Archaeological Sites: a New Sectioning Technique. *Science* 199: 530-531.

Brainerd, G.W.
1939 An Illustrated Key for the Identification of Mammal Bones. *Ohio State Archaeological and Historical Quaterly* 48(4): 324-328.

Brickman, B.L.
1984 *Sexual Dimorphism of the Cranium Prior to Puberty*. M.A. Thesis, Department of Anthropology, New York University.

Brinkhuizen, D.C. and A.T. Clason
1986 *Fish and Archaeology: Studies in Osteometry, Taphonomy, Seasonality and Fishing Methods*. B.A.R., International Series 294.

Brothwell, D.R.
1981 *Digging Up Bones*. 3rd edition. New York: Cornell University Press.

Brown, G.
1902 _Dentition as Indicative of the Age of the Animals of the Farm_. Revised Edition. Royal Agricultural Society of England.

Brown, C.L. and C.E. Gustafson
1979 A Key to Postcranial Skeletal Remains of Cattle/Bison, Elk, and Horse. _Reports of Investigation_, No. 57. Laboratory of Anthropology, Washington State University, Pullman.

Brown, J.C. and G.I. Twigg
1969 Studies on the Pelvis in British Muridae and Cricetidae (Rodentia). _J. Zool., Lond._ 158: 81-132.

Bruni, A.C. and U. Zimmerl
1951 _Anatomia degli Animali Domestici_, vol. 1 Milano: Casa Editrice Dottor Francesco Vallardi.

Bruns, M., T. Amorosi and T.H. McGovern
1985 An Archaeofauna from Cross Geos, Clibberswick on Unst, Shetland Islands: Preliminary Report. Unpublished Ms. on file at the Bioarchaeological Laboratory, Department of Anthropology, Hunter College (City University of New York).

Bull, G. and S. Payne
1982 Tooth Eruption and Epiphyseal Fusion in Pigs and Wild Boar. In _Ageing and Sexing Animal Bones from Archaeological Sites_. B. Wilson, C. Grigson, and S. Payne, eds. pp. 55-71. B.A.R., Brit. Series 109.

Bullock, D. and J. Rackham
1982 Epiphyseal Fusion and Tooth Eruption of Feral Goats from Moffatdale, Dumfries and Galloway, Scotland. In _Ageing and Sexing Animal Bones from Archaeological Sites_. B. Wilson, C. Grigson, and S. Payne, eds. pp. 73-80. B.A.R., Brit. Series 109.

Burt, W.H. and R.P. Grossenheider
1976 A Field Guide to the Mammals: Field Marks of all North American Species Found North of Mexico. 3rd edition. Boston: Mifflin Company.

Butenbieck, V.E. and H. Wissdorf
1982 Contribution to the Designation of the Headbones of the Turkey (Meleagris gallopavo) with Regard to the NAA (1979). Zoological Journal of Anatomy. 107: 153-184.

Butzer, K.W.
1975 The Ecological Approach to Archaeology: Are We Really Trying. American Antiquity 40(1): 106-111.

Cannon, D.Y.
1987 Marine Fish Osteology: A Manual for Archaeologists. Department of Anthropology, Simon Fraser University, Burnaby, B.C., Canada V5A 1S6.

Cart[illegible] H.H.
1976 Vertebrae of the Larger Mammals of Western Europe. Ossa 3/4: 109-127.

Carlsen, O.
1987 Dental Morphology. Copenhagen; Denmark: Munksgaard.

Casteel, R.W.
1972 A Key on Scales to the Families of Native California Freshwater Fishes. Proceedings of California Academy of Science 39(7): 75-86.

1973 The Scales of the Native Freshwater Fishes of Washington. Northwest Science 47(7): 320-238.

Casteel, R.W.

1974a Identification of the Species of Pacific Salmon (genus Oncorhynchus) Native to North America Based Upon Otolith. Copeia 2: 305-311.

1974b On the Remains of Fish Scales from Archaeological Sites. American Antiquity 39: 557-581.

1976a Fish Remains in Archaeology: and Paleoenvironmental Studies. New York: Academic Press.

1976b Incremental Growth Zones in Mammals and Their Archaeological Value. Papers of the Kroeber Anthropological Society (Berkeley) 47: 1-27.

Caughly, G.

1965 Horn Rings and Tooth Eruption as Criteria of Age in the Himalayan Thar, Hemitagus iemlahicus. National Zoological Journal of Science 8: 333-351.

Chaplin, R.E.

1971 The Study of Animal Bones from Archaeological Sites. New York: Seminar Press.

Cheverud, J.M.

1981 Epiphyseal Union and Dental Eruption in Macacca mulatta. Amercan Journal of Physical Anthropology 56: 157-167.

Clothier, C.R.

1950 A Key to Some Southern California Fishes Based on Vertebral Characters. California Department of Fish and Game, Fish Bulletin 79.

Clutton-Brock, J.

1981 Domesticated Animals from Early Times. Austin: University of Texas Press.

Cohn, S.A.
1957 Development of the Molar Teeth in the Albino Mouse. _American Journal of Anatomy_ 101: 295-320.

Cohen, A. and D. Serjeantson
1986 _A Maunal for the Identification of Bird Bones from Archaeological Sites_. London: Privately Published.

Cornwall, I.W.
1956 _Bones for the Archaeologists_. New York: The Macmillian Company.

Cowles, G.
1987 A Guide to the Identification of Bird Bones from Britian and Europe. In preparation.

Coy, J.P., R.T. Jones and K.A. Turner
1982 Absolute Ageing of Cattle from Teeth Sections and its Relevance to Archaeology. In _Ageing and Sexing Animal Bones From Archaeological Sites_. B. Wilson, C. Grigson, and S. Payne, eds. pp. 127-140. B.A.R., Brit. Series 109.

Crabtree, P.J.
1982 Worked Bone from Dun Ailine, Co. Kildare, Ireland. _MASCA Journal_ 2(1): 6-7.

Curgy, J.-J.
1965 Appartion et Soudres des Points D'ossification des Membres chez les Mammiferees. _Memories du Museum National D'historie Naturelle_, serie A, Zoologie, tome XXXII, fascicale 3. Paris.

Davis, S.J.M.
1987 _The Archaeology of Animals_. New Haven: Yale University Press.

Dawson, A.B.
1925 The Age Order of Epiphyseal Union in the Long Bones of the Albino Rat. Anatomical Record 31: 1-10.

1927 Further Studies on the Epiphyses of the Albino Rat Skeleton, with Special Reference to the Vertebral Column, Ribs, Sternum, and Hip Gridle. Anat. record 34: 351.

1934 Further Studies on Epiphyseal Union in the Skeleton of the Rat. Anat. Record 60: 84.

1935 The Sequence of Epipihyseal Union in the Skeleton of the Mouse, with Special Reference to the Phenomenon of "Lapsed Union". Anat. Record 63: 93.

Degerbol, M.
1970 The Urus (Bos primigenius Bojanus) and Neolithic Domesticated Cattle (Bos taurus domesticus Linne) in Denmark. K. danske Vidensk. Selsk. Skr. 17: 1-117.

Deniz, E. and S. Payne
1982 Eruption and Wear in the Mandibular Dentition as a Guide to Ageing Turkish Angora Goats. In Ageing and Sexing Animal Bones from Archaeological Sites. B. Wilson, C. Grigson, and S. Payne, eds. pp. 155-207. B.A.R., Brit. Series 109.

Domning, D.P.
1983 Marching Teeth of the Manatee. Natural History No. 5, pp. 8-11.

Driver, J.C.
1984 Zooarchaeological Analysis of Raw-Material Selection by a Saxon Artisan. Journal of Field Archaeology 11(4): 397-403.

Ellenberger, N.H. and H. Dittrich
1959 *An Atlas of Animal Anatomy*. New York: Dover Publications.

Emslie, S.D.
1982 Osteological Identification of Long-Eared and Short-Eared Owls. *American Antiquity* 47(1): 155-157.

Erickson, J.A. and W.G. Seliger
1969 Efficient Sectioning of Incisors for Estimating Ages of Mule Deer. *J. Wildl. Mgmt.* 33: 384-388.

Evans, R.R.
n.d. Preliminary Report on Limitations of the Dental Cementum Aging Technique for Moose.

Flannery, K.V.
1969 The Animal Bones. In *Prehistory and Human Ecology of the Deh Luran Plain*. F. Hole, K.V. Flannery, and J.A. Neely, eds. Ann Arbor: Memiors of the Museum of Anthropology, University of Michigan, No. 1.

Flower, W.H.
1885 *An Introduction to the Osteology of the Mammalia*. London: Macmillan Company.

Frison, G.C. and D.J. Stanford
1982 *The Agate Basin: A Record of the Paleoindian Occupation of the Northwestern High Plains*. New York: Academic Press.

Frison, G.C., M. Wilson and D.J. Wilson
1976 Fossil Bison and Artifacts from an Early Alithermal Period Arroyo Trap in Wyoming. *American Antiquity* 41(1): 28-57.

Frost, K.J.
1981 Descriptive Key to the Otoliths of Gadid Fishes of the Bering, Chukchi and Beaufort Seas. *Arctic* 34(1): 55-59.

Gaffney, E.S.
1979 Comparative Cranial Morphology of Recent and Fossil Turtles. Bulletin of the American Museum of Natural History 162(2).

Getty, R.
1975 Anatomy of the Domestic Animals. 5th edition. Philadelphia: W.B. saunders Company.

Gilbert, F.F.
1966 Aging White-tailed Deer by Annuli in the Cementum of the First Incisor. J. Wildl. Mgmt. 30(1): 200-202.

Gilbert, B.M.
1980 Mammalian Osteology. Laramie, Wyoming: B. Miles Gilbert, Publisher.

Gilbert, B.M., L.D. Smith and H.G. Savage
1981 Avian Osteology. Laramie Wyoming: B. Miles Gilbert, Publisher.

Glass, B.P.
1951 A Key to the Skulls of North American Mammals. Minneapolis.

Gorden, B.H.C.
1982 Tooth Sectioning as an Archaeological Tool. Canadian Studies Report, No.14e.

Gorden, C.C. and J.E. Buikstra
1981 Soil pH, Bone Preseervation, and Sampling Bias at Mortuary Sites. American Antiquity 46(3): 566-571.

Grant, A.
1975 The Use of Tooth Wear as a Guide to the Age of Domestic Animals - A Brief Explanation. In Excavation at Portchester Castle I: Roman.. B. Cunliffe. London: Society of Antiquaries Research Report, No. 32.

Gaŕnt, A.
1978 Dental Attrition and its Relevance to Age. In Research Problems in Zooarchaeology. D.R. Brothwell, K.D. Thomas, S. Clutton-Brock, eds. pp. 103-107. London: Institute of Archaeology, Occasional Publications, No. 3.

1982 The Use of Tooth Wear as a Guide to the Age of Domestic Ungulates. In Ageing and Sexing Animal Bones from Archaeological Sites. B. Wilson, C. Grigson, and S. Payne, eds. pp. 91-109. B.A.R., Brit. Series 109.

Grau, G.A., G.C. Sanderson and J.P. Rogers
1970 Age Determination of Raccoons. J. Wildl. Mgmt. 34: 364-372.

Grayson, D.K.
1977 On the Holocene History of Some Northern Great Basin Lagomorphs. Journal of Mammalogy 58: 507-513.

Grayson, D.K.
1978 Minimum Numbers and Sampling Sizes in Vertebrate Faunal Analysis. American Antiquity 43: 53-65.

1979 On the Quantification of Vertebrate Archaeofaunas, In Advances in Archaeological Method and Theory, vol. 2. M.B. Schiffer, ed. pp. 199-237. New York: Academic Press.

1984 Quantitative Zooarchaeology: Topics in the Analysis of Archaeological Faunas. New York: Academic Press.

Gregory, W.K.
1933 Fish Skulls: A Study of the Evolution of Natural Mechanisms. The American Philosophical Society vol. XXIII, Article II. Philadelphia.

Grigson, C.
1974 The Craniology and Relationship of Four Species of _Bos_ 1. Basic Craniology: _Bos taurus_ L. and its Absolute Size. _J. Arch. Sci._ 1: 353-379.

1975 The Craniology and Relationship of Four Species of _Bos_ 2. Basic Craniology: _Bos taurus_ L. Proportions and Angles. _J. Arch. Sci._ 2: 109-128.

1976 The Craniology and Relationship of Four Species of _Bos_ 3. Basic Craniology: _Bos taurus_ L. Sagittal Profiles and Other Non-Measuresurable Characters. _J. Arch. Sci._ 3: 115-136.

1978 The Craniology and Relationship of Four Species of _Bos_ 4. The Relationship between _Bos primigenius_ Boji and _Bos taurus_ L. and its Implications for Phylogeny of the Domestic Breeds. _J. Arch. Sci._ 5: 123-152.

1982 Sex and Age Determination of Some Bones and Teeth of Domestic Cattle: Review of the Literature. In _Ageing and Sexing Animal Bones from Archaeological Sites_. B. Wilson, C. Grigson, and S. Payne, eds. pp. 7-25. B.A.R., Brit. Series 109.

Grue H.
1976 Non-Seasonal Incremental Lines in Tooth Cementum of Domestic Dogs (_Canis familiaris_ L.) of Known Age. _Danish Review of Game Biology_ 10(2): 2-8.

Grue, H. and B. Jensen
1973 Annular Structure in Canine Tooth Cementum in Red Foxes (_Vulpes vulpes_ L. of Known Age. _Danish Review of Game Biology_ 8(7).

Grue, H. and B. Jensen
1976 Annual Cementum Structures in Canine Teeth in Arcic Foxes (*Alopex alopex* L.) from Greenland and Denmark. *Danish Review of Game Biology* 10(3).

Habermehl, K.-H.
1961 *Die Alterbestimmung bei Haustieren. Pelztieren und beim jagdbaren Wildtieren*. Berlin and Hamurg: Paul Parey.

1975 *Die Alterbestimmung bei Haustieren und Labortieren*. Berlin: Z. Auflage Verlag Paul Parey.

Hall, E.R.
1981 *The Mammals of North America*, vols. I-II. New York: John Wiley and Sons.

Hargarve, L.L.
1970 *Mexican Macaws: Comparative Osteology and Survey of Remains from the Southwest*. Tucson, Arizona: The University of Arizona Press.

Hargrave, L.L. and S.D. Emslie
1979 Osteological Identification of the Sandill Crane versus the Turkey. *American Antiquity* 44(2): 295-299.

Harkonen, T.
1986 *Guide to the Otoliths of the Boney Fishes of the Northeast Atlantic*. Denmark: Danibu ApS. (Biological Consultants, Henningsens Alle 58, Dk - 2900 Hellerup, Denmark).

Harrington, R.W.
1955 The Osteocranium of the American Cyprinid Fish, *Notropis bifrenatus*, with an Annotated Synonymy of the Telost Skull Bones. *Copeia* 267-290.

Hesse, B. and P. Wapnish
1985 *Animal Bone Archaeology: from Objects to Analysis*. Manuals on Archaeology 5. Washington: Taraxacum.

Higgs, E.S.
1972 *Papers in Economic Prehistory*. Cambridge: Cambridge University Press.

1975 *Paleoeconomy*. Cambridge: Cambridge University Press.

Higham, C.F.W.
1967 Stock Rearing as a Cultural Factor in Prehistoric Europe. *Proceedings of the Prehistoric Society* 33: 84-106.

Hildebrand, M.
1949 A Means to Distinguish Skulls of Coyotes and Dogs. *Journal of Mammalogy* 3(2): 167-171.

1954 Comparative Morphology of the Body Skeleton in Recent Canidae. *University of California: Publication in Zoology* 52(5): 399-470.

1955 Skeletal Differences Between Deer, Sheep, and Goats. Davis: *California Fish and Game Commission*, pp. 327-346.

Hillson, S.
1986 *Teeth*. New York: Cambridge University Press.

Holman, J.A.
1961 Osteology of Living and Fossil New World Quails. (Aves, Galliformes). *Bulletin of the Florida State Museum* 6(2).

Howard, W.E.
1949 A Means to Distinguish Skulls of Coyotes and Domestic Dogs. *Journal of Mammalogy* 30(2): 169-171.

Johnson, M.L.
1933 The Time and Order of Appearance of Ossification Centers in the Albino Mouse. *American Journal of Anatomy* 52: 241-271.

Jollie, M.
1984 Development of the Head Skeleton and Pectoral Girdle of Salmon, with a Note on the Scales. Canadian Journal of Zoology 62(9): 1757-1778.

Jones, A.K.G.
1982 Bulk Sieving and the Recovery of Fish Remains from Urban Archaeological Sites. In Environmental Archaeology in Urban Context. A.R. Hall and H.K. Kenward, eds. pp. 79-86. Council for British Archaeology (CBA) Report.

Kasper, J.C.
1980 Skeletal Identification of California Sea Lions and Harbor Seals for Archaeologists. San Diego Museum of Man: Ethnic Technology Notes, No. 17.

Kay, M.
1974 Dental Annuli Age Determination on White-tailed Deer from Archaeological Sites. Plains Anthropologist 19: 224-227.

Kazakov, R.V., E.A. Dorofeeva, V.V. Kozlov and S.A. Ll'enkova
1982 Use of Osteological Characters for Identification of Reciprocal Hybrids Between Atlantic Salmon, Salmo salar, and Brown Trout Samo trutta. Journal of Icthology (Voprosy Ikhtiologii) 22(4): 165-170.

Keiss, R.E.
1969 Comparison of Eruption Wear Patterns and Cementum Annuli as Age Criteria in Elk. J. Wildl. Mgmt. 33: 175-180.

Kenward, H. and C. Engleman
1985 Rapid Scanning of Urban Archaeological Deposits for Insect Remains. Circaea 3(3): 163-172.

King, J.E.
1983 Seals of the World. 2nd Edition. New York: Cornell University Press.

Klebanova, E.A. and G.A. Kleveza1
1966 (Stratification of the Periostal Zone of Long Bones as a Criterion for Determining the Age of Mammals). In Russian Zool. Zh. 45: 406-413. National Lending Library, Russian Translation No. 3647.

Klein, R.G. and K. Cruz-Uribe
1984 The Analysis of Animal Bones from Archaeological Sites. Chicago: The University of Chicago Press.

Klein, R.G., K. Wallwardem and C. Wolf
1983 The Calculation and Interpretation of Ungulate Age Profiles from Dental Crown Height. In Hunter-Gatherer Economy in Prehistory: A European Perspective. G. Bailey, ed. pp. 47-57. New York: Cambridge University Press.

Kler, O.
1923 Proceedings of the First All Russian Congress on Zoology, Anatomy and Histology in Petrograd. pp. 194-195.

1927 Ruski Zoologicheski Zhurnal (Russian Zoological Journal). 7(4): 114-133.

Klevezal, G.A.
1970 Retrospective Evalution of Individual Features of the Growth of Mammals by the Layered Structure of Dentin and Bone. Academy of Science 1(4), U.S.S.R.. Translated 1973 from Russian for the Department of the Secretary of State, Translation Bureau Multilingual Services Division, Canada.

Klevezal, G.A. and S.E. Kleinberg
1969 Age Determination of Mammals from Annual Layers in Teeth and Bones. Academy of Sciences, U.S.S.R., Translated 1969 from Russian for the Department of the Interior and the National Science Foundation, U.S. Department of Commerce, Clearinghouse for Fedral Scientific and Technical information, Springfield, Virginia.

Koch, T, ed.
1973 Locomotion System: Apparatus locomotorius. In Anatomy of the Chicken and Domestic Birds. pp. 6-159. Iowa: Iowa State University Press.

Kroon, H.M.
1929 Die Lehre von Altersbestimmung bei den Haustieren. 2nd edition. Hanover.

Langenau, E.E.
1972 Application and Limitations of Aging White-tailed Deer by Annuli in the Cementum of the First Lower Incisor. Research and Development Report No. 272. Michigan: Department of National Resources.

Lavocat, R.
1966 Faunes et Flores Prehistoriques De L'Europe Occidentale. Tome III, Editions. Paris: N. BouBee et C ie.

Lawlor, T.E.
1979 Handbook to the Orders and Families of Living Mammals. Eureka, California: Eureka Priniting Company, Inc..

Lawrence, B.
1951 Mammals Found at the Awatouri Site, Part I, Post Cranial Skeletal Characters of Deer, Pronghorn, and Sheep-Goat with Notes on Bos and Bison, Part II. Papers of the Peabody Museum of Archaeology and Ethnology 35(3). Cambridge Mass.: Harvard University.

Laws, R.M.
1952 A New Method of Age Determination for Mammals. Nature 169: 972-973.

Legge, A.J.
1981 Aspects of Cattle Husbandry. In Farming Practice in British Prehistory. R. Mercer, ed. pp. 169-181. Edinburgh: University Press.

Lesbre, M.F.X.
1897 Contribution a L'etude de L'ossification du Squellette des Mammiferes Domestique. Annales de la Societe D' agriculture Science et Industrie de Lyon 5: 1-106.

Levine, M.A.
1982 The Use of Crown Height Measurements and Eruption-Wear Sequences to Age Horse Teeth. In Ageing and Sexing Animal Bones from Archaeological Sites. B. Wilson, S. Payne, and C. Grigson, eds. pp. 223-250. B.A.R. Brit. Series 109.

1983 Mortality Models and the Interpretation of Horse Population Structure. In Hunter-Gatherer Economy in Prehistory: A European Perspective. G. Baily, ed. pp. 23-46. New York: Cambridge University Press.

Levitan, B.
1983 Reducing the work-load: sub-sampling animal bone assemblages. Circaea 1(1): 7-12.

Lewin, R.
1983 Archaeological Analysis Gets Some Teeth. Science 221: 446-447.

Linhart, S.B. and F.F. Knowlton
1967 Determining Age of Coyotes by Tooth Cementum Layers. J. Wildl. Mgmt. 31: 362-365.

Lockard, G.R.
1972 Further Studies of Dental Annuli for Ageing White-tailed Deer. J. Wildl. Mgmt. 36: 46-55.

Lodge, G.A. and G.E. Lammimg
1968 Growth and Development of Mammals. New York: Plenum Press.

Low, W.A. and I. McT. Cowan
1963 Age Determination of Deer by Annular Structure of Dental Cementum. J. Wildl. Mgmt. 27(3): 466-471.

Lowe, V.P.W.
1967 Teeth as Indicators of Age with Special Reference to Red Deer (Cervus elephas) of Known Age from Rhum. J. Zool., Lond. 152: 137-153.

Madsen, R.M.
1967 Age Determination of Wildlife: a Bibliography. Washington D.C.: U.S. Department of the Interior Bibliography No.2.

Mayer, W.V. and S. Bernick
1963 Effect of Hibernation on Tooth Structure and Dental Caries. In Mechanisms of Hard Tissue. R.F. Sognnaes, ed. Washington: American Association for Advancement of Science.

Meadow, R.H.
1978 "Bonecode" -- A System of Numerical Coding for Faunal Data from Middle Eastern Sites. In Approaches to Faunal Analysis in the Middle East. R.H. Meadow, and M.A. Zeder, eds. pp. 169-186. Harvard University: Peabody Museum of Archaeology and Ethnology, Peabody Museum Bulletin 2.

Mead, J.G.
1975 Anatomy of the external nasal passages and facial complex in the Delphinidae (Mammalia: Cetacea). Smithsonian Contributions to Zoology 207.

McKusick, C.R.
1986 Southwest Indian Turkeys. Arizona: C.R. McKuisck. (Route #1, Box 35-D, Globe, Az. 85501).

McGovern, T.H. and T. Amorosi
1986 Morphological Markers for Distinguishing the Bones of Cattle and Caribou: A Zooarchaeological Manual. Unpublished Ms. on file at the Bioarchaeological Lab., Dept. of Anthropology, Hunter College, C.U.N.Y..

Morales, A. and K. Rosenlund
1979 Fish Bone Measurements: An Attempt to Standardize the Measuring of Fish Bone from Archaeological Sites. Copenhagen: Steenstrupia.

Morris, P.A.
1970 A Method for Determining the Absolute Age in the Hedgehog. Journal of Zoology 161: 277-281.

Morris, P.A.
1972 A Review of Mammalian Age Determination Methods. Mammal Review 2(3): 69-104.

Miller, W.C., and E.D.S. Roberston
1947 Practical Animal Husbandry. 5th edition. Edinburgh.

Mitchell, B.
1967 Growth Layers in the Cementum of the Teeth of Barren Ground Caribou. Canadian Journal of Zoology 41: 111-113.

Mujib, K.A.
1967 The Cranial Osteology of the Gadidae. Journal of the Fisheries Research Board of Canada 24: 1315-1375.

Myers, V.S. and M.A. Emmerson
1966 The Age and Manner of Epiphyseal Lines in Equine Limbs. pp. 21-39. Proceedings of the 12th Annual Convention of American Association of Equine Practitioners.

Navia, J.M. and A.J. Narkates
1980 Dental Research. In _The Laboratory Rat_. H.J. Baker, ed. New York: Academic Press.

Needham, A.E.
1964 _The Growth Process in Animals_. London: D. Van Nostrand Company, Inc.

Ness, A.R.
1967 A Measure of Asymmetry of the Skulls of Odontocete Whales. _Jounral of Zoology_ 153: 209-221.

Noddle, B.
1974 Ages of Epiphyseal Closure in Feral and Domestic Goats and Ages of Dental Eruption. _J. Arch. Sci._ 1: 195-204.

Noe-Nygaard, N.
1977 Butchering and Marrow Fracturing as a Taphonomic Factor in Archaeological Deposits. _Paleobiology_ 3: 218-237.

Norden, C.R.
1961 Comparative Osteology of Represenitive Salmonid Fishes, with Particular Reference to the Grayling _Thymallus arcticus_ and Its Phylogeny. _Journal of the Fisheries Research Board of Canada_ 18: 679-791.

O'Connor, T.P.
1985 On Quantifying Vertebrates - Some Sceptical Observations. _Circaea_ 3(1): 27-30.

Olsen, S.J.
1951 Post-Cranial Skeletal Characters of _Bison_ and _Bos_. _Papers of the Peabody Museum of Archaeology and Ethnology_ 35(3). Cambridge Mass.: Harvard University.

1961a The Relative Value of Fragmentary Mammalian Remains. _American Antiquity_ 26(4): 538-540.

1961b Problems of Mammal Skull Identification Due to Age Differences in the Dentition. _American Antiquity_ 27(2): 231-234.

Olsen, S.J.
1967 A Note in Regard to Osteological Field Manuals. American Antiquity 32: 321.

1968 Fish, Amphibian and Reptile Remains from Archaeological Sites: Part 1 -- Southeastern and Southwestern United States. Papers of the Peabody Museum of Archaeology and Ethnology 56(2). Cambridge, Mass. : Harvard University.

1979a Osteology for the Archaeologist: No. 3, The American Mastodon and the Wooly Mammoth: No. 4, North American Birds: Skulls and Mandibles: No. 5 North American Birds: The Postcranial Skeletons. Papers of the Peabody Museum of Archaeology and Ethnology 56. Cambridge, Mass.: Harvard University.

1980 Mammalian Remains from Archaeological Sites: Part 1 -- Southeastern and Southwestern United States. Papers of the Peabody Museum of Archaeology and Ethnology 56(1). Cambridge, Mass.: Harvard University.

1982 An Osteology of Some Maya Mammals. Papers of the Peabody Museum of Archaeology and Ethnology 73. Cambridge, Mass.: Harvard University.

Parker, S.
1988 Skeleton. Eyewitness Books. New York: Alfred A. Knopf.

Parker, W.K.
1873 On the Structure and Development of the Skull of the Salmon Salmo salar L.. Philosophical Transaction of the Royal Society 163: 95-145.

Parmalee, P.W., A.A. Paloumpis and N. Wilson
1972 Animals Utilized by Woodland Peoples Occupying the Apple Creek Site, Illinois. In Illinois Valley Archaeological Program, Research Papers, vol. 5. S. Struever, ed. Springfield Illinois: Illinois State Museum.

Payne, S.
1972a A Metrical Distinction Between Sheep and Goat Metacarpals. In The Domestication and Exploitation of Plants and Animals. P.J. Ucko, and G.W. Dimbleby, eds. pp. 295-305. Chicago: Duckworth.

1972b Partial Recovery and Sample Bias: The Results of Some Sieving Experiments. In Papers in Economic Prehistory. E.S. Higgs, ed. pp. 49-64. Cambridge: Cambridge University Press.

1972c On the Interpretation of Bone Samples from Archaeological Sites. In Papers in Economic Prehistory. E.S. Higgs, ed. pp. 65-81. Cambridge: Cambridge University Press.

1973 Kill-off Patterns in Sheep and Goats: the Mandibles from Asvan Kale. Anatolian Studies 23: 281-303.

1975 Partial Recovery and Sample Bias. In Archaeozoological Studies. A.T. Clason, ed. New York: North Holland/American Elsevior.

Perkins, D.
1973 The Beginning of Animal Domestication in the Near East. American Journal of Archaeology 77(3): 279-282.

Peyer, B.
1968 Comparative Odontology. Chicago: University of Chicago Press.

Pichugin, M.Yu
1983 An Osteological Description of Southern Char of the Genus Salvelinus (Salomindae) from the My and Bol'shaya Iska River Populations (Amur Lagoon). Journal of Icthyology 23(4): 27-38.

Prummel, W.
1987a Atlas for the Identification of foetal Skeletal Elements of Cattle, Horse, Sheep and Pig. Part I. Archaeozoologica 1(1): 23-30.

1987b Atlas for the Identification of foetal Skeletal Elements of Cattle, Horse, Sheep and Pig. Part II. Archaeozoologica 1(2): 11-41.

Ranson, A.B.
1966 Determining Age of White-tailed Deer from Layers in Cementum of Molars. J. Wildl. Mgmt. 30(1): 197-199.

Reiland, S.
1978 Growth and Skeletal Development of the Pig. Acata Radiologica Suppl. 358: 15-22.

Reynolds, S.H.
1913 The Vertebrate Skeleton. Cambridge: At The Uinversity Press.

Ritchie, W.A.
1980 The Archaeology of New York State. New York: Harbor Hill Books.

Robinette, H.R. and H.J. Stains
1970 Comparative Study of the Calcanea of the Pinnipedia. Journal of Mammalogy 51(3): 527-541.

Romer, A.S.
1949 The Vertebrate Body. Philadelphia: Saunders.

Romer, A.S.
1956 Osteology of the Reptiles. Chicago: University of Chicago Press.

Rooney, J.R.
1963 In Equine Medicine and Surgery. J.F. Bone, E.J. Catcott, A.A. Gabel, L.E. Johnson, and W.F. Riley, eds. pp. 407-409. Wheaton, Illinois: American Veterinary Publication, Inc.

Rudge, M.R.
1972 Horns as Indicators of Age in Goats (Capra hircus L.). NZ. Journal Science 15(2): 255-263.

Russell, D., T. Amorosi and T.H. McGovern
1986 An Archaeofauna from Storaborg, Southern Iceland. Unpublished Ms. on file at the Department of Archaeology, National Museum Iceland and the Bioarchaeological Lab., Dept. of Anthropology, Hunter College, C.U.N.Y..

Ryder, M.L.
1969 Animal Bones in Archaeology. Oxford and Edinburgh: Published for the Mammal Society by Blackwell Scientific Publications.

Sauer, P.R.
1966 Determining Sex of Black Bears from the Size of the Lower Canine Tooth. New York Fish and Game Journal 13(2): 140-145.

Saxon, A. and C.F.W. Higham
1968 Identification and Interpretation of Growth Rings in the Secondary Dental Cementum of Ovis aries L. Nature 219: 634-635.

Scheffer, V.B. and R.S. Peterson
1967 Growth Layers in Teeth of Suckling Fur Seals. Growth 31: 35-38.

Schmid, E.F.
1972 *Atlas of Animal Bones*. New York: Elsevier Publishing Company.

Schmidt, G.
1960 *Epiphysen und Apophysen in der Rontgenlogischen Darstellung in den Vorder und Hinterextremiten der Fohlen*. Hanover: Inaugural Disseration.

Schneider, F.
1958 *Die Rinder des Latene - Oppidums Manching. Studien an vor - und fruhaeschichtlichen Tierresten Bayerns* 5. Munich: Kiefhaber, Kiefhaber and Elbl..

Schultz, L.P. and A.D. Weander
1935 A Review of the Cods of the Northeastern Pacific with Comparative Notes on Related Species *Copeia* 1935(3): 127-139.

Sergeant, D.E. and D.H. Pimlott
1959 Age Determination in Moose from Sectioned Incisor Teeth. *J. Wildl. Mgmt.* 23(3): 315-321.

Shipman, P., A. Walker and D. Bichell
1985 *The Human Skeleton*. Cambridge: Harvard University Press.

Sicker, H. and S.N. Bhaskar
1972 *Orban's Oral Histology and Embryology*. St. Louis: C.V. Mosely Company.

Silver, I.A.
1969 The Aging of Domestic Animals. In *Science in Archaeology*. D. Brothwell, and E. Higgs, eds. pp. 250-268. New York: Praeger Publishers.

Simkiss, K.
1975 *Bone and Biomineralization: Studies in Biology*, No. 53. Southhampton, Great Britian: Camelot Press, Ltd.

Sinclair, D.
1978 Human Growth After Birth, 3rd edition. New York: Oxford University Press.

Sisson, S. and J.D. Grossman
1966 The Anatomy of Domestic Animals, 4th. edition. London: W.B. Saunders Company.

Smith, G.S.
1979 Mammalian Zooarchaeology, Alaska: a Manual for Identifying and Analyzing Mammal bones from Archaeolgical Sites in Alaska. Anthropology and Historic Preservation, Cooperative Park Studies Unit, University of Alaska. Fairbanks: Alaska.

Smith, R.N.
1966 Observations on the Ossification of the Appendicular Skeleton of Sheep. Ph.D. Disseration, University of Bristol.

Speth, J.D.
1983 Bison Kills and Bone Counts; Decision Making by Ancient Hunters. Chicago: The University of Chicago Press.

Spiess, A.E.
1976 Genus Phoca: Metrical Analysis for Osteoarchaeology. Unpublished Ms..

1979 Reindeer and Caribou Hunters: An Archaeological Study. New York: Academic Press.

Spiess, A.E. and M.H. Hedden
1983 Kidder Point and Sears Island in Prehistory. Occassional Publications in Maine Archaeology Number Three. The Maine Historic Preservation Commission, Augusta, Maine.

Spinage, C.A.
1973 A Review of the Age Determination of Mammals by Means of Teeth, with Special Reference to Africa. East African Wildlife Journal 11: 165-187.

Stallibrass, S.
1982 The Use of Cement Layers for Absolute Ageing of Mammalian Teeth: A Selective Review of the Literature, With Suggestions for Further Studies and Alternative Application. In Ageing and Sexing Animal Bones from Archaeolgical Sites). B. Wilson, C. Grigson, and S. Payne, eds. pp. 109-126. B.A.R., Brit Series 109.

Steele, D.G. and C.A. Bramblett
1988 The Anatomy and Biology of the Human Skeleton. College Station: Texas A & M University Press.

Stewart, T.D.
1979 Essentials of Forensic Anthropology: Especially as Developed in the United States. Springfield, Illinois: Charles C. Thomas Publisher.

Streitferdt, U.K.
1972 Osteo-Archaeologische Untersuchung an Tierknochenfunden aus vier Romschan Statioen. In Suddeutschen Raum. Munich: Disseration.

Struever, S.
1968 Flotation Techniques for the Recovery of Small - Scale Archaeological Remains. American Antiquity 33(3): 353-362.

Styles, B.W.
1981 *Faunal Exploitation and Resource Selection: Early Late Woodland Subsistence in the Lower Illinois Valley*. Northwestern University Archaeological Program, Scientific Papers, No. 3 Evanston: Illinois.

Tchernavin, J.V.
1938 Changes in the Salmon Skull. *Transactions of the Zoological Society of London* 24: 103-185.

Thomas, D.H.
1969 Great Basin Hunting Patterns: a Quantitative Method for Treating Faunal Remains. *American Antiquity* 34(4): 392-401.

1972 Analysis of Faunal Remains. In *The Archaeology of Barrel Springs Site (NV-Pe-104), Pershing County, Nevada*. R.A. Cowan. Archaeological Research Facility, Department of Anthropology, University of California, Berkeley.

Tohara, S.
1950 Radiographical Studies on the Ossification of Leg-Bones of Horses. *Japanese Journal of Veterinary Science* 12: 1-12.

Torres, V.R.P., A.A. Enciso and E.G. Porres
1986 *The Osteology of South American Camelids*. Archaeological Research Tools, volume 3, Institute of Archaeology, University of California, Los Angeles.

Traquair, R.H.
1865 On the Asymmetry of the Pleuronectidae, As Elucidated by an Examination of the Skeleton in the Turbot, Halibut and Plaice. *Transactions of the Linnean Society of London* 25: 263-296.

Turner, A.
1984 Sub-Sampling Animal Bone Assemblages: Reducing the Work Load or Reducing the Information? *Ciraea* 2(2): 69-75.

Turner, J.C.
1977 Cemental Annulations as an Age Criterion in North American Sheep. J. Wildl. Mgmt. 41(2): 211-217.

Ubelaker, D.H.
1978 Human Skeletal remains: Excavation, Analysis, Interpretation. Manuals on Archaeology 2., Washington: Taraxacum.

van Nostrand, F.C. and A.B. Stehenson
1964 Age Determination for Beavers by Tooth Development. J. Wildl. Mgmt. 28: 430-434.

Vaughan, J.
1980 Bone Growth and Modeling. In Growth in Animals. T.L.J. Lawrence, ed. pp. 83-99. London: Butterworth.

Vladykov, V.D.
1962 Osteological Studies on Pacific Salmon of the Genus Oncorhynchus. Fisheries Research Board of Canada Bulletin No. 136. Ottawa.

von den Driesch, A.
1976 A Guide to the Measurememt of Animal Bones from Archaeological Sites. Peabody Museum Bulletin 1. Peabody Museum of Archaeology and Ethnology. Cambridge Mass.: Harvard University.

Walker, R.
1985 A Guide to Post-Cranial Bones of East African Animals. Hylochoerus Press.

Walker, W.F.
1982 A Study of the Cat. New York: Saunders College Publishing.

Warick, R. and P.L. Williams
1973 Gray's Anatomy. Philadelphia: W.B. Saunders Company. 35th, British Edition.

Watson, J.P.N.
1978 The Interpretation of Epiphyseal Fusion Data. In *Research Problems in Zooarchaeology*. D.R. Brothwell, K.D. Thomas, J. Clutton-Brock, eds. Institute of Archaeology, Occassional Publications, No. 3 London.

Wheat, J.B.
1972 *The Olsen-Chubbuck Site: a Paleo-Indian Bison Kill*. Society for American Archaeology Memoir, No. 26.

Wilson, B., C. Grigson, and S. Payne, eds.
1982 *Ageing and Sexing Animal Bones from Archaeological Sites*. B.A.R., Brit. Series 109.

Wilson, M.V.H. and P. Veilleux
1982 Comparative Osteology and Relationships of the Umbridae (Pisces: Salmoniformes). *Zoological Journal of the Linnean Society* 76: 321-352.

Wirtschafter, Z.T.
1960 *The Gensis of the Mouse Skeleton: a Laboratory Atlas*. Springfield, Illinois: C.C. Thomas.

Wolf-Heidegger, G.
1961 *Atlas der Systematischen Anatomie des Menschen. I. Skeleton Iuncturae Ossium Systema Musculorum*. Karger, Basel-New York.

Wolsan, M.
1982 A Comparative Analysis of the Ribs of Ungulates for Archaeozoological Purposes. *Acata Zool. Crancov*. 26(6): 167-228.

Woolfender, G.E.
1961 Post Cranial Osteology of the Waterfowl. *Bulletin of the Florida State Museum* 6(1).

Zeuner, F.E.
1963 *A History of Domesticated Animals*. London.

Ziegler, A.C.
1965 The Role of Faunal Remains in Archaeological Investigations. In *Symposium on Central California Archaeology*. F. Curtis, ed. pp. 47-75. Sacramento Anthropological Society Papers, No. 3. Sacramento.

Zietzschmann, O.
1943 *Ellenberger-Baum. Hanbuch der Vergleichenden Anatomie der Haustiere*, Berlin: Springer-Verlag.

Zietzschmann, O., and O. Krooling
1955 *Lehrbuch der Entwicklungs-Gesschicte der Haustiere*. 2nd edition. Berlin-Hamburg.

www.ingramcontent.com/pod-product-compliance
Ingram Content Group UK Ltd.
Pitfield, Milton Keynes, MK11 3LW, UK
UKHW061844041026
15591UKWH00015B/901

9 780860 546788